ERCOFTAC Series 29

Series Editors

ERCOFTAC (European Research Community on Flow, Turbulence and Combustion) was founded as an international association with scientific objectives in 1988. ERCOFTAC strongly promotes joint efforts of European research institutes and industries that are active in the field of flow, turbulence and combustion, in order to enhance the exchange of technical and scientific information on fundamental and applied research and design. Each year, ERCOFTAC organizes several meetings in the form of workshops, conferences and summer schools, where ERCOFTAC members and other researchers meet and exchange information. The *ERCOFTAC* series publishes the proceedings of ERCOFTAC meetings, which cover all aspects of fluid mechanics. The series comprises proceedings of conferences and workshops, and of textbooks presenting the material taught at summer schools. The series covers the entire domain of fluid mechanics, which includes physical modelling, computational fluid dynamics including grid generation and turbulence modelling, measuring techniques, flow visualization as applied to industrial flows, aerodynamics, combustion, geophysical and environmental flows, hydraulics, multi-phase flows, non-Newtonian flows, astrophysical flows, laminar, turbulent and transitional flows.

Indexed by SCOPUS, Google Scholar and SpringerLink.

Martin White · Tala El Samad ·
Ioannis Karathanassis · Abdulnaser Sayma ·
Matteo Pini · Alberto Guardone
Editors

Proceedings of the 4th International Seminar on Non-Ideal Compressible Fluid Dynamics for Propulsion and Power

Springer

Editors
Martin White
School of Engineering and Informatics
University of Sussex
Brighton, UK

Tala El Samad
School of Science & Technology
City, University of London
London, UK

Ioannis Karathanassis
School of Science & Technology
City, University of London
London, UK

Abdulnaser Sayma
School of Science & Technology
City, University of London
London, UK

Matteo Pini
Faculty of Aerospace Engineering
Delft University of Technology
Delft, Zuid-Holland, The Netherlands

Alberto Guardone
Department of Aerospace Science
and Technology
Politecnico di Milano
Milan, Italy

ISSN 1382-4309 ISSN 2215-1826 (electronic)
ERCOFTAC Series
ISBN 978-3-031-30935-9 ISBN 978-3-031-30936-6 (eBook)
https://doi.org/10.1007/978-3-031-30936-6

This Springer imprint is published by the registered company Springer Nature Switzerland AG
The registered company address is: Gewerbestrasse 11, 6330 Cham, Switzerland

Preface

The topic of non-ideal compressible fluid dynamics (NICFD) for propulsion and power deals with studying the fluid dynamics of compressible flows for which the ideal gas law does not apply, with an emphasis on reactive and non-reactive flows within power and propulsion systems. The study of NICFD effects finds application within industries that are critical to ensure the provision of clean and secure energy, which includes power generation, refrigeration, heat pumps and clean combustion amongst others.

To understand the field of NICFD, we can break the topic into its constituent parts. The term *non-ideal* relates to the fundamental thermodynamic behaviour of unconventional fluids where the behaviour cannot be predicted assuming the fluid behaves as an ideal gas. Such thermodynamic behaviour can be observed in flows in supercritical states, flows close to the saturation curve, flows close to the critical point and two-phase vapour–liquid flows.

Secondly, we have *compressible fluid dynamics* which relates to the study of fundamental fluid dynamic aspects such as compressible high-speed flows, shockwave formation, boundary layers, turbulence modelling, acoustics and phase change, amongst others. Combining this with non-ideal relates to understanding such fluid behaviour within thermodynamic regions where non-ideal effects are expected. Hand-in-hand with understanding these effects comes the need for computational fluid dynamic, experimental and measurement techniques, developed specifically for such flows.

Finally, *power & propulsion* relates to the application of this knowledge to practical engineering systems operating with non-ideal fluids. This relates to a range of technologies that are becoming, or are likely to become, increasingly important to ensure the provision of clean and secure energy, alongside efficient propulsion. This includes power cycles such as organic Rankine cycles and supercritical carbon dioxide power cycles, refrigeration systems, heat pumps, clean combustion, rocket engines, alongside subsequent system components such as turbomachinery, heat exchangers and combustors.

With this in mind, the seminar themes range from theoretical foundations, to advanced numerical and experimental practices, and to applications. The seminar provides an exciting platform to bring together researchers and scientists who are pioneering theoretical, numerical and experimental advancements in order to share, learn and discuss the latest insights in this field. The key themes of the conference include the following areas: experiments; fundamentals; numerical methods; optimisation and uncertainty quantification; critical and supercritical flows; turbulence and mixing; multi-component fluid flows; applications in ORC power systems; applications in supercritical CO_2 power systems; steam turbines; cryogenic flows; condensing flows in nozzles; cavitating flows and super- and trans-critical fluids in space propulsion.

The biannual NICFD seminar was established in 2016 due to the growing interest in NICFD effects, particularly stemming from advances in propulsion and power applications. The seminar aims to bring together researchers working in the field to discuss the latest advancements in the field. In 2019, the ERCOFTAC Special Interest Group on Non-Ideal Compressible Fluid Dynamics (SIG-49) was founded (SIG-49), which was setup to further promote the exchange of scientific information and to encourage and consolidate the interaction between NICFD researchers and professionals.

This volume contains the proceedings from NICFD 2022: The 4th International Seminar on Non-Ideal Compressible Fluid Dynamics for Propulsion & Power, which was held during 3–4 November 2022, and hosted by the Thermo-Fluids Research Centre at City, University of London, UK. Further details of the conference can be found here: NICFD2022. The published proceedings are composed of 23 papers reporting on the latest developments in the thematic areas of: Fundamentals; Numerical Modelling and Methods; Multi-phase Flows; Reacting Flows and Experiments.

The conference organisers are extremely grateful to everybody that helped make NICFD 2022 a success. This includes all the authors of the submitted papers, as well as the reviewers and members of the scientific committee, who together ensured a high-quality of the contributions collected here, alongside the accompanying technical presentations. Thanks also go to the City Events team, the session chairs and the student helpers who ensured that the event ran smoothly.

Martin White
Chair

Tala El Samad
Ioannis Karathanassis
Abdulnaser Sayma
Matteo Pini
Conference Co-chairs

Alberto Guardone
Founding Chair

Contents

Fundamentals

Adaptive Simulations of Cylindrical Shock Waves in Polytropic van der Waal Gas

Barbara Re(✉), Alessandro Franceschini, and Alberto Guardone

Department of Aerospace Science and Technology, Politecnico di Milano, Milano, Italy
{barbara.re,alberto.guardone}@polimi.it,
alessandro1.franceschini@mail.polimi.it

Abstract. The propagation of converging cylindrical shock waves, collapsing at the axis of symmetry, in a non-ideal gas is investigated by using innovative interpolation-free mesh adaptation techniques. The high pressure, temperature, and energy that can be reached close to the focus point call for thermodynamic models able to take into account the non-ideal gas effects. A distinguishing feature of converging shock waves is the non-constant propagation speed, which makes the flow field behind the shock intrinsically unsteady. To efficiently simulate this configuration, it is fundamental to adapt the computational domain as time evolves. Hence, at each time step, we modify the grid spacing through node insertion, deletion, or relocation according to the current position of the flow features. In this work, any interpolation of the solution between the original and the adapted grids is avoided thanks to a peculiar strategy able to describe mesh adaptation within the arbitrary Lagrangian-Eulerian framework. The adaptive simulation framework, equipped with the polytropic van der Waals equation of state, is assessed first in dilute conditions, where it is possible to compare numerical results with theoretical predictions. In particular, we compute the Guderley-like self-similar solution describing shocks of different intensities propagating in the siloxane MM. Then, it is used to verify the possibility to reach higher energy density when the shock is initiated in the NICFD regime. Finally, we investigate the interaction between the converging shock wave and an arc-shaped obstacle, which is a fundamental phase of the so-called reshaping process, useful to increase the stability of converging shocks.

Keywords: cylindrical shock waves · van der Waals model · mesh adaptation · finite volume scheme

1 Introduction

Converging cylindrical shock waves, collapsing at the axis of symmetry, have been investigated in several theoretical, numerical, and experimental campaigns since the first half of last century [5,7]. Differently from the planar case, converging shocks are intrinsically unsteady phenomena, because the speed of the

M. White et al. (Eds.): NICFD 2022, ERCOFTAC 29, pp. 3–12, 2023.
https://doi.org/10.1007/978-3-031-30936-6_1

shock increases as the radius decreases. Hence, during the propagation towards the focus point, the shock intensity increases, reaching particularly high values of fluid temperature and pressure close to the focus point. The possibility to reach such a high-energy condition has prompted many research activities, motivated by the potential applications in different fields, spanning from medical treatments, e.g., lithotripsy for renal calculi, to power generation, for instance, for the ignition of the nuclear fusion process [1]. Unfortunately, the stability of these flow structures, defined as the capability of the converging wave to approach the cylindrical shape damping out disturbances during the propagation, is of great concern, because small perturbations can make the propagation front deviate from the cylindrical shape [7]. The stability decreases also as the shock Mach number increases. A possible strategy to prevent the onset of instabilities is the so-called *shock-reshaping* process, which turns a cylindrical shock into a prismatic one through controlled reflections. To minimize the possible losses that would reduce the shock intensity, these reflections are generated by lenticular obstacles, i.e., symmetric aerodynamic profiles with sharp leading and trailing edge [6].

The aim of this work is the assessment of a numerical simulation tool for 2D cylindrical converging shocks, without and with obstacles, in non-ideal compressible fluid dynamic regimes, where, for the same pressure jump, higher temperature and energy are reached. The proposed CFD tool performs unsteady inviscid flow simulations using a mesh adaptation strategy, which refines and makes the grid coarser following the shock evolution. Notably, this adaptation is performed within the arbitrary Lagrangian-Eulerian (ALE) framework, so that no interpolation of the solution from the original to the adapted grid is required [2,8]. The siloxane MM is the chosen working fluid for the tests here presented and its thermodynamic behavior is described by the polytropic van der Waals model.

This work represents the first step of a wider project, whose long-term goal is the study of how the obstacle geometry and the fluid dynamic behavior affect the intensity and the stability of the converging shock waves.

2 The Numerical Method

Assuming the flow as inviscid, i.e., neglecting thermal conductivity and viscosity, the fluid flow behavior is described by the unsteady compressible Euler equations. To close the system, the van der Waals equation of state (EOS) is considered:

$$p = \frac{RT}{v - b} - \frac{a}{v^2} \tag{1}$$

where p is the pressure, T the temperature, v the specific volume, R the gas constant, and a and b are the fluid-specific parameters of the van der Waals equation that account for non-ideal gas effects about covolume and intermolecular attractive forces. To a first approximation and to compare numerical results with theoretical predictions, the polytropic assumption is made. So, the value of the specific heat at constant volume c_v is computed through the Aly-Lee equation [3] at $T = 300$ K, and it is assumed to be constant during the simulation, even though the high temperatures and the low densities that can be reached at the focus point could jeopardize the hypothesis of polytropic gas [12].

2.1 The Unsteady Adaptive Finite Volume Scheme

The governing equations are integrated in space over an unstructured, triangular mesh that discretizes the domain Ω, through a node-centered, edge-based finite volume scheme. The equations are discretized within an ALE framework that allows the finite volumes to move and deform independently from the flow velocity. Thanks to a peculiar three-step procedure, also the topology changes due to mesh adaptation—such as node insertion, node deletion, and edge swapping—are described within a standard ALE formulation [8]. A backward-finite-difference implicit scheme with a dual time-step technique is used as time integrator.

The unsteady adaptive discretization strategy is well-suited for the simulation of the converging shock waves because the mesh can be modified to follow moving flow features: as shown in Fig. 1, the minimum grid spacing is reached only close to the shock wave, increasing the accuracy while keeping the number of grid nodes under control. The adaptation process is driven by criteria based on the derivatives of the flow variables: where large variations are detected, the local grid spacing is reduced, whereas it is enlarged where the solution is smooth, without jeopardizing solution accuracy [9,10]. The adaptation criterion is a weighted combination of the magnitude of the Hessian of the Mach number and the gradient of the density, with weights 0.8 and 0.2, respectively. The same strategy can be used also to follow arbitrarily large displacements of the boundary domain [11], but in this work we consider only a domain with fixed boundaries.

Due to the axial symmetry of the numerical domain, the computational cost is reduced by considering only a slice of the 2D cylinder section. The computational domain, shown in Fig. 1, is thus defined by three boundaries: the lower and the upper walls that merge at the focus point, which are modeled as inviscid walls, and the external boundary shaped as an arc of a circle, where non-reflecting boundary conditions are imposed.

The simulations are initialized with null velocity, and a jump in pressure and density at a certain radial distance (R_0) from the focus point, hence with a radial version of the Riemann problem. The post-shock density is computed according to the Rankine-Hugoniot adiabatic jump relations for a perfect gas, as in [13]. After the diaphragm rupture, a shock wave and a contact discontinuity propagate inward, while a rarefaction fan propagates outward. We remind that, unlike the planar case, the internal, i.e., post-shock, state is time-dependent because the shock intensity increases while converging.

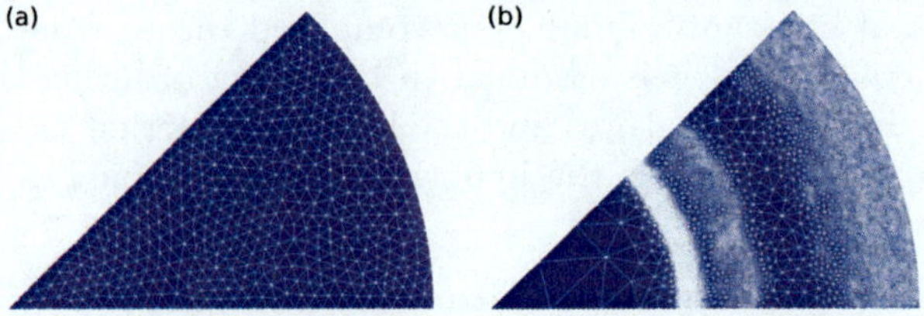

Fig. 1. Computational domain: a) initial grid, with a uniform grid spacing $h_0 = 0.01$ m; b) example of an adapted grid obtained during the simulation. The radial length of the mesh is 0.31 m. During adaptation, the imposed minimum edge size is $h_{\min} = 0.001$ m

2.2 Characterization of Converging Shock Waves

The evolution of cylindrical shock waves can be approximated according to Guderley's law [5]:

$$\frac{R_s}{R_0} = \left(1 - \frac{t}{\tilde{\tau}}\right)^{\alpha}, \tag{2}$$

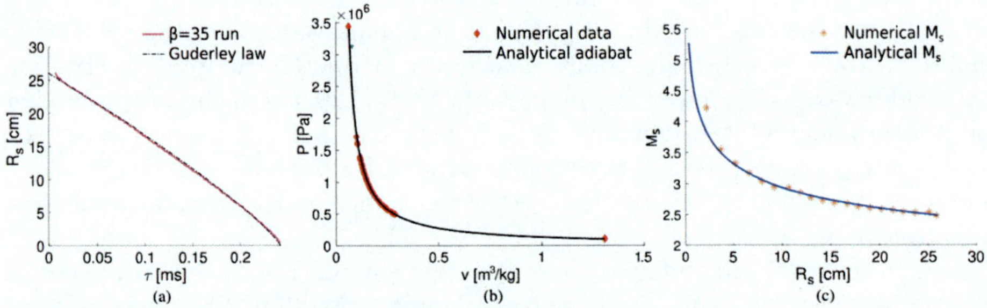

Fig. 2. Comparison of numerical data and theoretical predictions for the MM in dilute conditions, for the $\beta = 35$ case.

which expresses the shock position R_s as function of the time t. R_0 is the initial shock radius and $\tilde{\tau}$ the total focusing time, namely the time required by the shock to reach the focus point. Finally, α is the self-similarity exponent, which is a function of shock geometry (planar, cylindrical, or spherical) and thermodynamics of the gas. For the ideal gas, the theoretical value of α for cylindrical shocks is 0.834, while in this work, the numerical value according to the van der Waals EOS is computed through a non-linear least squares approach with numerical data.

A further theoretical result useful for validation is the post-shock pressure along the Hugoniot adiabatic curve, according to the van der Waals model, which can be defined in the p-v plane as

$$p_2 = p_2(p_1, v_1, v_2) = \left[\frac{1}{\delta_\gamma}\left(p_1 + \frac{a}{v_1^2}\right)(v_2 - b) - \frac{1}{2}p_1(v_2 - v_1) + \frac{a}{v_2\delta_\gamma}\left(\frac{b}{v_2} - 1\right)\right] \Big/ \left[\left(\frac{1}{2} + \frac{1}{\delta_\gamma}\right)v_2 - \left(\frac{v_1}{2} + \frac{b}{\delta_\gamma}\right)\right], \tag{3}$$

where $\delta_\gamma = R/c_v$, and the subscripts 1 and 2 refer to the pre- and post-shock state, respectively.

Finally, the monotonic behavior of the shock front intensity is evaluated by computing the shock Mach number as a function of the shock radius. To do so, Guderley's law is derived with respect to time as:

$$M_S^G(t_k) = \frac{1}{c_1} \cdot \left|\frac{\partial R_s(t_k)}{\partial t}\right|_{t_k} = \frac{R_s \cdot \alpha}{c_1 \cdot (\tilde{\tau} - t_k)}, \tag{4}$$

where c_1 is the pre-shock speed of sound and t_k is the k-th time step.

From the numerical results, we compute the shock position R_s in the radial direction in a post-process step by using an ad-hoc algorithm, devised following the three criteria proposed by Vignati and Guardone [13]. We evaluate the post-shock state through a similar strategy, but, due to the fact the shock wave is represented over more than one cell due to numerical viscosity, it is extracted at a position $R_e \geq R_s$, and R_e is computed by solving a local maximum problem of the second order derivative of the pressure in the proximity of R_s [4].

3 Results

Simulations of converging shock waves with siloxane MM are now presented. Different values of the pressure jump $\beta = p_2/p_1$ across the diaphragm of the numerical Riemann problem are tested. In the first case, we consider $\beta = 35$, with $p_1 = 0.0516\ p_c$ and density $\rho_1 = 0.0039\ \rho_c$, namely in dilute gas conditions (the subscript c indicates critical variables), where a fair agreement with the perfect gas model is expected. The diaphragm is located at $R_0 = 26$ cm, whereas the domain length in the radial direction is 31 cm. The numerical results and the theoretical predictions are compared in Fig. 2. The shock position, in panel (a), is compared with the theoretical Guderley's law with the ideal gas self-similar exponent $\alpha = 0.834$. The fitted value of α results equal to 0.852, deviating about 2% from the ideal gas model case, in these dilute conditions. In Fig. 2(b), post-shock states computed from the numerical simulation are plotted along with the shock adiabatic curve in the $p - v$ plane, observing a good agreement. Starting from the pivot state, namely the one at the highest specific volume, throughout its propagation, the converging shock wave progressively increases the fluid post-shock pressure, eventually attaining values in the order of GPa. The latest comparison for this validation case is the shock Mach number as a

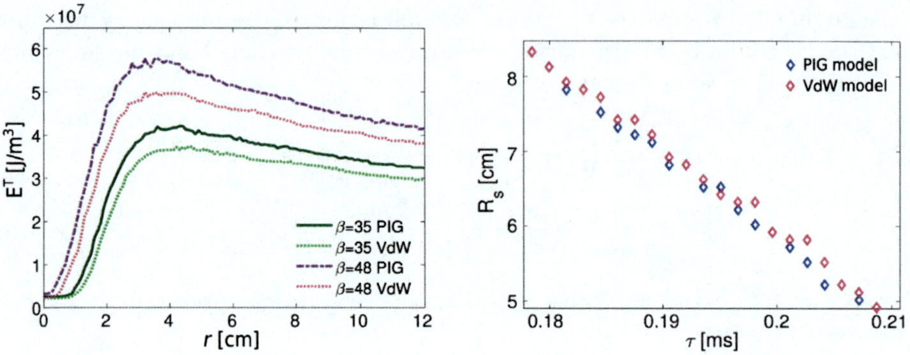

Fig. 3. Comparison between the results for the MM siloxane in dilute conditions, modelled with ideal and van der Waals EOS. Left: total energy close to the focus point; right: shock position close to focus point.

function of shock radius, through Eq. (4), where, again, a good agreement is achieved and the Mach number of the shock front monotonically increases as the shock propagates. More specifically, from Fig. 2(c), it can be seen that the region close to the focus point is characterized by a higher slope of the curve, meaning that the last instants of the convergence process are contributing the most to the increase of fluid energy. This is the reason behind the importance of keeping as stable as possible the shock front during the implosion process in practical applications.

Now, the importance and influence of the selected thermal EOS to model the fluid behavior are assessed, considering the same numerical setup for each test. Tested values of pressure jump are $\beta = 35$ and $\beta = 48$. Relevant fluid dynamics variables are sampled from probes located along a radial coordinate r. Figure 3 (left) shows the differences in the total energy density, at two different time instants, one for each value of β. The differences in energy are important, especially once the converging shock has reached the focus point. Pressure values are lower with van der Waal EOS than with the perfect gas because attractive forces reduce fluid pressure at a given thermodynamic state. Consequently, lower pressure means lower temperature and lower total energy E^T. Moreover, results with the van der Waals model are characterized by larger values of shock radius at a given time, which reflect eventually into Guderley's law. Indeed, looking at the region close to the focus point, reported in Fig. 3 (right), the ideal gas confirms its tendency of overestimating the shock wave speed to the van der Waals model, because, at a given time step, lower values of shock radius are determined by a higher shock wave speed.

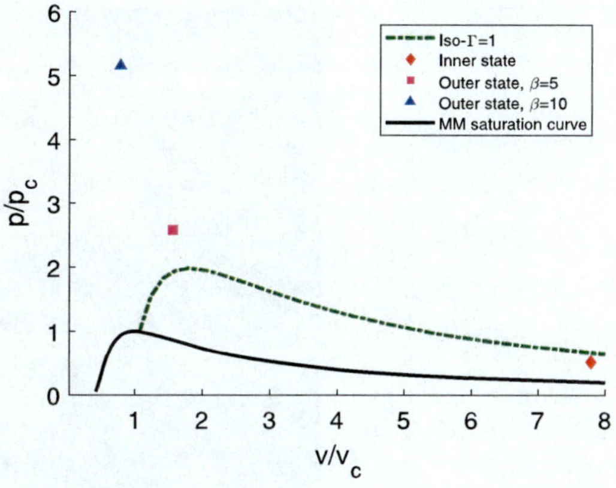

Fig. 4. Initial Riemann problem states for the tests with siloxane MM in NICFD conditions. The saturation curve is drawn in black, along with the isoline at $\Gamma = 1$ that delimits the NICFD region. The red marker denotes the initial, pre-shock, state. The blue and the violet markers depict, respectively, the post-shock state corresponding to $\beta = 5$, and $\beta = 10$, respectively.

Converging shock waves are then simulated in the non-ideal regime, which is bounded by $\Gamma = 1$ iso-line and is characterized by classical gas dynamics features, but the speed of sound dependence on the density is the opposite to the ideal regime. The initial conditions, displayed in Fig. 4, are $p_1 = 0.5157\ p_c$ and $\rho_1 = 0.1284\ \rho_c$, and two values of pressure jumps are simulated, $\beta = 5$ and $\beta = 10$. Shock position in time is depicted in Fig. 5 and it qualitatively agrees with the behavior predicted by Guderley's law. Moreover, the extrapolated power law exponents α computed for the two different β values show a variation lower than 1%, as proof of the validity of the implemented framework.

To evaluate if the compression factor attained in the proximity of the focus point is enhanced due to initial dense working conditions, the behavior of the pressure coefficient trace, defined as $c_p = \frac{p}{p_1}$, is investigated by computing the trace at the first probe (located at $r = 0.001$ m), shown in Fig. 6. As expected, starting from the same pre-shock conditions, higher values of β are associated with lower total focusing time, since the shock wave converges faster. In fact, the trace peak for $\beta = 10$ is reached at an earlier time. Moreover, doubling the pressure jump β, the maximum pressure coefficient value increases from $c_p^{\max} = 42.4$ to $c_p^{\max} = 106.4$, so it increases more than twice. This super-linear proportionality of the pressure coefficient on the pressure jump has been observed only in the non-ideal regime. Hence, working in dense conditions, with an initial condition close to the saturation curve, could lead easier to higher gas energy states at the focus point, which is an aspect of primary importance for practical applications of cylindrical converging shock waves.

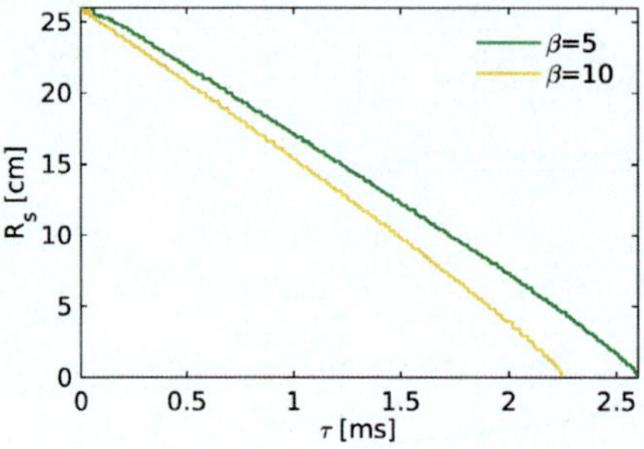

Fig. 5. Guderley's law from numerical simulation of MM siloxane in NICFD conditions.

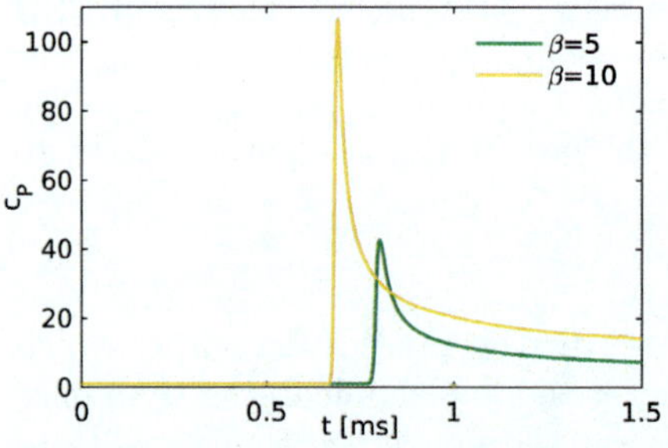

Fig. 6. Pressure coefficient traces in NICFD conditions at $r = 0.001$ m.

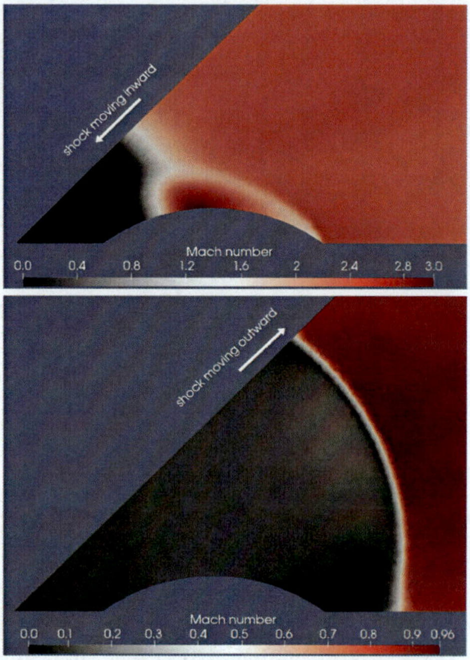

Fig. 7. Mach number contour plots for the test with the obstacle, in dense condition, at a time instant before (top) and after (bottom) the focus time.

Finally, a numerical simulation of a cylindrical converging shock wave that interacts with an obstacle is performed. The obstacle is an arc-shaped profile located in the proximity of the focus point, which generates a truly 2D flow field. Figure 7 (left) shows the Mach reflection caused by the shock-obstacle interaction. This flow structure, composed by the incident shock, the reflected shock and the third shock that travels parallel to the surface called Mach stem, is required to obtain cylindrical shock reshaping, regardless of the reflection type occurring in correspondence with the obstacle leading edge [13]. The robustness of the developed framework for adaptive 2D simulations of cylindrical shock waves has been proven through the computation of the flow field behavior also after the focus point is reached. Figure 7 (right) shows the Mach field when the shock wave has already been reflected at the focus point and passed the obstacle.

4 Conclusions

For the first time, 2D interpolation-free adaptive simulations of cylindrical converging shock waves have been carried out, assessing the validity of an innovative unsteady solver for the ALE formulation of the Euler equations over adaptive grids in NICFD regimes. The siloxane MM has been modeled through the van der Waals EOS, accounting for non-ideal gas effects. Several initial jump conditions for the radial Riemann problem have been tested, along with the effects of the thermodynamic model. The ideal gas model proved to overestimate the propagation velocity of the converging shock, even in dilute conditions. The simulations in the NICFD regime demonstrated that working in such conditions is efficient to obtain a higher energy state of the gas in close proximity to the focus point. Finally, the ultimate purpose of this work has been about the numerical simulation of a converging shock wave that eventually interacts with an obstacle, undergoing a reshaping process. In practical applications, reshaping into polygonal shocks is exploited since the latter are more stable. Results showed how the numerical method adopted is suitable for this type of problem. The Mach reflection pattern leading to the shock wave reshaping has been simulated. Moreover, the robustness of the method permits to represent properly the flow field once the converging shock wave has reached the focus point and is bounced back, moving outwards the computational domain.

References

1. Clark, D., Tabak, M.: A self-similar isochoric implosion for fast ignition. Nucl. Fusion **47**, 1147–1156 (2007). https://doi.org/10.1088/0029-5515/47/9/011
2. Colombo, S., Re, B.: An ALE residual distribution scheme for the unsteady Euler equations over triangular grids with local mesh adaptation. Comput. Fluids **239**, 105414 (2022). https://doi.org/10.1016/j.compfluid.2022.105414
3. Colonna, P., Nannan, N.R., Guardone, A., Lemmon, E.W.: Multiparameter equations of state for selected siloxanes. Fluid Phase Equilib. **244**, 193–211 (2006). https://doi.org/10.1016/j.fluid.2006.04.015
4. Franceschini, A.: Cylindrical shock waves in high molecular complexity van der Waals fluid. Master's thesis, Politecnico di Milano, July 2022
5. Guderley, K.G.: Stark kugelige und zylindrische verdichtungsstöße in der nähe des kugelmittelpunktes bzw. der zylinderachse. Luftfahrtforschung **19**, 302 (1942)
6. Kjellander, M., Tillmark, N., Apazidis, N.: Thermal radiation from a converging shock implosion. Phys. Fluids **22**(4), 046102 (2010). https://doi.org/10.1063/1.3392769
7. Perry, R.W., Kantrowitz, A.: The production and stability of converging shock waves. J. Appl. Phys. **22**, 878–886 (1951). https://doi.org/10.1063/1.1700067
8. Re, B., Dobrzynski, C., Guardone, A.: An interpolation-free ALE scheme for unsteady inviscid flows computations with large boundary displacements over three-dimensional adaptive grids. J. Comput. Phys. **340**, 26–54 (2017). https://doi.org/10.1016/j.jcp.2017.03.034
9. Re, B., Dobrzynski, C., Guardone, A.: Assessment of grid adaptation criteria for steady, two-dimensional, inviscid flows in non-ideal compressible fluids. Appl. Math. Comput. **319**, 337–354 (2018). https://doi.org/10.1016/j.amc.2017.03.049
10. Re, B., Guardone, A.: An adaptive ALE scheme for non-ideal compressible fluid dynamics over dynamic unstructured meshes. Shock Waves **29**, 73–99 (2019). https://doi.org/10.1007/s00193-018-0840-2

11. Re, B., Guardone, A., Dobrzynski, C.: An adaptive conservative ALE approach to deal with large boundary displacements in three-dimensional inviscid simulations. In: 55th AIAA Aerospace Sciences Meeting, Grapevine, Texas, 9–13 January 2017 (2017). https://doi.org/10.2514/6.2017-1945
12. Vignati, F.: Dynamics of cylindrical converging shock waves interacting with circular-arc obstacles. Ph.D. thesis, Politecnico di Milano (2015)
13. Vignati, F., Guardone, A.: Dynamics of cylindrical converging shock waves interacting with aerodynamic obstacle arrays. Phys. Fluids **27**, 066101 (2015). https://doi.org/10.1063/1.4921680

Non-ideal Compressible Flows in Radial Equilibrium

Paolo Gajoni and Alberto Guardone(✉)

Politecnico di Milano, Milan, Italy
{paolo.gajoni,alberto.guardone}@polimi.it

Abstract. Compressible flows at radial equilibrium, *i.e.* θ-independent flows, are investigated in the ideal, dilute-gas regime and in the non-ideal regime close to the liquid-vapour saturation curve and the critical point. A differential relation for the Mach number dependency on the radius is derived for both ideal and non-ideal conditions. For ideal flows, the relation is integrated analytically.

For flows of low molecular complexity fluids, such as diatomic nitrogen or carbon dioxide, the Mach number is a monotonically decreasing function of the radius of curvature, with the polytropic exponent γ being the only fluid-dependent parameter. In non-ideal conditions, the Mach number profile depends also on the total thermodynamic conditions of the fluid. For high molecular complexity fluids, such as toluene and MM (hexamethyldisiloxane), a non-monotone Mach profile is uncovered for non-ideal flows in supersonic conditions. For Bethe-Zel'dovich-Thompson fluids, the non-monotone behaviour is possible also in subsonic conditions.

Keywords: Non-ideal Compressible Fluid Dynamics (NICFD) · non-ideal flows · radial equilibrium

1 Introduction

Flows of molecularly complex fluids in the neighbourhood of the liquid-vapour saturation curve and the critical point significantly depart from the ideal-gas behaviour typical of dilute thermodynamic states. Thompson [14] firstly noted the crucial role of the so-called fundamental derivative of gas dynamics,

$$\Gamma = \frac{v^3}{2c^2}\left(\frac{\partial^2 P}{\partial v^2}\right)_s = 1 + \frac{c}{v}\left(\frac{\partial c}{\partial P}\right)_s, \tag{1}$$

in outlining the dynamic behaviour of compressible flows in non-ideal conditions. Different gasdynamic regimes can be defined, based on the value of Γ. Flows developing through thermodynamic states featuring $\Gamma > 1$ are said to evolve in the ideal gasdynamic regime, since the usual behaviour of ideal gases is observed. By contrast, one speaks of non-ideal regime, if the flow evolution possibly experiences values of $\Gamma < 1$, or non-classical regime, when negative values of Γ are reached. The most unconventional phenomena include, for $\Gamma < 1$, the Mach number decrease in steady supersonic

M. White et al. (Eds.): NICFD 2022, ERCOFTAC 29, pp. 13–21, 2023.
https://doi.org/10.1007/978-3-031-30936-6_2

nozzles and around rarefactive ramps [4,5,8,13] and the increase of the Mach number across oblique shock waves [16], whereas expansion shock waves and split waves are characteristic of the non-classical regime [12,15]. State-of-the-art thermodynamic models [3,10] predict values of $\Gamma < 1$ in the vapour-phase region close to saturation for fluids with high molecular complexity [1], or even $\Gamma < 0$ for the so-called Bethe-Zel'dovich-Thompson (BZT) fluids. Unfortunately, no experimental evidence of the occurrence of $\Gamma < 0$ is available, yet [7,11].

2 Scope of Research

Two-dimensional compressible flows at radial equilibrium, *i.e.* θ-independent flows, assuming a polar representation of the flowfield shown in Fig. 1, are here investigated. Formally, they represent the flow within an annulus, in which the curvature is constant with θ, but they can also be illustrative of the behaviour of flows subjected to a strong curvature. The subject of this paper is in fact inspired by the behaviour of a flow within a turbine cascade, which is repeatedly deviated through alternating rotors and stators. The curvature prescribed by the passage between two adjacent blades of an Organic Rankine Cycle (ORC) turbine disk [2], for example, can be seen in Fig. 2.

The research, however, does not deal directly with turbomachine flows, but it just aims at describing simple two-dimensional flows, which are mainly of theoretical interest. The purpose is in fact to derive a differential relation for the Mach number dependency on the radius of curvature, for flows at equilibrium, in both ideal and non-ideal conditions. For ideal flows, the relation, henceforth called radial equilibrium equation, is integrated analytically.

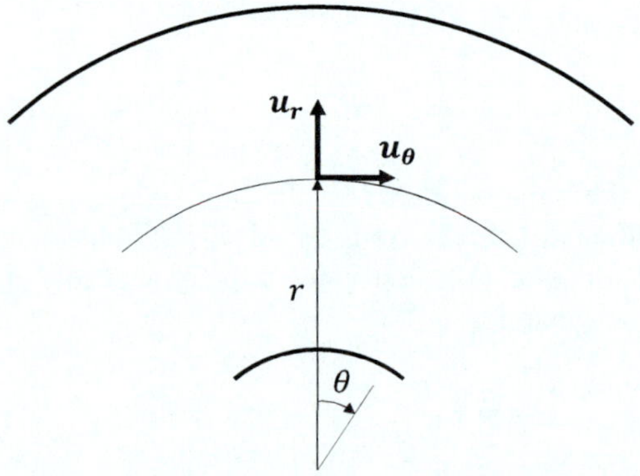

Fig. 1. Radial equilibrium flow representation.

Fig. 2. Example of an ORC turbine disk [9].

3 Compressible Flows at Radial Equilibrium

In an attempt to obtain a simple, and possibly closed-form, solution, several simplifying hypotheses are taken into account. A steady flow of a single-phase monocomponent fluid is investigated, under the assumptions of null heat transfer and viscous effects. The flow is therefore homototalenthalpic (h_t = const. $= \bar{h}_t$) and homoentropic (s = const. $= \bar{s}$).

The compressible Euler equations in polar coordinates are taken as starting point for the derivation of the radial equilibrium equation. Exploiting the above hypotheses, the continuity and momentum equations lead to the intuitive definition of the pressure gradient established due to the curvature,

$$\frac{dP}{dr} = \rho\frac{u_\theta^2}{r} = \rho\frac{u^2}{r} \tag{2}$$

where r represents the radial coordinate and $u_\theta = u$ is the flow velocity, pointing in the θ-direction, already shown in Fig. 1. Through the definitions of the speed of sound c and the Mach number $M = u/c$, an equivalent expression for the density gradient can be obtained:

$$\frac{d\rho}{dr} = \left(\frac{\partial\rho}{\partial P}\right)_s \frac{dP}{dr} = \frac{1}{c^2}\frac{dP}{dr} = \rho\frac{M^2}{r}. \tag{3}$$

The specification of a thermodynamic model finally yields the expression for the Mach variation along the radius. The most general formulation is achieved exploiting sophisticated models able to describe gases behaviour even in non-ideal conditions. In this context, a non-dimensional measure of the Mach derivative with the density is introduced [4] as

$$J = \frac{\rho}{M}\frac{dM}{d\rho} = 1 - \Gamma - \frac{1}{M^2}. \tag{4}$$

The Mach number variation with the radius is therefore computed:

$$\frac{dM}{dr} = \frac{dM}{d\rho}\frac{d\rho}{dr} = \frac{M}{\rho}\left(1 - \Gamma - \frac{1}{M^2}\right)\rho\frac{M^2}{r} = \frac{M}{r}\left[(1-\Gamma)M^2 - 1\right]. \tag{5}$$

The above equation is then recast in non-dimensional form by defining a dimensionless radial coordinate $\tilde{r} = r/r_i$, where r_i represents the internal radius of the channel where the radial equilibrium is established.

The final expression reads

$$\frac{dM}{d\tilde{r}} = -\frac{M}{\tilde{r}}\left[1 + (\Gamma - 1)M^2\right]. \tag{6}$$

The possibility to write the radial equilibrium equation as a function of $\tilde{r}$ proves that the same solution is valid for all possible values of the internal radius of curvature and the only parameter that matters is the width of the curve, *i.e.* the external radius r_e. Equation (6) can eventually be modified exploiting Eq. (4), yielding

$$\frac{dM}{d\tilde{r}} = \frac{M^3}{\tilde{r}}J. \tag{7}$$

Therefore, for thermodynamic conditions featuring negative values of J, the Mach number decreases towards the external radius. By contrast, M increases towards $\tilde{r}_e$ when $J > 0$.

Starting from Eq. (6), an easier expression, valid in the dilute-gas regime, can be derived. For an ideal polytropic gas, the fundamental derivative of gas dynamics simply reduces to the constant value $\Gamma = (\gamma + 1)/2 > 1$, being $\gamma > 1$ the polytropic exponent. Thus, the radial equilibrium equation for an ideal gas reads

$$\frac{dM}{d\tilde{r}} = -\frac{M}{\tilde{r}}\left(1 + \frac{\gamma - 1}{2}M^2\right), \tag{8}$$

where γ appears as the only fluid-dependent parameter.

The above equation can be integrated analytically yielding

$$M(\tilde{r}) = \frac{M_i}{\sqrt{\left(1 + \frac{\gamma-1}{2}M_i^2\right)\tilde{r}^2 - \frac{\gamma-1}{2}M_i^2}}, \tag{9}$$

where M_i is the Mach number at the internal radius $\tilde{r}_i$, by definition equal to 1, chosen as initial condition for the integration. Varying M_i, for a selected fluid, all possible radial equilibrium solutions are computed, for different mass flow rates.

An analytical integration is instead not possible in non-ideal conditions, *i.e.* of Eq. (6), since the value of Γ depends on the specific thermodynamic state. Within this work, therefore, the Dormand-Prince method [6], belonging to the Runge-Kutta family of ODE solvers, has been selected for the integration of Eq. (3), rewritten in non-dimensional form as

$$\frac{d\rho}{d\tilde{r}} = \rho\frac{M^2}{\tilde{r}}. \tag{10}$$

The integration of Eq. (6) is indeed numerically less stable than the integration of the equation above, since the Mach number profile can be non-monotone with the radius, whereas the density always increases towards the external radius.

Being the entropy constant and equal to $\bar{s}$, the differential equation is written as an ODE, expressing the Mach number as a function of the density only as

$$M = \frac{u}{c} = \frac{\sqrt{2(\bar{h}_t - h(\rho, \bar{s}))}}{c(\rho, \bar{s})} = M(\rho), \tag{11}$$

where the definition of the total enthalpy $\bar{h}_t$ has been used and $h(\rho, \bar{s})$ and $c(\rho, \bar{s})$ are computed from the REFPROP library [10], which contains state-of-the-art Equations of State. The density at the internal radius ρ_i, computed from M_i, is imposed as initial condition and the integration proceeds for increasing values of the radius. The Mach profile $M(\tilde{r})$ is finally recovered from $\rho(\tilde{r})$ exploiting Eq. (11).

4 Results

The radial equilibrium profiles, computed as previously explained, are now presented for ideal and non-ideal conditions. Suitable fluids and thermodynamic states are selected, aiming to show their influence on the solution.

4.1 Ideal Gas

Diatomic nitrogen N_2 and carbon dioxide CO_2 are compared in the context of ideal polytropic gases. They feature different values of the polytropic exponent, namely $\gamma = 1.4$ for N_2 and $\gamma = 1.29$ for CO_2. Four values of M_i (0.5, 1, 1.5, 2) are taken into account. Figure 3 shows the solutions for a channel with external radius $r_e = 5\,r_i$.

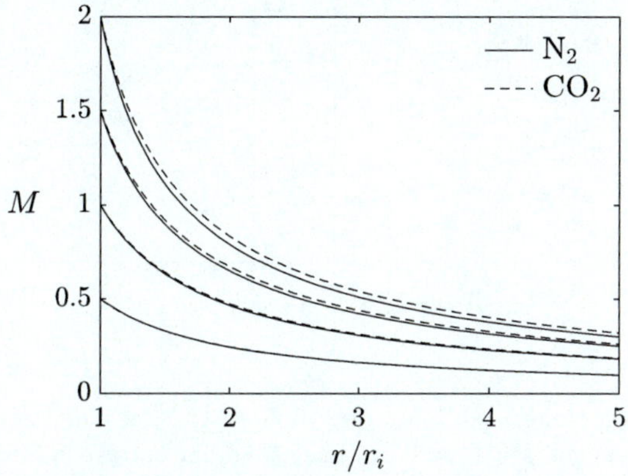

Fig. 3. Mach distribution along $\tilde{r} = r/r_i$ for an ideal flow at radial equilibrium. Comparison between N_2 and CO_2, with different values of M_i.

In all cases, the Mach number is larger at the internal radius and reduces monotonically towards the external one. The decrease is much faster in the inner part of the channel, in particular when supersonic conditions are imposed at $\tilde{r}_i$. For low Mach number flows, the γ-dependence is negligible, since the two curves are almost overlapping, whereas their difference increases with M_i. This is due to the fact that, reducing the Mach number, the compressibility of the flow also decreases, so that thermodynamics, here represented by the polytropic exponent, has less and less influence on the solution.

4.2 Non-ideal Gas

The dynamic behaviour of compressible flows in non-ideal conditions strongly depends on the fluid molecular complexity. Carbon dioxide is again considered as low molecular complexity fluid, for which only a quantitative departure from the ideal-gas behaviour is expected. A qualitatively different result is instead forecasted for the siloxanes MM (hexamethyldisiloxane, $C_6H_{18}OSi_2$), representative of high molecular complexity fluids, and D6 (dodecamethylcyclohexasiloxane, $C_{12}H_{36}O_6Si_6$), which is a BZT fluid. Moreover, in non-ideal conditions, the solution is determined also by total thermodynamic states, which, by contrast, have no influence in the ideal, dilute-gas regime. At this purpose, two values of the total pressure are tested, namely $P_t = 0.5\,P_c$ and $P_t = 2\,P_c$, being P_c the critical pressure, and, for each of them, four values of the total temperature T_t.

Figure 4 shows the radial equilibrium profiles obtained for CO_2. The Mach number at the internal radius is set to $M_i = 0.5$, since larger values could lead to entry the two-phase region, for the total states considered.

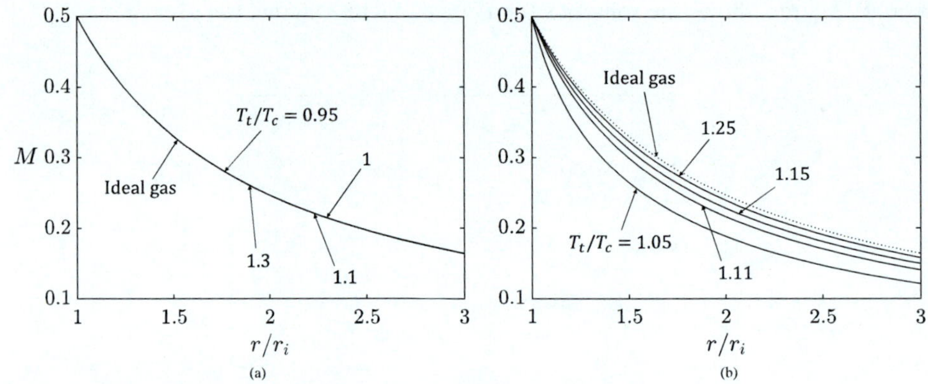

Fig. 4. Mach distribution along $\tilde{r} = r/r_i$, for a flow of CO_2 at radial equilibrium, with total pressure $P_t/P_c = 0.5$ (a) and $P_t/P_c = 2$ (b). Each solid line corresponds to a different value of the total temperature T_t, while dotted lines are obtained from the CO_2 ideal-gas model.

The ideal-gas solution is also superimposed for a direct comparison. With low total pressure, *i.e.* $P_t = 0.5\,P_c$, all the Mach profiles are essentially coincident with the ideal-gas solution, even for thermodynamic states very close to saturation. Considering instead $P_t = 2\,P_c$, the curves deviate more from the ideal one, in particular for low values of T_t, which lead to thermodynamic states closer to the critical point. Anyway, the ideal-gas-like behaviour is qualitatively retrieved, with the Mach number monotonically decreasing with the radius.

Dealing with MM, for a better comprehension of the thermodynamic states involved, the P-v diagram is also reported in Fig. 5, with the axes non-dimensionalized by means of critical pressure P_c and volume v_c. The presence of a thermodynamic region featuring non-ideal gasdynamic behaviour is clearly shown by the Γ isolines.

For the study of the radial equilibrium, a larger Mach number at the internal radius ($M_i = 1.75$) is considered, since the non-ideal behaviour appears in supersonic conditions. The Mach evolution along the radius is shown in Fig. 6, again superimposed to the ideal solution. The latter is found computing the polytropic exponent in the ideal-gas limit as

$$\gamma = \frac{c_p\,(T_c,\, P \to 0)}{c_v\,(T_c,\, P \to 0)}, \tag{12}$$

being c_p and c_v respectively the constant pressure and volume specific heats.

With total pressure $P_t = 0.5\,P_c$, all the states involved display values of the fundamental derivative of gas dynamics lower than 1. Nevertheless, the Mach profile deviates only quantitatively from the ideal model, with a departure which increases approaching the saturation curve. The non-monotonicity of the Mach number appears instead when the total pressure $P_t = 2\,P_c$ is considered. In this case, in fact, it is possible to reach states featuring lower values of Γ. Considering low total temperatures T_t, in the supersonic regime, a Mach number increase with the radius is observable. With a larger T_t, however, this behaviour disappears and the radial equilibrium profile approaches the ideal one.

Finally, for D6, the main result consists in the possibility to achieve a non-monotone Mach variation with the radius also in subsonic conditions. This is feasible due to the presence of thermodynamic states with negative values of Γ. A thermodynamic diagram displaying the Mach number evolution as a function of the density along several isentropes is reported in Fig. 7. A small region presenting values of $J > 0$ in subsonic conditions is indeed found.

The total conditions previously analyzed for the other fluids lead to results close to the ones shown in Fig. 6 for MM. Therefore, suitable total states, namely $P_t/P_c = 1.1171$ and $T_t/T_c = 1.0094$, are chosen to compute the Mach number profile presented in Fig. 8, which clearly shows the non-classical behaviour typical of BZT fluids.

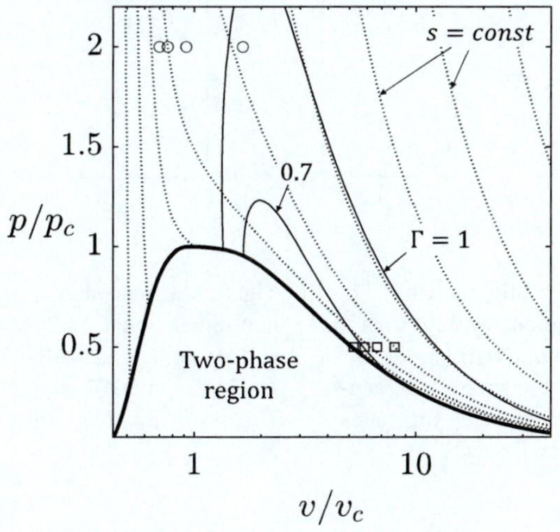

Fig. 5. Thermodynamic diagram for MM showing the states used for the radial equilibrium study: □ total conditions for Fig. 6a; ○ for Fig. 6b. Solid lines represent Γ isolines, while dotted lines are isentropes.

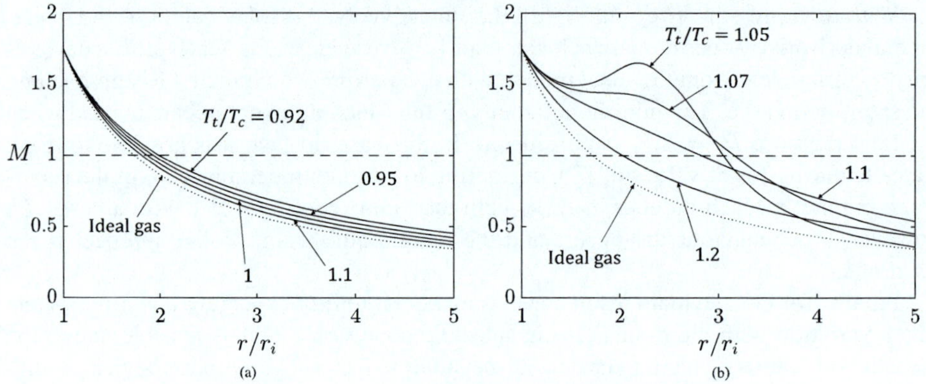

Fig. 6. Mach distribution along $\tilde{r} = r/r_i$, for a flow of MM at radial equilibrium, with total pressure $P_t/P_c = 0.5$ (a) and $P_t/P_c = 2$ (b). Each solid line corresponds to a different value of the total temperature T_t, while dotted lines are obtained from the MM ideal-gas model.

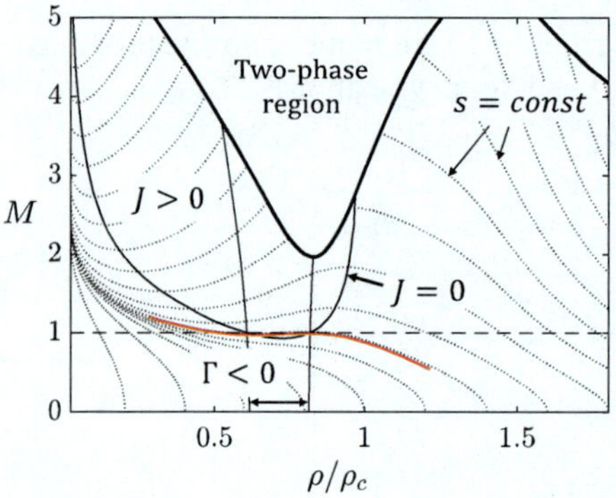

Fig. 7. Thermodynamic diagram for D6 showing the Mach number evolution as a function of ρ/ρ_c, obtained fixing $h_t = h(1.02\,P_c, v_c)$. Dotted lines represent isentropes. The red curve represents the states involved in Fig. 8.

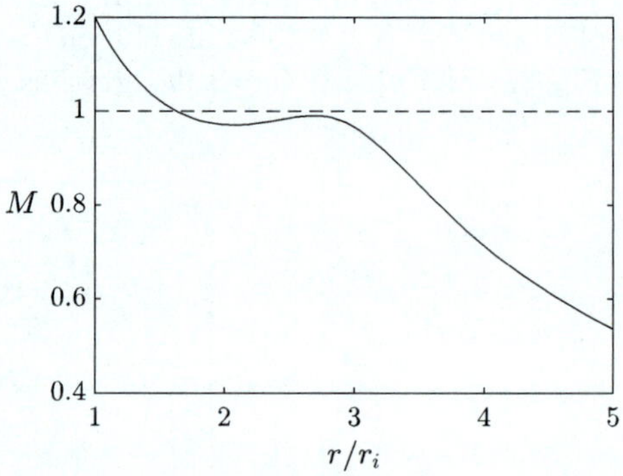

Fig. 8. Mach number distribution along the non-dimensional radius $\tilde{r} = r/r_i$, for a flow of D6 at radial equilibrium with $P_t/P_c = 1.1171$ and $T_t/T_c = 1.0094$. The states involved are also shown in red in Fig. 7

5 Conclusions

An original formulation for the Mach number dependency on the radius of curvature has been derived for compressible flows at radial equilibrium. The ordinary differential equation has been derived for a generic fluid, through the definition of the fundamental derivative of gas dynamics Γ, and then specialized for a polytropic ideal gas. In the latter case, the equation has been integrated analytically, obtaining a monotonically decreasing profile of the Mach number with the radius.

For thermodynamic states close to the liquid-vapour saturation curve and the critical point, however, the gas dynamic behaviour departs from the ideal-gas solution. Low molecular complexity fluids qualitatively behave as ideal gases and only a quantitative departure has been found. By contrast, high molecular complexity fluids can exhibit, in supersonic conditions, a non-monotone evolution of the Mach number with the radius. For BZT fluids, the non-monotone behaviour is also possible in subsonic regime.

These kinds of solutions, which formally represent flows within an annulus, are not of immediate practical applicability and they are of mainly theoretical interest. However, they can be a starting point for understanding and studying more complex flows as they can be an indication of the behaviour of ideal and non-ideal gases experiencing curvature effects.

References

1. Colonna, P., Guardone, A.: Molecular interpretation of nonclassical gas dynamics of dense vapors under the van der Waals model. Phys. Fluids **18**(5), 056101 (2006)
2. Colonna, P., Harinck, J., Rebay, S., Guardone, A.: Real-gas effects in organic Rankine cycle turbine nozzles. J. Propul. Power **24**(2), 282–294 (2008)
3. Colonna, P., van der Stelt, T., Guardone, A.: Fluidprop (version 3.0): a program for the estimation of thermophysical properties of fluids. Asimptote, Delft, The Netherlands (2012). http://www.fluidprop.com
4. Cramer, M., Best, L.: Steady, isentropic flows of dense gases. Phys. Fluids A Fluid Dyn. **3**(1), 219–226 (1991)
5. Cramer, M., Crickenberger, A.: Prandtl-Meyer function for dense gases. AIAA J. **30**(2), 561–564 (1992)
6. Dormand, J.R., Prince, P.J.: A family of embedded Runge-Kutta formulae. J. Comput. Appl. Math. **6**(1), 19–26 (1980)
7. Fergason, S., Guardone, A., Argrow, B.: Construction and validation of a dense gas shock tube. J. Thermophys. Heat Transfer **17**(3), 326–333 (2003)
8. Harinck, J., Guardone, A., Colonna, P.: The influence of molecular complexity on expanding flows of ideal and dense gases. Phys. Fluids **21**(8), 086101 (2009)
9. Klonowicz, P., Lampart, P., et al.: Optimization of an axial turbine for a small scale orc waste heat recovery system. Energy **205**, 118059 (2020)
10. Lemmon, E., Huber, M.L., McLinden, M.O., et al.: NIST standard reference database 23: reference fluid thermodynamic and transport properties-REFPROP, version 9.1 (2007)
11. Mathijssen, T., et al.: The flexible asymmetric shock tube (fast): a Ludwieg tube facility for wave propagation measurements in high-temperature Vapours of organic fluids. Exp. Fluids **56**(10), 1–12 (2015)
12. Menikoff, R., Plohr, B.J.: The Riemann problem for fluid flow of real materials. Rev. Mod. Phys. **61**(1), 75 (1989)
13. Romei, A., Vimercati, D., Persico, G., Guardone, A.: Non-ideal compressible flows in supersonic turbine cascades. J. Fluid Mech. **882**, A12 (2020)
14. Thompson, P.A.: A fundamental derivative in gasdynamics. Phys. Fluids **14**(9), 1843–1849 (1971)
15. Thompson, P.A., Lambrakis, K.: Negative shock waves. J. Fluid Mech. **60**(1), 187–208 (1973)
16. Vimercati, D., Gori, G., Guardone, A.: Non-ideal oblique shock waves. J. Fluid Mech. **847**, 266–285 (2018)

Data-Driven Regression of Thermodynamic Models in Entropic Form

Matteo Pini(✉), Andrea Giuffre', Alessandro Cappiello, Matteo Majer, and Evert Bunschoten

Propulsion and Power, Delft University of Technology, Delft, The Netherlands
m.pini@tudelft.nl

Abstract. Modeling non-ideal compressible flows in the context of computational fluid-dynamics (CFD) requires the calculation of thermodynamic state properties at each step of the iterative solution process. To this purpose, the use of a built-in fundamental equation of state (EoS) in entropic form, i.e., $s = s(e, \rho)$, can be particularly cost-effective, as all state properties can be explicitly calculated from the conservative variables of the flow solver. This approach can be especially advantageous for massively parallel computations, in which look-up table (LuT) methods can become prohibitively expensive in terms of memory usage. The goal of this research is to: i) develop a fundamental relation based on the entropy potential; ii) create a data-driven model of entropy and its first and second-order derivatives, expressed as a function of density and internal energy; iii) test the performance of the data-driven thermodynamic model on a CFD case study. Notably, two Multi-Layer Perceptron (MLP) models are trained on a synthetic dataset comprising 500k thermodynamic state points, obtained by means of the Span-Wagner EoS. The thermodynamic properties are calculated by differentiating the fundamental equation, thus ensuring thermodynamic consistency. Conversely, thermodynamic stability is properly enforced during the regression process. Albeit the method is applicable to the development of equation of state models for arbitrary fluids and thermodynamic conditions, the present work only considers siloxane MM in the single phase region. The MLP model is implemented in the open-source SU2 software [8] and is used for the numerical simulation of non-ideal compressible flows in a planar converging-diverging nozzle. Finally, the accuracy and the computational performance of the data-driven thermodynamic model are assessed by comparing the resulting flow field, the wall time and the memory requirements with those obtained with direct calls to a cubic EoS, and with a LuT method.

Keywords: data-driven modeling · thermodynamic properties · computational fluid dynamics

1 Introduction

Flows of fluids not obeying to the perfect or the ideal gas law are encountered in many relevant engineering applications. In the field of propulsion and power, examples of these applications include unconventional turbines and compressors operating

M. White et al. (Eds.): NICFD 2022, ERCOFTAC 29, pp. 22–32, 2023.
https://doi.org/10.1007/978-3-031-30936-6_3

with dense vapors for power generation and thermal management systems [10], fuel injectors of gas turbines and regenerative cooling nozzles of rocket engines [1,2], in which the fluid is at supercritical conditions, and compact heat-exchangers for waste heat recovery systems, whereby the fluid can be either a non-ideal compressible liquid or a supercritical gas [4].

The accurate modeling of the thermodynamic properties of non-ideal fluid flows is usually obtained by resorting to multi-parameter equations of state (MEoS) in the Helmholtz free-energy form [5,21]. Models in this form for a variety of pure fluids and mixtures are implemented in popular thermodynamic software packages like RefProp [13], FluidProp [6], and CoolProp [3]. However, in the context of CFD, the computational overhead associated with direct calls to MEoS is typically excessive, especially if design optimization [19] or high-fidelity analyses [11,17] are involved.

Methods for the efficient calculation of thermodynamic properties can be divided in two classes: methods based on the interpolation of tabulated property values, commonly referred to as look-up table methods, and data-driven methods, in which thermodynamic models are constructed through a proper fitting of the fluid property values obtained with arbitrarily complex EoS. For instance, in [18], consistent thermodynamic models in the Helmholtz free-energy form satisfying the Maxwell relations were trained using an Artificial Neural Network (ANN), showing high-accuracy in the predicted fluid properties. Similarly, in [22], the authors compared the performance obtained by training a Multi-Layer Perceptron (MLP) and a Gaussian Process (GP) to replicate the statistical associating fluid theory (SAFT-VR) EoS for pure fluids, and concluded that the ANN outperforms the GP model.

For CFD simulations, LuT methods have been proven to be efficient and accurate, but data-driven methods provide advantages in terms of robustness and reduction of memory usage [12], thus making them particularly suitable for massively parallel computations. Longmire and Banuti [14,15] exploited tiny neural networks to compute the thermodynamic properties of CO_2 as a function of temperature, in the supercritical region. The objective was to perform viscous simulations of non-ideal flows in non-adiabatic laminary boundary layers, using the incompressible Navier-Stokes solver of SU2 [7]. Results show that the computational overhead while using the ANN model was about 20% higher compared to when using constant properties. In [16], the authors adopted an ANN to compute the thermodynamic and transport properties of multi-component mixtures for large-eddy simulations of turbulent mixing flows, and one-dimensional simulations of diffusive flames. As outcome of their study, the authors measured a speed-up of 1.5 and 2.3 times, respectively, compared to the baseline cases. Moreover, the authors stated that, for test cases involving many chemical species, the use of ANNs enables a reduction of memory usage of up to 5 orders of magnitude with respect to LuT methods. However, the data-driven regression technique described in the paper heavily relies on the injection of boundary information into the training dataset, making it application-specific. In addition, the average prediction errors for the primitive variables reported in the paper are in the range of 1–9%. This level of accuracy could lead to large discrepancies in the prediction of local flow phenomena involving large spatial gradients, e.g., shock-waves.

The objective of this work is the development of an accurate data-driven equation of state, tailored to compressible flow simulations of pure fluids and mixtures of fixed composition. The model is formulated in terms of entropy potential, i.e., $s = s(e, \rho)$. The desired level of accuracy and computational efficiency is attained by means of a semi-consistent formulation: a MLP model is trained on a dataset comprising 500k values of entropy and its first and second-order derivatives with respect to e and ρ, calculated with the Span-Wagner multi-parameter EoS implemented in [13]. It follows that, for a given thermodynamic state, s and its derivatives are computed with the MLP, while the primary and secondary thermodynamic properties are explicitly computed using their analytical definition based on the entropy potential. The semi-consistent MLP model has been implemented within the compressible Navier-Stokes solver of SU2 [8], and the resulting numerical model has been applied to perform simulations of non-ideal flows of siloxane MM in a converging-diverging nozzle.

2 Methodology

2.1 Fundamental Relation in Entropic Form

A thermodynamic model based on the entropy potential is highly suited for CFD applications featuring non-ideal flows, since its natural variables, i.e., e and ρ, are those used to retrieve the fluid properties in a flow solver. The proposed thermodynamic model is of general validity, i.e., it is applicable to any pure fluid and mixture of fixed composition, regardless of the thermodynamic state, both in the single and in the two-phase region. However, in the present work, its use is restricted to siloxane MM in the single phase region. Equations (1) to (4) provide the analytical expressions of the primary thermodynamic properties used by CFD compressible flow solvers, namely temperature, T, pressure, p, enthalpy, h, and speed of sound, c. Such relations have been derived from fundamental thermodynamic equations, by applying mathematical rules of multivariable differential calculus.

$$T = \left(\frac{\partial s}{\partial e}\right)_\rho^{-1} \quad (1) \qquad p = -\rho^2\, T \left(\frac{\partial s}{\partial \rho}\right)_e \quad (2) \qquad h = e - \rho \left(\frac{\partial s}{\partial \rho}\right)_e \left(\frac{\partial s}{\partial e}\right)_\rho^{-1} \quad (3)$$

$$\begin{aligned} c^2 = \left(\frac{\partial p}{\partial \rho}\right)_e - \left(\frac{\partial p}{\partial e}\right)_\rho \left(\frac{\partial s}{\partial \rho}\right)_e \left(\frac{\partial s}{\partial e}\right)_\rho^{-1} = \\ -\rho \left(\frac{\partial s}{\partial e}\right)_\rho^{-1} \cdot \left\{ \left(\frac{\partial s}{\partial \rho}\right)_e \cdot \left[2 - \rho \left(\frac{\partial s}{\partial e}\right)_\rho^{-1} \cdot \left(\frac{\partial^2 s}{\partial \rho \partial e}\right)\right] + \rho \left(\frac{\partial^2 s}{\partial \rho^2}\right) + \right. \\ \left. -\rho \left[- \left(\frac{\partial s}{\partial e}\right)_\rho^{-1} \cdot \left(\frac{\partial^2 s}{\partial e^2}\right) \cdot \left(\frac{\partial s}{\partial \rho}\right)_e + \left(\frac{\partial^2 s}{\partial \rho \partial e}\right)\right] \cdot \left(\frac{\partial s}{\partial \rho}\right)_e \cdot \left(\frac{\partial s}{\partial e}\right)_\rho^{-1} \right\} \end{aligned} \quad (4)$$

Secondary thermodynamic variables, which are needed to compute inviscid and viscous fluxes, and Jacobians, namely, $\left(\frac{\partial T}{\partial e}\right)_\rho$, $\left(\frac{\partial T}{\partial \rho}\right)_e$, $\left(\frac{\partial p}{\partial e}\right)_\rho$, $\left(\frac{\partial p}{\partial \rho}\right)_e$, are obtained by partial differentiation of Eqs. (1, 2), and are given by Eqs. (5) to (8).

$$\left(\frac{\partial T}{\partial e}\right)_\rho = -\left(\frac{\partial s}{\partial e}\right)_\rho^{-2} \cdot \left(\frac{\partial^2 s}{\partial e^2}\right) \quad (5) \qquad \left(\frac{\partial T}{\partial \rho}\right)_e = -\left(\frac{\partial s}{\partial e}\right)_\rho^{-2} \cdot \left(\frac{\partial^2 s}{\partial \rho \partial e}\right) \quad (6)$$

$$\left(\frac{\partial p}{\partial e}\right)_\rho = -\rho^2 \left(\frac{\partial s}{\partial e}\right)_\rho^{-1} \cdot \left\{-\left(\frac{\partial s}{\partial e}\right)_\rho^{-1} \cdot \left(\frac{\partial^2 s}{\partial e^2}\right) \cdot \left(\frac{\partial s}{\partial \rho}\right)_e + \left(\frac{\partial^2 s}{\partial \rho \partial e}\right)\right\} \tag{7}$$

$$\left(\frac{\partial p}{\partial \rho}\right)_e = -\rho \left(\frac{\partial s}{\partial e}\right)_\rho^{-1} \cdot \left\{\left(\frac{\partial s}{\partial \rho}\right)_e \cdot \left[2 - \rho \left(\frac{\partial s}{\partial e}\right)_\rho^{-1} \cdot \left(\frac{\partial^2 s}{\partial \rho \partial e}\right)\right] + \rho \left(\frac{\partial^2 s}{\partial \rho^2}\right)\right\} \tag{8}$$

Therefore, given the values of internal energy and density, all the necessary thermodynamic properties can be computed by resorting to the entropy potential and its first and second-order partial derivatives. Thermodynamic consistency is thus inherently guaranteed. Moreover, at each iteration of the flow solver, the thermodynamic state can be updated by means of explicit calculations, differently from what happens with other thermodynamic potentials, e.g., Helmholtz free energy, $a = a(T, \rho)$, for which the internal energy is retrieved by iteratively solving a non-linear equation.

The correctness of the analytical relations (4)–(8) has been assessed by comparing the thermodynamic property values obtained computing each right-hand side term of the equations through the CoolProp library [3] against those calculated by directly computing T, p, h, c^2, $\left(\frac{\partial T}{\partial e}\right)_\rho$, $\left(\frac{\partial T}{\partial \rho}\right)_e$, $\left(\frac{\partial p}{\partial e}\right)_\rho$, $\left(\frac{\partial p}{\partial \rho}\right)_e$ via direct EoS calls. For all properties, the deviations were, on average, in the order of $\mathcal{O}(10^{-13})$.

2.2 Data-Driven Modeling of Thermodynamic Properties

The development of the data-driven EoS model for the prescribed case study, i.e., siloxane MM in the vapor and supercritical regions, is documented in this section. A MLP, i.e., a feedforward ANN featuring one or multiple fully connected hidden layers, has been selected to accomplish this task. The reason is twofold. First, the computational overhead associated with the evaluation of the data-driven model is of primary importance for the target application. In general terms, the computational cost of a MLP model scales with the total number of neurons, and is lower than the cost of evaluating, for instance, a model based on a GP or a Support Vector Machine (SVM). On the other hand, an MLP model requires a larger amount of training data to reach the same level of accuracy of a GP or a SVM model. However, in the present work the dataset is synthetically generated by resorting to the MEoS implemented in [13], thus the availability of data is not a limiting factor.

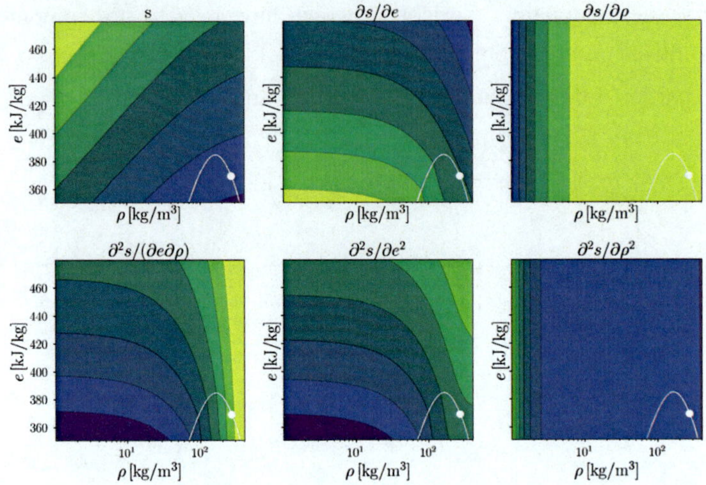

Fig. 1. Trends of the entropy potential, its first and its second-order derivatives with respect to density and internal energy. The white line identifies the saturation curve, while the white dot corresponds to the critical point. The range of density has been extended up to $405\ \mathrm{kg/m^3}$ for visualization purposes.

The ranges of variation of the input features, i.e., density and internal energy, selected to create the dataset are ρ = 1.0–295 $\mathrm{kgm^{-3}}$ and e = 350.6–479.8 $\mathrm{kJkg^{-1}}$, respectively. The trends of the labels, i.e., entropy potential and its first and second-order derivatives, are displayed in Fig. 1. Two observations can be drawn from such trends. On the one hand, the first and second-order partial derivatives of s with respect to ρ are characterized by large gradients at low values of density. As a result, training a MLP model over a wide range of density requires the use of a deeper network architecture, featuring a higher neuron count, thus an increased computational cost. To overcome this issue, while retaining a high level of accuracy, the dataset has been split in two parts, i.e., $\rho \leq 10\ \mathrm{kgm^{-3}}$, $\rho > 10\ \mathrm{kgm^{-3}}$, and a different MLP model has been trained over each portion of the dataset. On the other hand, the labels do not show large discontinuities across the saturation curve. As a consequence, training a MLP model over a dataset comprising both the single and the two-phase region should not involve additional complexity. Nevertheless, the development of data-driven EoS models for thermodynamic processes within the two-phase region is out of the scope of the present work.

As outcome of a sensitivity analysis, a dataset comprising 250k samples for each portion of the thermodynamic plane, i.e., $\rho \leq 10\ \mathrm{kgm^{-3}}$, $\rho > 10\ \mathrm{kgm^{-3}}$, has been selected as optimal trade-off between prediction accuracy and computational cost associated to training. To improve the performance of the MLP model, input features normalization and logarithmic transformation of the labels characterized by a highly skewed distribution, i.e., $\partial s/\partial \rho$ and $\partial^2 s/\partial \rho^2$, has been applied prior to training. Upon pre-processing, the dataset has been split into training, development, and test sets, counting 450k, 25k, and 25k samples, respectively.

In order to enhance the performance of the MLP models, a search for the optimal set of hyper-parameters was conducted by resorting to the latin-hypercube sampling, featuring 100 samples for each portion of the dataset. The design space comprises: the number of layers L, the number of neurons per layer $n^{[l]}$, the activation function, the learning rate α, and the mini-batch size. The performance of the different MLP architectures is measured in terms of computational cost at evaluation time $C(\Theta)$ and accuracy, i.e., loss $L(\Theta) = \frac{1}{n}\sum_{i=1}^{n}(\hat{y}_i - y_i)^2$ evaluated on the development set at the last epoch of training. The results are shown in Fig. 2. In the figures, the three different MLP architectures used for the CFD simulations are highlighted. The code used to train, test, and optimize the MLP architectures can be downloaded at[1].

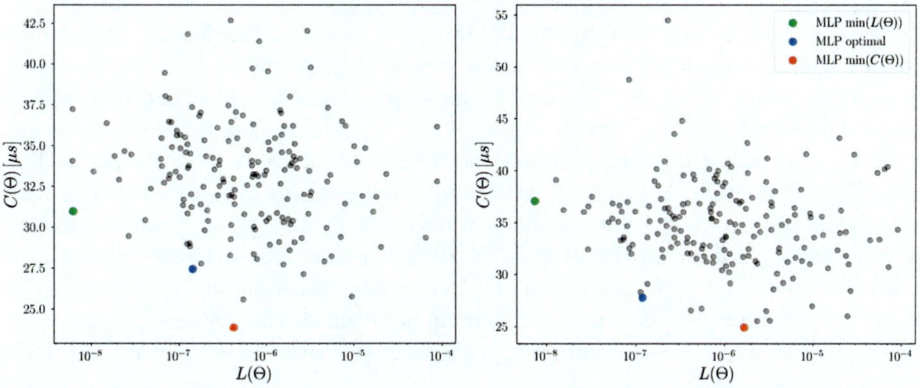

Fig. 2. Computational cost at evaluation time vs. loss evaluated on the development set for the MLP architectures selected for the hyper-parameters search. The results obtained for $\rho \leq 10$ kgm^{-3} are shown on the left, while the performance of the MLPs trained over the portion of the dataset featuring $\rho > 10$ kgm^{-3} are displayed on the right. The colored points highlight the MLP models selected for the CFD simulations.

Table 1. Set of hyper-parameters associated to the MLP architectures selected for the CFD analysis. For each hyper-parameter, the first entry corresponds to the model trained on the $\rho \leq 10$ kgm^{-3} dataset, whereas the second entry refers to the model trained on the $\rho > 10$ kgm^{-3} dataset.

Label	L	$n^{[1]}$	$n^{[2]}$	$n^{[3]}$	α	Activation	Mini-batch size
MLP optimal	2–2	90–19	16–51	0–0	$10^{-3.111}$–$10^{-4.019}$	swish - tanh	128–32
MLP min($C(\Theta)$)	2–2	58–49	20–44	0–0	$10^{-4.737}$–$10^{-3.359}$	tanh - selu	32–512
MLP min($L(\Theta)$)	3–3	90–29	50–70	9–66	$10^{-4.072}$–$10^{-3.844}$	sigmoid - gelu	32–16

3 Results

The computational performance of the MLP model was evaluated against direct calls to the MEoS implemented in [13]. The computational cost associated to the evaluation

[1] https://github.com/Propulsion-Power-TU-Delft/Deep.

of the thermodynamic state of siloxane MM on 250k samples of density and internal energy was measured for both the MLP model and the low-level C++ interface of CoolProp [3]. With reference to Fig. 2 and Table 1, the MLP architecture selected for this analysis was the MLP optimal. The benchmark has been performed on a workstation equipped with an 11th Gen Intel(R) Core(TM) i7-11700KF processor, featuring 8 cores and 16 threads, 32 GB of RAM, and a clock speed of 3.60 GHz. The evaluation of the entire dataset required 3.992×10^3 ms for the MLP, and 1.125×10^3 ms when resorting to the low-level interface of CoolProp. In other words, the data-driven model is approximately 2.8 times faster than the low-level interface of CoolProp.

3.1 CFD Analysis

In order to further assess the accuracy and the performance of the data-driven thermodynamic model, inviscid simulations of the supersonic flow within a planar De-Laval nozzle have been carried out. The converging-diverging nozzle has been designed to operate with siloxane MM as working fluid, at inlet conditions close to the critical point. At the inlet boundary, the total pressure and temperature have been set to 18.423 bar and 525 K, respectively. At the outlet boundary, a back-pressure equal to 10 bar has been prescribed, such to obtain a normal shock-wave in the diverging section of the nozzle. The computational domain, featuring 108k triangular elements, has been created with the unstructured mesh generator Gmsh [9]. To improve the resolution at the shock front and the convergence rate, local mesh refinement has been applied, by reducing the local cell size by a factor of 3. All simulations have been performed in two steps:

1. A maximum of 10k iterations using the ROE upwind scheme, featuring first-order spatial accuracy, and a Courant-Friedrichs-Lewy number (CFL) equal to 5;
2. A maximum of 30k iterations using the central scheme JST, featuring second and fourth order polynomial coefficients equal to 0.5 and 0.02, respectively, and a unitary CFL number.

Two different thermodynamic models have been used to verify the results obtained with the data-driven EoS: i) the polytropic Peng-Robinson (PPR) cubic EoS implemented in SU2; ii) an unstructured LuT method [20], resorting to a grid of about 250k thermodynamic state points, computed with the MEoS implemented in [13]. Figure 3 shows the contour of the Mach number obtained with the optimal MLP architecture, alongside with the relative percentage deviation computed with respect to the solution obtained with the LuT method. The maximum and minimum relative deviations are located at the shock-wave front, and are in the order of 5%, whereas the average relative deviations computed for the Mach, density, and pressure fields are 0.47%, 0.05%, and 0.03%, respectively. To further assess the discrepancy between the flow solutions computed with the different thermodynamic models, the Mach number at the nozzle center-line is displayed in Fig. 4. The excellent agreement between the Mach trend obtained with the MLP, the PPR and the LuT methods corroborate the validity of the novel data-driven thermodynamic model.

The ultimate goal of this study is to assess the suitability of data-driven thermodynamic models for massively parallel CFD calculations, both in terms of memory requirements and computational cost. To this purpose, the memory usage and

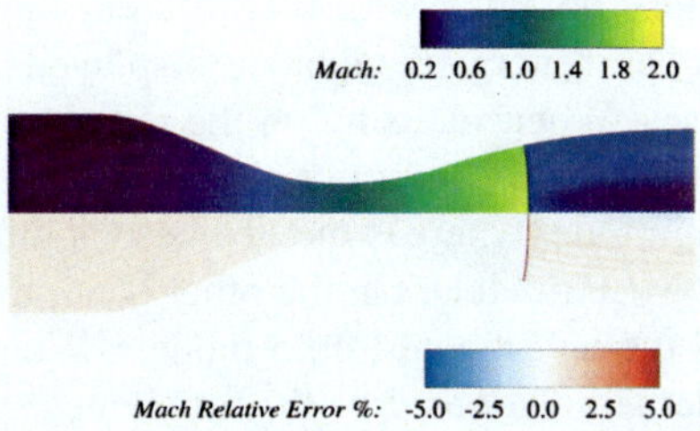

Fig. 3. Nozzle flow field. Upper half: Mach contours obtained with the optimal MLP architecture; Lower half: Mach number relative percentage deviation with respect to the *LuT* solution.

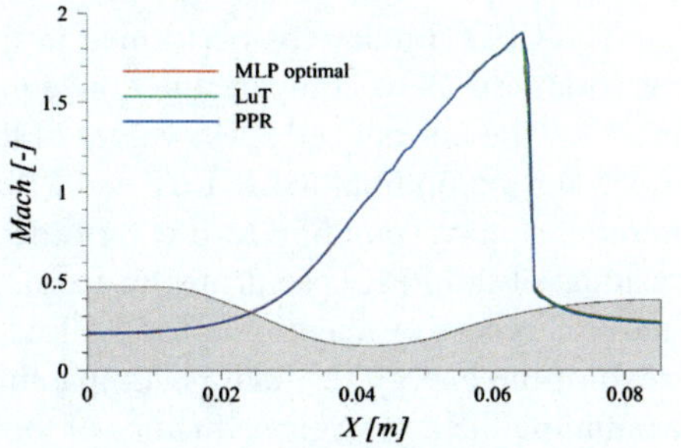

Fig. 4. Comparison of the Mach number trends computed along the nozzle center-line. The nozzle geometry is displayed in grey.

Table 2. Overview of the memory usage and the time/iteration obtained with the different fluid models.

Label	Memory [GB]	Time/Iteration [s]
LuT	9.0	2.754e−2
PPR	0.4	1.767e−2
MLP optimal	0.4	1.513e−1
MLP $\min(C(\Theta))$	0.4	1.882e−1
MLP $\min(L(\Theta))$	0.4	8.491e−1

the time per iteration of the flow solver have been recorded for each of the method described above. The benchmark provided in Table 2 has been conducted on a workstation equipped with an Intel(R) Xeon(R) Gold 5220R, featuring 24 cores, 48 threads, 256 GB of RAM, and a base and maximum clock speed of 2.2 and 4.0 GHz, respectively. The comparison shows a remarkable difference in terms of memory usage, with the data-driven model being comparable to the simpler cubic EoS, while the LuT method requires about 22.5 times more RAM. In terms of computational cost, the fastest of the MLP models is 5.5 times slower than the LuT method, and more than 8.5 times slower than the PPR. According to Fig. 2, the architecture labelled as MLP $\min(C(\Theta))$ should have provided the lowest computational cost. However, the model labelled as MLP optimal has been found to be the least computational intensive, when integrated within the flow solver. This discrepancy can be attributed to the sub-optimal implementation of the MLP model within SU2, as compared to the one featured in the state-of-the-art platform for machine learning that has been used to create such models, i.e., TensorFlow. The optimization of the MLP implementation within SU2 will be the objective of future research, with the purpose of reducing the computational overhead as compared to the LuT method.

The CFD simulations performed in this study have been run for a fixed number of iterations, so as to compare the convergence history obtained with the different fluid models. The normalized RMS values of the density and energy residuals is displayed in Fig. 5 for the optimal MLP, LuT and PPR test cases. When focusing on the initial 10k iterations, corresponding to the 1st order simulations, one can notice that the density residual of the PPR case drops by 5 orders of magnitude, whereas the decrease is limited to 3 orders of magnitude for the LuT and the MLP models. On the other hand, the trend of the energy residual is comparable for all the investigated fluid models. When examining the convergence history of the 2nd order simulations, i.e., the following 30k iterations, the LuT test case shows a reduction of the normalized residuals of 1.5 orders of magnitude, followed by an oscillatory trend. Conversely, both the PPR and the MLP cases are characterized by a smooth trend of the residuals. Such behavior is expected, since both fluid models are based on a continuous mathematical function, which does not involve any interpolation. As a result, the higher computational overhead associated with the MLP model can be partially compensated by its superior convergence properties, as compared to LuT methods.

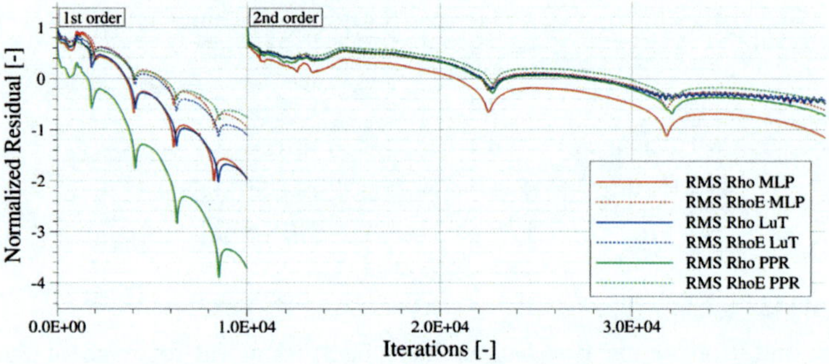

Fig. 5. Normalized density and energy residuals for the 1st and 2nd order simulations performed with the three investigated fluid models. The case featuring the PPR EoS reached the lowest absolute density residual.

4 Conclusions and Future Works

The research documented in this paper demonstrated that data-driven equations of state models based on the entropy potential are a valid alternative over look-up table (LuT) methods for non-ideal compressible flow simulations. Different artificial neural networks of the type of multi-layer perceptron (MLP) have been generated to compute the thermodynamic properties of siloxane MM in the vapor and supercritical region. The data-driven models have been implemented in the open-source SU2 software to perform a simulation of a shock-induced non-ideal flow within a planar converging-diverging nozzle. Comparisons between simulation results obtained with the MLP and with the LuT and a polytropic Peng-Robinson model have been made in terms of accuracy and computational performance. Results computed with MLP and LuT are in excellent quantitative agreement, but the CFD simulations carried out with the LuT are approximately 5 times faster than those run with the MLP. Conversely, the use of the MLP is 20 times less memory intensive, and shows better convergence properties. Future work will target the optimization of the MLP implementation within the flow solver of SU2 for massively parallel computations of non-ideal compressible flows in turbomachinery applications.

References

1. Banuti, D.: Why we need to care about supercritical and non-ideal injection. In: APS Division of Fluid Dynamics Meeting Abstracts, p. T01.008, AA. University of New Mexico, January 2021
2. Banuti, D.T.: A thermodynamic look at injection in aerospace propulsion systems. In: AIAA Scitech 2020 Forum, Number 0 in AIAA SciTech Forum. American Institute of Aeronautics and Astronautics, 15 July 2022
3. Bell, I.H., Wronski, J., Quoilin, S., Lemort, V.: Pure and pseudo-pure fluid thermophysical property evaluation and the open-source thermophysical property library CoolProp. Ind. Eng. Chem. Res. **53**(6), 2498–2508 (2014)
4. Brun, K., Friedman, P., Dennis, R.: Fundamentals and Applications of Supercritical Carbon Dioxide (sCO_2) Based Power Cycles. Woodhead Publishing (2017)
5. Colonna, P., Nannan, N.R., Guardone, A., Lemmon, E.W.: Multiparameter equations of state for selected siloxanes. Fluid Phase Equilib. **244**(2), 193–211 (2006)
6. Colonna, P., van der Stelt, T.P., Guardone, A.: FluidProp (version 3.0): a program for the estimation of thermophysical properties of fluids. Asimptote, Delft, The Netherlands (2012). http://www.fluidprop.com
7. Economon, T.D.: Simulation and adjoint-based design for variable density incompressible flows with heat transfer. AIAA J. **58**(2), 757–769 (2022)
8. Economon, T.D., Palacios, F., Copeland, S.R., Lukaczyk, T.W., Alonso, J.J.: SU2: an open-source suite for multiphysics simulation and design. AIAA J. **54**(3), 828–846 (2015). https://doi.org/10.2514/1.J053813
9. Geuzaine, C., Remacle, J.F.: Gmsh: a 3-D finite element mesh generator with built-in pre- and post-processing facilities. Int. J. Numer. Methods Eng. **79**(11), 1309–1331 (2009)
10. Giuffre, A., Colonna, P., Pini, M.: The effect of size and working fluid on the multi-objective design of high-speed centrifugal compressors. Int. J. Refrig **143**, 43–56 (2022)
11. Hoarau, J.-C., Cinnella, P., Gloerfelt, X.: Large eddy simulations of strongly non-ideal compressible flows through a transonic cascade. Energies **14**(3), 772 (2021)
12. Ihme, M., Schmitt, C., Pitsch, H.: Optimal artificial neural networks and tabulation methods for chemistry representation in LES of a bluff-body swirl-stabilized flame. Proc. Combust. Inst. **32**(1), 1527–1535 (2009)
13. Lemmon, E.W., Bell, I.H., Huber, M.L., McLinden, M.O.: NIST standard reference database 23: reference fluid thermodynamic and transport properties-REFPROP, version 10.0. National Institute of Standards and Technology (2018)
14. Longmire, N., Banuti, D.T.: Onset of heat transfer deterioration caused by pseudo-boiling in CO_2 laminar boundary layers. Int. J. Heat Mass Transf. **193**, 122957 (2022)
15. Longmire, N.P., Banuti, D.: Extension of SU2 using neural networks for thermo-fluids modeling. In: AIAA Propulsion and Energy 2021 Forum, Number 0 in AIAA Propulsion and Energy Forum. American Institute of Aeronautics and Astronautics, 10 July 2022
16. Milan, P.J., Hickey, J.-P., Wang, X., Yang, V.: Deep-learning accelerated calculation of real-fluid properties in numerical simulation of complex flowfields. J. Comput. Phys. **444**, 110567 (2021)

17. Przytarski, P.J., Wheeler, A.P.S.: Accurate prediction of loss using high fidelity methods. J. Turbomach. **143**(3), 031008 (2022)
18. Rosenberger, D., Barros, K., Germann, T.C., Lubbers, N.: Machine learning of consistent thermodynamic models using automatic differentiation. Phys. Rev. E **105**(4), 045301 (2022)
19. Rubino, A., Colonna, P., Pini, M.: Adjoint-based unsteady optimization of turbomachinery operating with nonideal compressible flows. J. Propul. Power **37**(6), 910–918 (2022)
20. Rubino, A., Pini, M., Kosec, M., Vitale, S., Colonna, P.: A look-up table method based on unstructured grids and its application to non-ideal compressible fluid dynamic simulations. J. Comput. Sci. **28**, 70–77 (2018)
21. Span, R., Wagner, W.: Equations of state for technical applications. II. Results for nonpolar fluids. Int. J. Thermophys. **24**(1), 41–109 (2003). https://doi.org/10.1023/A:1022310214958
22. Zhu, K., Müller, E.A.: Generating a machine-learned equation of state for fluid properties. J. Phys. Chem. B **124**(39), 8628–8639 (2020)

Numerical Modelling and Methods

Direct Numerical Simulation of Wall-Bounded Turbulence at High-Pressure Transcritical Conditions

Marc Bernades(✉), Francesco Capuano, and Lluís Jofre

Universitat Politècnica de Catalunya, Barcelona, Spain
{marc.bernades,francesco.capuano,lluis.jofre}@upc.edu

Abstract. Supercritical fluids are commonly utilized in energy conversion and propulsion applications due to the higher power and efficiencies they provide. Their increased performance is connected to the thermophysical properties they exhibit around the pseudo-boiling region, in which density is relatively large while transport coefficients are similar to those of a gas. Consequently, higher levels of turbulence intensity can be achieved, resulting in mixing and heat transfer enhancements with respect to fluids operating at atmospheric conditions. However, supercritical fluids turbulence is a research field still in its infancy, and, thus, requires to be carefully investigated and further characterized. In this regard, this work analyzes supercritical wall-bounded turbulence by computing direct numerical simulations of high-pressure N_2 at transcritical temperature conditions imposed by a temperature difference between the bottom and top walls.

Keywords: High pressure · Real-gas thermodynamics · Supercritical fluids · Transcritical conditions · Turbulence

1 Introduction

Supercritical fluids are used in a wide range of engineering applications, like for example in gas turbines, supercritical water-cooled reactors and liquid rocket engines. They operate within high-pressure thermodynamic spaces in which intermolecular forces and finite packing volume effects become important. In this regard, it is important to distinguish between supercritical gas-like and liquid-like fluids separated by the pseudo-boiling line [12,13]: (i) a supercritical liquid-like fluid is one whose density is large, and whose transport coefficients behave similar to a liquid; (ii) the density of supercritical gas-like fluids is smaller, and their transport coefficients vary similar to gases. This set of thermophysical characteristics presents very interesting properties that can be leveraged, for instance, to achieve turbulent regimes in microfluidic devices [4]. Generally, in most macroscale energy applications, for example related to power and heat transfer [13,16,31,32], turbulence is a key mechanism for achieving high levels of performance and efficiency due to the notable increase in mixing and transfer rates that it provides [28].

M. White et al. (Eds.): NICFD 2022, ERCOFTAC 29, pp. 35–42, 2023.
https://doi.org/10.1007/978-3-031-30936-6_4

The inherent physics complexity of supercritical fluids across the pseudo-boiling region typically leads to poor prediction capabilities of Reynolds-averaged Navier-Stokes (RANS) simulation approaches [32]. Consequently, the coupled hydrodynamic and thermal behavior of high-pressure transcritical turbulent flows, for example in terms of enhancement/deterioration of heat transfer, cannot be captured correctly. In this regard, scale-resolving methodologies based on direct numerical simulation (DNS) and large eddy simulation (LES) are typically required to properly characterize and/or predict high-pressure transcritical turbulent flows. Selected examples of scale-resolving studies include (i) DNS of supercritical heat transfer in pipe flows by Bae et al. [1,2], (ii) DNS of transcritical flows close to the critical point by Sengupta et al. [25], (iii) LES and DNS of transcritical channel flows by Doehring & Adams [9] and Ma et al. [19], respectively, and (iv) DNS of transcritical turbulent boundary layers by Kawai [14].

Nonetheless, scale-resolving computational studies of high-pressure transcritical flows are significantly challenging due to the appearance of spurious pressure oscillations and amplification of aliasing errors [18], mostly as a result of the large density gradients across the pseudo-boiling region. Consequently, straightforward discretizations of the compressible equations of supercritical fluid motion typically turn out to be unsuitable. For instance, the commonly used conservative (*divergence*) formulation of the Navier-Stokes equations requires to be stabilized, either by filtering the conservative variables [17] or by a proper blending with upwind-biased methods such as HLLC [29] or WENO schemes [26]. However, these approaches are known to add exceedingly high levels of artificial dissipation, thus suppressing part of the turbulent spectrum. In this regard, the computational study presented in this work is based on a novel discretization approach grounded on extending a stable and non-dissipative formulation of the kinetic-energy preserving scheme introduced by Kennedy & Gruber [15], and later shown to be energy-preserving by Pirozzoli [22] and Coppola et al. [8] (referred to as KGP), to high-pressure transcritical flow problems [3].

Therefore, the objectives of this work are twofold: (i) demonstrate the suitability of a novel stable energy-preserving scheme tailored for real-gas compressible turbulence, and (ii) characterize the physics of high-pressure transcritical wall-bounded turbulence by means of studying a canonical channel flow setup. To that end, the work is organized as follows. First, in Sect. 2, the flow physics modeling of supercritical fluids is described together with the numerical scheme utilized. Next, in Sect. 3, the computational results are presented and analyzed. Finally, the work is concluded and future directions are proposed in Sect. 4.

2 Flow Physics and Numerical Modeling Framework

The framework utilized for studying supercritical fluids turbulence in terms of (i) equations of fluid motion, (ii) real-gas thermodynamics, (iii) high-pressure transport coefficients, and (iv) numerical scheme is described below.

2.1 Equations of Fluid Motion

The turbulent flow motion of supercritical fluids is described by the following set of conservation equations of mass, momentum, and total energy

$$\frac{\partial \rho}{\partial t} + \nabla \cdot (\rho \mathbf{v}) = 0, \tag{1}$$

$$\frac{\partial (\rho \mathbf{v})}{\partial t} + \nabla \cdot (\rho \mathbf{v}\mathbf{v}) = -\nabla P + \nabla \cdot \boldsymbol{\tau}, \tag{2}$$

$$\frac{\partial (\rho E)}{\partial t} + \nabla \cdot (\rho \mathbf{v} E) = -\nabla \cdot \mathbf{q} - \nabla \cdot (P\mathbf{v}) + \nabla \cdot (\boldsymbol{\tau} \cdot \mathbf{v}), \tag{3}$$

where ρ is the density, $\mathbf{v}$ is the velocity vector, P is the pressure, $\boldsymbol{\tau}$ is the viscous stress tensor for Newtonian fluids, E is the total energy, and $\mathbf{q}$ is the Fourier conduction heat flux.

2.2 Real-Gas Thermodynamics

The thermodynamic space of solutions for the state variables pressure P, temperature T, and density ρ of a single substance is described by an equation of state. One popular choice for systems at high pressures, which is used in this study, is the Peng-Robinson equation of state [21] written as

$$P = \frac{R_u T}{\bar{v} - b} - \frac{a}{\bar{v}^2 + 2b\bar{v} - b^2}, \tag{4}$$

with R_u the universal gas constant, $\bar{v} = W/\rho$ the molar volume, and W the molecular weight. The coefficients a and b take into account real-gas effects related to attractive forces and finite packing volume, respectively, and depend on the critical temperatures T_c, critical pressures P_c, and acentric factors ω. They are defined as

$$a = 0.457\frac{(\mathrm{R_u} T_c)^2}{P_c}\left[1 + c\left(1 - \sqrt{T/T_c}\right)\right]^2 \quad \text{and} \quad b = 0.078\frac{\mathrm{R_u} T_c}{P_c}, \tag{5}$$

where coefficient c is provided by

$$c = \begin{cases} 0.380 + 1.485\omega - 0.164\omega^2 + 0.017\omega^3 & \text{if } \omega > 0.49, \\ 0.375 + 1.542\omega - 0.270\omega^2 & \text{otherwise.} \end{cases} \tag{6}$$

The Peng-Robinson real-gas equation of state needs to be supplemented with the corresponding high-pressure thermodynamic variables based on departure functions [24] calculated as a difference between two states. In particular, their usefulness is to transform thermodynamic variables from ideal-gas conditions (low pressure - only temperature dependant) to supercritical conditions (high pressure). The ideal-gas parts are calculated by means of the NASA 7-coefficient polynomial [5], while the analytical departure expressions to high pressures are derived from the Peng-Robinson equation of state as detailed in Jofre & Urzay [13].

2.3 High-Pressure Transport Coefficients

The high pressures involved in the analyses conducted in this work prevent the use of simple relations for the calculation of the dynamic viscosity μ and thermal conductivity κ. In this regard, standard methods for computing these coefficients for Newtonian fluids are based on the correlation expressions proposed by Chung *et al.* [6,7]. These correlation expressions are mainly function of critical temperature T_c and density ρ_c, molecular weight W, acentric factor ω, association factor κ_a and dipole moment $\mathcal{M}$, and the NASA 7-coefficient polynomial [5]; further details can be found in dedicated works, like for example [13,23].

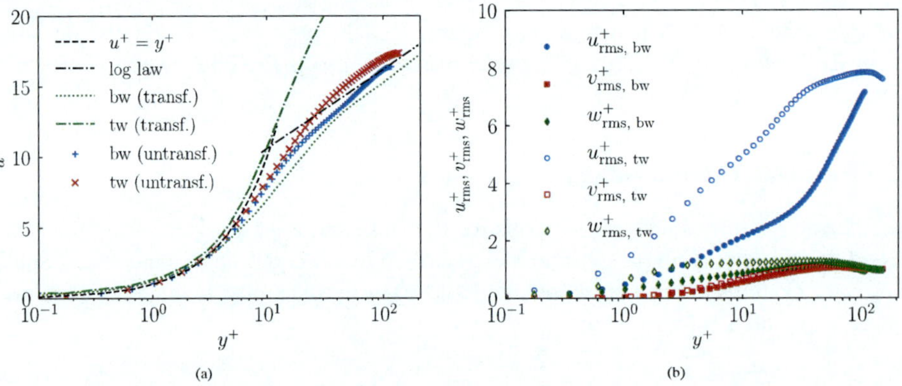

Fig. 1. Time-averaged streamwise velocity u^+ (a) and Favre-averaged rms velocity fluctuations u^+_{rms}, v^+_{rms} and w^+_{rms} (b) along the y^+ direction for the bottom and top walls.

2.4 Numerical Scheme

The equations of fluid motion introduced in Sect. 2.1 are numerically solved by adopting a standard semi-discretization procedure, i.e., they are firstly discretized in space and then integrated in time. In particular, spatial differential operators are treated using second-order centered finite-differences, and time-integration is performed by means of a third-order strong-stability preserving (SSP) Runge-Kutta explicit scheme [10]. As introduced in Sect. 1, the discretization of the convection terms is based on a stable kinetic-energy preserving scheme initially assessed in a previous work [3]. Specifically, the KGP splitting is applied to the continuity and momentum equations, and to the transport of enthalpy. By doing so, the method globally preserves kinetic energy by convection, and locally and globally preserves mass, momentum, total and internal energy [8].

3 Computational Results

High-pressure transcritical turbulent channel flow is studied by means of DNS based on the flow physics framework described in Sect. 2 utilizing the in-house MPI+OpenACC compressible flow solver RHEA [11]. The problem setup and discussion of flow statistics are described below.

3.1 Problem Setup

The turbulent channel flow problem is a reference experiment widely used in the computing community to validate and analyze wall-bounded turbulent flows [27]. To study and characterize high-pressure supercritical fluid turbulence, the fluid selected for this work is N_2 whose critical pressure and temperature are $P_c = 3.4\,\text{MPa}$ and $T_c = 126.2\,\text{K}$. The fluid system is at a supercritical bulk pressure of $P_b/P_c = 2$ and confined between bottom (bw) and top (tw) isothermal walls, separated at a distance $H = 2\delta$ with $\delta = 100\,\mu\text{m}$ the channel half-height, at $T_{bw}/T_c = 0.75$ and $T_{tw}/T_c = 1.5$, respectively. This problem setup imposes the fluid to undergo a transcritical trajectory by operating within a thermodynamic region across the pseudo-boiling line. The friction Reynolds number selected at the bottom wall is $Re_{\tau,bw} = \rho_{bw} u_{\tau,bw} \delta / \mu_{bw} = 100$ to ensure fully-developed turbulent flow conditions [4], with parameters corresponding to dynamic viscosity $\mu_{bw} = 1.6 \cdot 10^{-4}\,\text{Pa} \cdot \text{s}$, density $\rho_{bw} = 839.4\,\text{kg/m}^3$, and friction velocity $u_{\tau,bw} = 1.9 \cdot 10^{-1}\,\text{m/s}$. The mass flow rate in the streamwise direction is imposed through a body force controlled by a feedback loop where a proportional controller (gain $k_p = 0.1$) aims to reduce the difference between the desired ($Re_{\tau,bw} = 100$) and measured (numerical) $Re_{\tau,bw}$ values.

The computational domain is $4\pi\delta \times 2\delta \times 4/3\pi\delta$ in the streamwise (x), wall-normal (y), and spanwise (z) directions, respectively. The streamwise and spanwise boundaries are set periodic, and no-slip conditions are imposed on the horizontal boundaries (x-z planes). The grid is uniform in the streamwise and spanwise directions with resolutions in wall units (based on bw values) equal to $\Delta x^+ = 9.8$ and $\Delta z^+ = 3.3$, and stretched toward the walls in the vertical direction with the first grid point at $y^+ = y u_{\tau,bw}/\nu_{bw} \approx 0.1$ and with sizes in the range $0.2 \lesssim \Delta y^+ \lesssim 2.3$. Thus, this grid arrangement corresponds to a DNS of size $128 \times 128 \times 128$ grid points. The simulation strategy starts from a linear velocity profile with random fluctuations as proposed by Nelson & Fringer [20], which is advanced in time with CFL $= 0.3$ to reach turbulent steady-state conditions after approximately 5 flow-through-time (FTT) units; based on the bulk velocity u_b and the length of the channel $L_x = 4\pi\delta$, a FTT is defined as $t_b = L_x/u_b \sim \delta/u_\tau = 5.2 \cdot 10^{-4}$ s. In this regard, flow statistics are collected for roughly 10 FTTs once steady-state conditions are achieved.

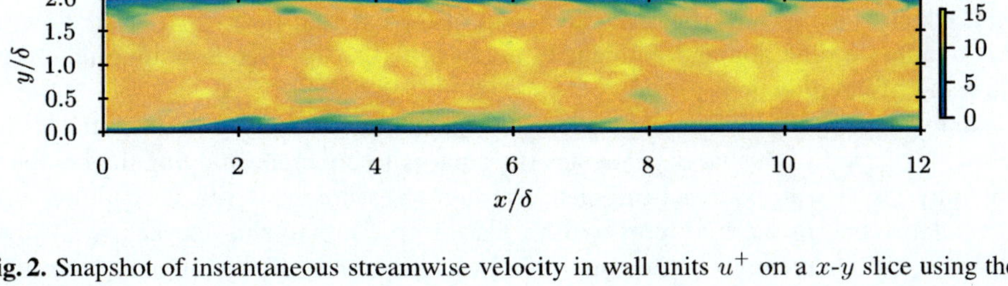

Fig. 2. Snapshot of instantaneous streamwise velocity in wall units u^+ on a x-y slice using the KGP scheme.

3.2 Flow Statistics

The time-averaged mean streamwise velocity u^+ and root-mean-squared (rms) velocity fluctuations $u^+_{\rm rms}$, $v^+_{\rm rms}$, $w^+_{\rm rms}$ along the wall-normal direction y^+ in viscous units for the bottom and top walls provided by RHEA are depicted in Fig. 1. Two main results can be inferred from the plots. First, as shown in Fig. 1(a), the shape of the time-averaged u^+ profiles tend to follow the topology of turbulent boundary layers for both walls. However, the profiles do not exactly collapse to the curves (black lines) corresponding to the "law of the wall" for an incompressible flat plate. An interesting observation is that the profiles neither collapse to the standard law when utilizing the state-of-the-art transformation (green lines) proposed by Trettel & Larsson [30], which was developed for non-isothermal compressible boundary layers, due to the presence of a sustained baroclinic instability close to the top wall. Second, focusing on the Favre-averaged r.m.s. velocity fluctuations shown in Fig. 1(b), turbulence intensity is significantly different between the top (gas-like) and bottom (liquid-like) walls. In particular, the top wall presents larger fluctuations in the viscous sublayer and buffer region, while turbulence intensity grows following a concave parabola from the bottom wall to the center of the channel. An equivalent, although significantly less pronounced, behavior is observed for the fluctuations in the wall-normal velocity, whereas they virtually collapse to the same curve for the spanwise direction. Finally, a snapshot of the instantaneous streamwise velocity in wall units u^+ on a x-y slice is displayed in Fig. 2 to provide qualitative information of the wall-bounded turbulence for $Re_{\tau,bw} = 100$ computed by RHEA using the KGP scheme.

4 Conclusions and Future Work

This work reports DNS results of a differentially heated channel flow at transcritical thermodynamic conditions. Of note, simulations were obtained using a recently proposed energy-preserving numerical scheme, and were completely free of any form of artificial dissipation. Preliminary analysis of the flow fields shows that, even though the first-order flow statistics follow the topology of turbulent boundary layers, they (i) do not collapse exactly to the standard "law of the wall", and (ii) the streamwise velocity fluctuations differ significantly between the bottom and top walls. These differences are mainly driven by a baroclinic instability introducing vorticity in the vicinity of the top wall, which is generated from the interaction of the external force driving the flow in the streamwise direction and the density gradient across the pseudo-boiling region in the wall-normal direction.

In this regard, the interesting and rich flow phenomena observed in the initial set of results presented in this work motivates the authors to continue studying the complex flow physics of wall-bounded turbulence at high-pressure transcritical conditions. In particular, focus in the mid-term will be placed on (i) analyzing the near-wall flow behavior, (ii) characterizing the baroclinic instability, and (iii) investigating the flow structures.

Acknowledgements. This work is funded by the European Union (ERC, SCRAMBLE, 101040379). Views and opinions expressed are however those of the authors only and do not necessarily reflect those of the European Union or the European Research Council. Neither the European Union nor the granting authority can be held responsible for them. The authors gratefully acknowledge the Formació de Professorat Universitari scholarship (FPU-UPC R.D 103/2019) of the Universitat Politècnica de Catalunya - BarcelonaTech (UPC) (Spain), the SRG (2021-SGR-01045) program of the Generalitat de Catalunya (Spain), the Beatriz Galindo program (Distinguished Researcher, BGP18/00026) of the Ministerio de Educación y Formación Profesional (Spain), and the computer resources at FinisTerrae III and the technical support provided by CESGA (RES-IM-2023-1-0005). Francesco Capuano is a Serra Húnter fellow.

References

1. Bae, J.H., Yoo, J.Y., Choi, H.: Direct numerical simulation of turbulent supercritical flows with heat transfer. Phys. Fluids **17**, 105104 (2005)
2. Bae, J.H., Yoo, J.Y., McEligot, D.M.: Direct numerical simulation of heated CO_2 flows at supercritical pressure in a vertical annulus at Re = 8900. Phys. Fluids **20**, 055108 (2008)
3. Bernades, M., Capuano, F., Trias, F.X., Jofre, L.: Energy-preserving stable computations of high-pressure supercritical fluids turbulence. In: 9th European Congress on Computational Methods in Applied Sciences and Engineering (ECCOMAS), pp. 1–12 (2022)
4. Bernades, M., Jofre, L.: Thermophysical analysis of microconfined turbulent flow regimes at supercritical fluid conditions in heat transfer applications. J. Heat Transfer **144**, 082501 (2022)
5. Burcat, A., Ruscic, B.: Third millennium ideal gas and condensed phase thermochemical database for combustion with updates from active thermochemical tables. Technical report, Argonne National Laboratory (2005)
6. Chung, T.H., Ajlan, M., Lee, L.L., Starling, K.E.: Generalized multiparameter correlation for nonpolar and polar fluid transport properties. Ind. Eng. Chem. Res. **27**, 671–679 (1988)
7. Chung, T.H., Lee, L.L., Starling, K.E.: Applications of kinetic gas theories and multiparameter correlation for prediction of dilute gas viscosity and thermal conductivity. Ind. Eng. Chem. Fundam. **23**, 8–13 (1984)
8. Coppola, G., Capuano, F., Pirozzoli, S., de Luca, L.: Numerically stable formulations of convective terms for turbulent compressible flows. J. Comput. Phys. **382**, 86–104 (2019)
9. Doehring, A., Kaller, T., Schmidt, S.J., Adams, N.A.: Large-eddy simulation of turbulent channel flow at transcritical states. Int. J. Heat Fluid Flow **89**, 108781 (2021)
10. Gottlieb, S., Shu, C.-W., Tadmor, E.: Strong stability-preserving high-order time discretization methods. SIAM Rev. **43**, 89–112 (2001)
11. Jofre, L., Oyarzún, G.: RHEA - an open-source Reproducible and Hybrid-architecture flow solver Engineered for Academia (2020)
12. Jofre, L., Urzay, J.: A characteristic length scale for density gradients in supercritical monocomponent flows near pseudoboiling. In: Annual Research Briefs, pp. 277–282. Center for Turbulence Research, Stanford University (2020)
13. Jofre, L., Urzay, J.: Transcritical diffuse-interface hydrodynamics of propellants in high-pressure combustors of chemical propulsion systems. Prog. Energy Combust. Sci. **82**, 100877 (2021)
14. Kawai, S.: Heated transcritical and unheated non-transcritical turbulent boundary layers at supercritical pressures. J. Fluid Mech. **865**, 563–601 (2019)
15. Kennedy, C.A., Gruber, A.: Reduced aliasing formulations of the convective terms within the Navier-Stokes equations for a compressible fluid. J. Comput. Phys. **227**(3), 1676–1700 (2008)

16. Knez, Z., Markocic, E., Leitgeb, M., Primozic, M., Knez, M., Skerget, M.: Industrial applications of supercritical fluids: a review. J. Energy **77**, 235–243 (2014)
17. Larsson, J., Lele, S., Moin, P.: Effect of numerical dissipation on the predicted spectra for compressible turbulence. In: Annual Research Briefs, pp. 47–52. Center for Turbulence Research, Stanford University (2007)
18. Ma, P.C., Lv, Y., Ihme, M.: An entropy-stable hybrid scheme for simulations of transcritical real-fluid flows. J. Comput. Phys. **340**, 330–357 (2017)
19. Ma, P.C., Yang, X.I.A., Ihme, M.: Structure of wall-bounded flows at transcritical conditions. Phys. Rev. Fluids **3**, 034609 (2018)
20. Nelson, K.S., Fringer, O.B.: Reducing spin-up time for simulations of turbulent channel flow. Phys. Fluids **29**, 105101 (2017)
21. Peng, D.Y., Robinson, D.B.: A new two-constant equation of state. Ind. Eng. Chem. Fundam. **15**, 59–64 (1976)
22. Pirozzoli, S.: Generalized conservative approximations of split convective derivative operators. J. Comput. Phys. **229**, 7180–7190 (2010)
23. Poling, B.E., Prausnitz, J.M., O'Connell, J.P.: Properties of Gases and Liquids, 5th edn. McGraw Hill, New York (2001)
24. Reynolds, W.C., Colonna, P.: Thermodynamics: Fundamentals and Engineering Applications, 1st edn. Cambridge University Press, Cambridge (2019)
25. Sengupta, U., Nemati, H., Boersma, B.J., Pecnik, R.: Fully compressible low-Mach number simulations of carbon-dioxide at supercritical pressures and trans-critical temperatures. Flow Turbul. Combust. **99**, 909–931 (2017). https://doi.org/10.1007/s10494-017-9872-4
26. Shu, C.W.: High order ENO and WENO schemes for computational fluid dynamics. In: Barth, T.J., Deconinck, H. (eds.) High-Order Methods for Computational Physics. LNCSE, vol. 9, pp. 439–582. Springer, Heidelberg (1999). https://doi.org/10.1007/978-3-662-03882-6_5
27. Smits, A.J., McKeon, B.J., Marusic, I.: High-Reynolds number wall turbulence. Ann. Rev. Fluid Mech. **43**(3), 53–75 (2011)
28. Sreenivasan, K.R.: Turbulent mixing: a perspective. PNAS **116**, 18175–18183 (2019)
29. Toro, F.: Riemann Solvers and Numerical Methods for Fluid Dynamics, 3rd edn. Springer, Heidelberg (2009). https://doi.org/10.1007/b79761
30. Trettel, A., Larsson, J.: Mean velocity scaling for compressible wall turbulence with heat transfer. Phys. Fluids **28**, 026102 (2016)
31. Xie, G., Xu, X., Lei, X., Li, Z., Li, Y., Sunden, B.: Heat transfer behaviours of some supercritical fluids: a review. Chin. J. Aeronaut. **35**(1), 290–306 (2022)
32. Yoo, J.Y.: The turbulent flows of supercritical fluids with heat transfer. Ann. Rev. Fluid Mech. **45**, 495–525 (2013)

Development, Validation and Application of an ANN-Based Large Eddy Simulation Subgrid-Scale Turbulence Model for Dense Gas Flows

Alexis Giauque(✉), Aurélien Vadrot, and Christophe Corre

Univ Lyon, Ecole Centrale de Lyon, Univ Claude Bernard Lyon 1, INSA Lyon, LMFA, UMR 5509, AURA, 69130 Ecully, France
{alexis.giauque,aurelien.vadrot,christophe.corre}@ec-lyon.fr

Abstract. Using a DNS database (available to the NICFD research community) including homogeneous isotropic turbulence (HIT), mixing layers and channel flows, the authors have developed a novel LES model for the Subgrid-Scale (SGS) turbulent Reynolds tensor based on artificial neural networks (ANN) with optimized hyperparameters. Particular attention has been paid to ensure Galilean invariance properties are satisfied by the ANN-based model. To accommodate the large size of the database (over 500 million samples), the ANN is trained on a parallel CPU architecture. *A priori* performance reveals determination coefficients (r^2-score) larger than 0.9 can be achieved, yielding an accurate prediction of the SGS turbulent Reynolds tensor magnitude for flow conditions never met during training. *A posteriori* validation is performed by implementing the model into AVBP, a LES solver developed at CERFACS [10]. The computation of a dense gas mixing layer at a convective Mach number of 2.2 shows that the newly developed ANN-based model provides levels of accuracy comparable with or even better than those provided by existing models (Implicit LES, Sigma model [13]). To further assess the ANN-based model, a 3D ORC turbine configuration is computed and the numerical prediction is compared with measurements gathered by Baumgartner, Otter and Wheeler [1] at Cambridge University and presented at NICFD 2020. Preliminary results from this recently launched European PRACE project are presented.

Keywords: ORC · Large Eddy Simulation · SGS Modeling · Dense gas · Machine Learning

1 Introduction

Global warming due to the release of huge amounts of CO_2 in the atmosphere generated by the consumption of fossil fuel is now considered as one of the most important threat to the stability of modern society. To reduce the production of CO_2, renewable energies are currently being developed at an accelerated pace. Among the possible technological solutions available, the Organic Rankine Cycle has been proposed to harvest low to moderate temperature heat sources. ORC systems use, instead of water, organic

M. White et al. (Eds.): NICFD 2022, ERCOFTAC 29, pp. 43–52, 2023.
https://doi.org/10.1007/978-3-031-30936-6_5

fluids displaying low boiling temperatures and large heat capacities. Among those fluids, some, when in their gaseous form and in specific thermodynamic conditions, are called dense gases. Dense gases are characterized by a fundamental derivative Γ lower than unity. Following this definition, the speed of sound in these gases decreases with increasing density along isentropic lines. Numerous phenomena occur in flows of dense gases that are related to their specific thermodynamic nature: lower speeds of sound are reached and most importantly the strength of shockwaves is significantly reduced. This phenomenon, first evidenced numerically by Schnerr and Leidner in 1996 [17] and again shown by Cinnella and Congedo in 2005 [2], is at the origin of a renewed interest for dense gases in the context of ORC. In 2016, Dura Galiana et al. [4, 5] show that up to $2/3$ of the losses in turbine expanders comes from viscous effects in the wake. The trailing edge region where the wake starts is both the region of origin of shockwaves necessary for the flow to adapt to the back pressure, and a region displaying a large turbulence activity because of the merging boundary layers coming from the blade. A natural question in this context is that of the interaction between turbulence and thermodynamics in dense gas flows (more specifically in turbine expanders) and its impact on SGS turbulence modeling.

To the best of the authors' knowledge, this question has not yet been addressed and state of the art comparisons between CFD and experiments either rely on Euler equations with a focus on compressibility-induced features [9] or on Large Eddy simulations (LES) using existing subgrid scales models developed in the context of perfect gas flows [11]. To better understand the behavior of turbulence in dense gas flows, the authors have analyzed Direct Numerical Simulations (DNS) in academic configurations such as Homogeneous Isotropic Turbulence, Mixing Layer and Channel Flow. Numerous physical phenomena specific to the dense nature of the gas have been described such as the strongly modified statistics of shocklets in compressible turbulence, the decoupling of internal and kinetic energy in the mixing layer which has also been observed in the channel flow and confirms independent findings from [18]. Among all results, the most striking one, observed by Vadrot during his PhD [19] is the strongly modified growth rate (multiplied by a factor 2) of the dense gas mixing layer when compared to the perfect gas one.

The precise description of losses occurring in ORC turbines requires LES as a way to accurately capture turbulence in general and more specifically features such as boundary layer transition, shockwave/boundary layer interaction and wake turbulence dynamics. LES is now a proven approach in the understanding and fine tuning of complex systems in presence of turbulent phenomena. In this approach, the most energetic part of the turbulent spectrum is captured by the numerical solver and the remaining part has to be modeled using so called "subgrid" models. Most models have been developed in the context of perfect gases and almost entirely focus on the SubGrid-Scale Reynolds Stress Tensor (SGS RST).

Recent findings by the authors [8] show that (1) additional subgrid-scale terms should be considered in dense gases, (2) usual closure models for the SGS RST lack precision when applied to turbulent dense gas flows. Developing turbulence models for dense gas flows requires to take into account highly non-linear Equations of State (EoS) needed to accurately describe the dense gas thermodynamic behavior. This significantly complexifies the usual modeling strategies and leads the authors to propose instead the use of Artificial Neural Networks (ANN), since they are well adapted to the identification of correlations in a non-linear context. This paper presents the development, validation and application of an ANN-based Large Eddy Simulation Subgrid-Scale Turbulence model for dense gas flows. In Sect. 2, the methodology used for the development of the model is presented, with a focus on the Galilean invariance of the training data and on the optimization of hyperparameters of the neural network. In Sect. 3, a preliminary *a posteriori* validation of the model is presented to demonstrate the influence of the ANN-based model on (i) the growth rate of a dense gas supersonic mixing layer and (ii) the total pressure distributions experimentally measured in an annular stator configuration at Cambridge University.

2 Methodology: Development of an ANN Model for the Subgrid-Scale Reynolds Tensor in LES

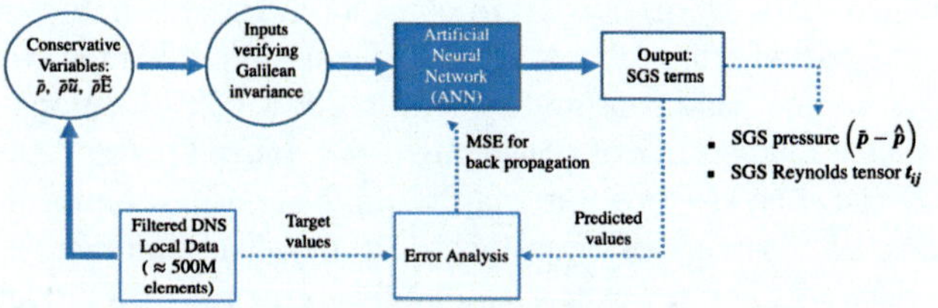

Fig. 1. Schematic representation of the ANN training process

The general description of the ANN training process is illustrated in Fig. 1. Starting from a DNS database of dense gas flows, local data are spatially filtered at a wavelength lying within the inertial range of turbulence for each case considered. In practice, a Gaussian filter is applied with a characteristic wavelength equal to roughly twenty times the initial DNS resolution. These filtered data are composed of all conservative ($\bar{\rho}$, $\bar{\rho}\tilde{u}_i$, $\bar{\rho}\tilde{E}$) and thermodynamic fields ($\hat{p}$, $\hat{T}$) along with their gradients ($\tilde{g}_{ij} = \frac{\partial \tilde{u}_i}{\partial u_j}$, $\nabla\hat{p}$, $\nabla\hat{T}$). $\bar{\phi}$ denotes the resolved large-scale component of a flow variable (ϕ). $\tilde{\phi} = \frac{\overline{\rho\phi}}{\bar{\rho}}$ denotes the Favre averaged of a flow variable ϕ while $\hat{\phi}$ denotes computable variables that are computed from conservative filtered fields (see [8] for a more detailed description). Components of the Turbulent Reynolds Tensor ($t_{ij} = \bar{\rho}(\widetilde{u_i u_j} - \tilde{u}_i\tilde{u}_j)$) are also stored for later use during training.

Among all the potentially available data when running a LES, actual input data are selected using a sensitivity analysis later described in this paper and preprocessed to satisfy Galilean invariance principles. Following usual practice, the database is then split into a training and a testing part. The training section of the database is used first to fit the parameters of the ANN (*i.e.* the weights and biases of each connection and neuron). The testing section is used next to assess the performance of the ANN and verify the absence of overfitting. To perform this analysis, the ANN prediction is compared to the Reynolds Tensor components target values available in the database.

2.1 Description of the Database

The methodology used in this study to develop a SGS model tailored for dense gas flows relies on Direct Numerical Simulations (DNS) computed by the authors in the course of the last five years. Figure 1 illustrates the academic dense gas flow configurations considered in this research project: Forced Homogeneous Isotropic Turbulence (Forced-HIT), Channel Flow and Mixing Layer. The reader interested in the details of the comparison of these flows with their perfect gas counterparts is referred to [6, 7] for the Forced-HIT, [8] for the channel and [20, 21] for the mixing layer. The code AVBP, developed at CERFACS, has been used for the development of the database and is also the code used to assess a-posteriori the properties of the developed SGS model. During the course of this entire study, the 3rd order in space and time convective scheme TTGC is used [3].

2.2 Choice of the Input Variables

The choice has been made from the start to consider as input variables only local data taken at the same location where the ANN is expected to predict the Reynolds Tensor components. This choice has been guided by CPU cost considerations in the context of High Performance Computing of dense gas flows. Indeed, other types of non-local neural networks exist such as Convolutional Neural Networks (non local in space) [16] or networks using LSTM neurons (non local in time). These attractive types of neurons and neural networks are however such that their cost for LES modeling over large discretization grids would eventually prevent their practical use.

Using therefore Gaussian-filtered local data only, the additional underlying principle guiding the choice of the ANN input variables is the Galilean invariance principle. This principle is based on a set of invariance properties including translation and rotation of the reference frame and uniform motion with respect to the reference frame. To make sure the ANN prediction satisfies the Galilean invariance principle, the input data should themselves be unchanged when the frame of reference is translated or when one adds a uniform translation motion to the database. Verifying the invariance with respect to the rotation of the frame of reference is more complex. Indeed, except for the HIT, the DNS constituting the database display the bulk flow direction as preferred direction. In order to verify the rotational invariance, each sample is randomly rotated before it is added to the database, with the distribution of rotation angles around the original x, y and z axis as flat as possible and ranging from 0 to 2π.

Actual input data used in this work have been initially chosen among all the data potentially available when running a LES. Based on the *a-priori* performance analysis of the ANN-based SGS model, the following set of 15 inputs has been eventually retained: y^+, $\bar{\rho}$, $\bar{T}$, $\tilde{g}_{ij} = \frac{\partial \tilde{u}_i}{\partial u_j}$, $\tilde{s}_{ij} = \frac{1}{2}(\tilde{g}_{ij} + \tilde{g}_{ji})$. Although strain rate tensor components $\tilde{s}_{ij}$ are strongly correlated with velocity gradients $\tilde{g}_{ij}$ and do not improve the *a-priori* performance, it has been observed that they improve the consistency of *a posteriori* ANN-based SGS model predictions.

2.3 Optimization of the Hyperparameters

The type of ANN used in this work is a multi-layer perceptron (a fully connected class of feedforward ANN). The hyperparameters of the ANN are of two main types. The first type is constitutive of the ANN in the sense that it also applies when the ANN is later used in *a posteriori* validation. Those type-1 hyperparameters are:

- the number of hidden layers,
- the number of neurons in each hidden layer,
- the activation function of each neuron.

The second type of hyperparameters is intrinsically linked to the training of the ANN. Those type-2 hyperparameters are:

- the size of the batches used to train the ANN,
- the optimization algorithm used for back-propagation,
- the measure used to assess the accuracy of the ANN,
- the regularization parameters (L_2 regularization in this work) to balance the amplitude of weights over the whole network and to avoid over-fitting.

The most popular methods used to tune an ANN are hyperparameter searches, which browse the hyperparameters space, testing a large number of hyperparameters combinations, for an optimized amount of computational time, in order to identify the one providing the best accuracy. The fundamental issue with these methods is the lack of information they provide about the sensitivity of the ANN to the choice of its hyperparameters: they are often seen as black-box tools which only yield the most effective ANN, with no information provided about the most influential hyperparameters or the most relevant range for each of them. Even though this information is actually stored in the search results, it is rarely exploited.

In this work, in addition to hyperparameter searches, an Hilbert-Schmidt Independence Criterion (HSIC) analysis is performed following the proposal made in [14]. HSIC evaluates in a first step the probability distribution of hyperparameters among tested combinations of hyperparameters randomly selected by the hyperparameters search method, denoted $P(\text{'hyperparameter'})$. In a second step, HSIC computes the probability distribution for an ANN of being among the best decile networks (the 10% of the best ANN), denoted $P(\text{'hyperparameter'}|Z)$, where Z is a random variable which is equal to one if the ANN is among the best decile. Two types of results are eventually obtained: a classification of hyperparameters by rank of importance (Fig. 2a)) and the optimal choice for a given hyperparameter (Fig. 2b)). In the given example, HSIC allows to identify the significant influence of the optimization algorithm used for back-propagation and recommends the use of the ADAbelief method for optimal ANN performance.

2.4 Analysis of the Training Process

The training process uses the complete training database, combining HIT, channel and mixing layer data with preprocessing to enforce Galilean invariance, over several epochs to optimize the weights and biases of the ANN until the optimization process

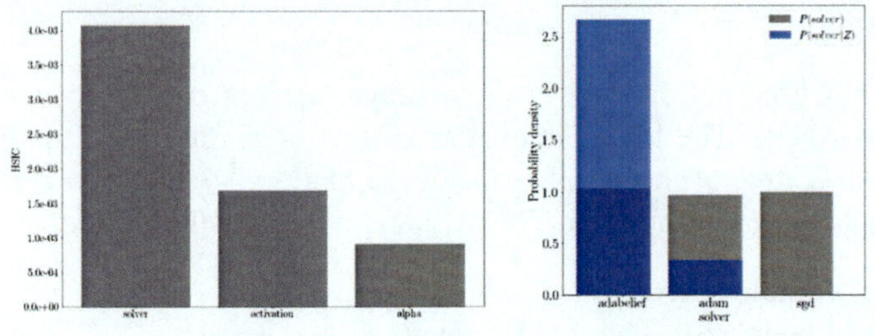

a) b)

Fig. 2. a) Classification of HSIC and b) Probability density function among solver choices.

converges and the accuracy of the model reaches an asymptotic value. Figure 3 illustrates this process for the three off-diagonal terms of the Turbulent Reynolds tensor (t_{xy}, t_{xz} and t_{yz}). One observes that for the testing database, values of the r^2-score larger than 0.9 are reached after 20 to 30 epochs. To illustrate this level of correlation between the filtered DNS data and the ANN model, Fig. 4 shows a side-by-side comparison of t_{xy} for the mixing layer at $M_c = 2.2$ from the filtered DNS (Fig. 4a)) and the ANN prediction (Fig. 4b)). $M_c = \frac{\Delta u}{c_1+c_2}$ is the convective Mach number (Δu is the differential speed between upper and lower parts of domain and c_1 and c_2 are the corresponding sound speeds). Note that the prediction is computed for a temporal solution which was not included in the training database. It can be observed in Fig. 4 that the local nature of the ANN model does not preclude the accurate reproduction, both in size and amplitude, of the t_{xy} peak regions.

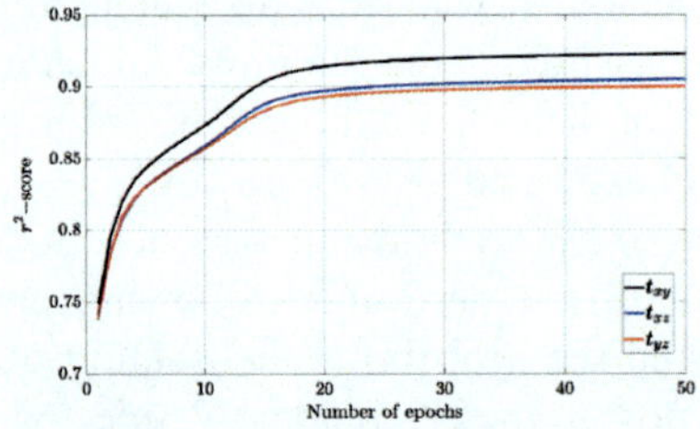

Fig. 3. Evolution of the r^2–score computed for the three off-diagonal terms of the Turbulent Reynolds tensor over the testing database as a function of the number of epochs.

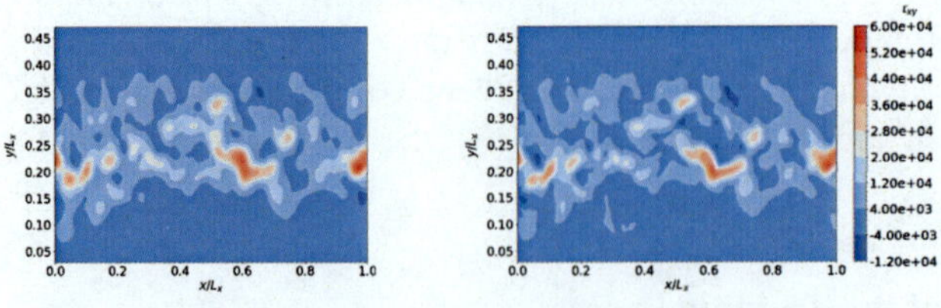

a) b)

Fig. 4. Comparison of t_{xy} between the (a) true values and (b) the predicted ones.

3 Results: Validation of the ANN-Based Turbulent Closure Model

3.1 Mixing Layer at Convective Mach Number of 2.2

The ANN model tuned in the previous *a priori* analysis bas been implemented in the LES solver AVBP developed at CERFACS, as an alternative to state-of-the-art SGS models available in this solver. The mixing layer case at $M_c = 2.2$ is considered as a first validation test-case. Four different meshes are designed to assess the accuracy of the ANN-based model as a function of the grid resolution. For this numerical experiment, the following spatial resolutions are considered: $\Delta/\Delta_{DNS} = 4$, 8, 16 and 32. The first two resolutions are too close to the DNS to be reachable by practical LES. The two coarsest are representative of LES designed to have the turbulent kinetic energy spectrum cut in the inertial regime for large enough Reynolds numbers. Figure 5a) compares the reference DNS temporal evolution of the mixing layer thickness with the LES simulations using the grid $\Delta/\Delta_{DNS} = 32$ and (i) the ANN model, (ii) the Sigma model and (iii) implicit LES. Because of the coarse grid resolution, the initial solution starts for all LES with a mixing layer thickness larger than in the DNS. The evolution of the mixing layer thickness is initially very similar between implicit LES and the ANN model but both strategies start to depart from each other after a non-dimensional time τ larger than 4000 and eventually the ANN model better recovers the growth rate of the DNS. The Sigma model at first predicts a larger mixing layer but quickly saturates and strongly under-predicts the growth rate later in time. Note that the comparison of the growth rates is meaningful when each computation reaches a linear regime, that is for $\tau \in [6000, 7000]$.

Figure 5b) directly compares the mixing layer growth rates reached in the self-similar regime by each model as a function of the space resolution. It can be seen that for this $M_c = 2.2$ dense gas mixing layer the ANN model consistently better predicts the mixing layer thickness for all four spatial resolutions considered, with an error of 10% maximum with respect to the DNS.

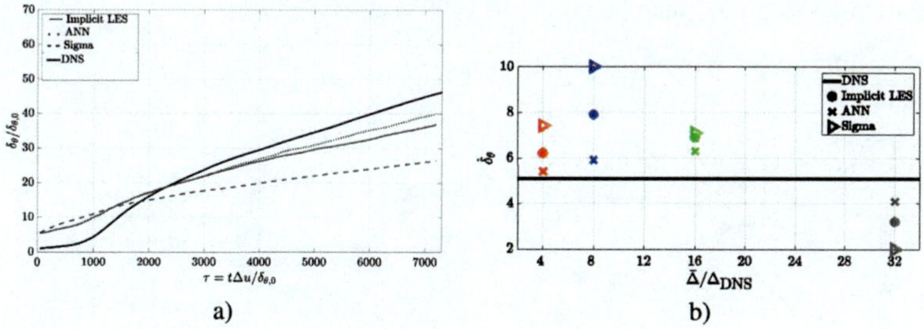

Fig. 5. a) Temporal evolution of the momentum thickness for a mixing layer at $M_c = 2.2$. Comparison is made between DNS and three LES at $\Delta/\Delta_{DNS} = 32$. b) Comparison of growth rates for DNS and three LES as a function of the resolution (Δ/Δ_{DNS}).

3.2 Annular Stator Case

In order to further assess the accuracy of the ANN-based model for flow configurations representative of industrial applications, it is expected to be applied to the simulation of the annular stator case tested in Cambridge University [1] and presented at the last NICFD conference [15]. More precisely, a comparison is sought between available measurements and LES simulations performed using both the novel model and a state-of-the-art algebraic model (WALE model [12]). Figure 6(a) displays the supersonic annular stator of chord 1cm which leads to Reynolds numbers ($Re = 500000$) that are tractable by LES (see Fig. 6(b) for a typical LES flow field visualization). Another interesting feature of this experiment is the relatively large amount of data available thanks to static wall pressure measurements along the shroud and total pressure measurements in the wake. Those can be used to discriminate numerical results obtained through different grid resolutions and turbulence models (see Fig. 7). To properly investigate the effect of the ANN model on the LES results, three different grid resolutions are considered: from a coarse grid with 46 million cells and a wall-law to a fine grid with 450 million cells allowing a wall-resolved calculation. Computations are performed on each grid using the classical WALE subgrid scale model or the novel ANN model. Figure 7 presents preliminary results for the total pressure ratio in the wake obtained with the WALE model and compared to experimental measurements. A more detailed comparison will be proposed and discussed at the NICFD 2022 Conference.

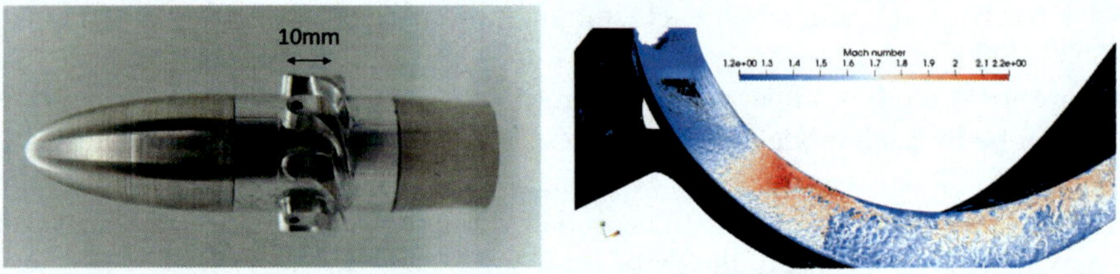

Fig. 6. a) Picture of the annular cascade [15] and b) Flow field visualization at the extrados of Q-criterion colored with the Mach number (WL-400M case).

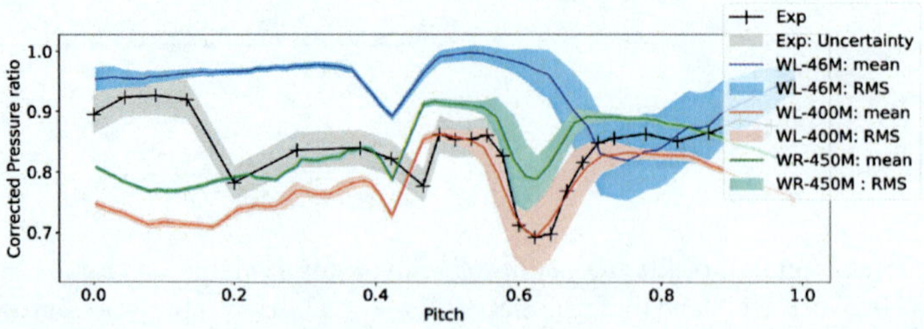

Fig. 7. Evolution of the total pressure ratio as a function of the pitch angle for three different meshes using an algebraic subgrid model (Wale model [12]). Comparison with experimental measurements [15].

4 Conclusions and Future Works

A new modeling strategy has been developed for the SGS turbulent Reynolds tensor using supervised machine learning tools. The optimization and the training of an ANN using a rich database including several dense gas turbulent flow configurations enabled to obtain an effective SGS model capable to provide reliable results for cases which were not encountered during the training phase. The *a posteriori* validation has been initiated by performing LES of a dense gas mixing layer at $M_c = 2.2$ using four grid sizes and three turbulent closure strategies: the ANN model was found to better predict growth rates for all grid sizes. Further results regarding the application of the ANN model in the context of an annular stator case are still being gathered and will be presented during the NICFD conference.

Acknowledgments. This work is supported by the JCJC ANR EDGES project, grant #ANR−17−CE06−0014−01 of the French Agence Nationale de la Recherche. Simulations have been carried out using HPC resources at CINES under the project grant #A0102A07564. We also acknowledge PRACE for awarding us access to Joliot-Curie at GENCI@CEA, France. The authors would like to thank the CFD team at CERFACS for their support and for giving access to their solver AVBP. The authors would like to thank the PMCS2I team at Ecole Centrale de Lyon.

References

1. Baumgärtner, D., Otter, J.J., Wheeler, A.P.: The effect of isentropic exponent on transonic turbine performance. J. Turbomach. **142**(8), 081007 (2020)
2. Cinnella, P., Congedo, P.M.: Aerodynamic performance of transonic Bethe-Zel'dovich-Thompson flows past an airfoil. AIAA J. **43**(2), 370–378 (2005)
3. Colin, O., Rudgyard, M.: Development of high-order Taylor-Galerkin schemes for LES. J. Comput. Phys. **162**(2), 338–371 (2000)
4. Durá Galiana, F.J., Wheeler, A.P., Ong, J.: A study of trailing-edge losses in organic rankine cycle turbines. J. Turbomach. **138**(12), 121003 (2016)
5. Galiana, F.D., Wheeler, A., Ong, J., Ventura, C.M.: The effect of dense gas dynamics on loss in ORC transonic turbines. In: Journal of Physics: Conference Series, vol. 821, p. 012021. IOP Publishing (2017)
6. Giauque, A., Corre, C., Menghetti, M.: Direct numerical simulations of homogeneous isotropic turbulence in a dense gas. In: Journal of Physics: Conference Series, vol. 821, p. 012017. IOP Publishing (2017)
7. Giauque, A., Corre, C., Vadrot, A.: Direct numerical simulations of forced homogeneous isotropic turbulence in a dense gas. J. Turbul. **21**(3), 186–208 (2020)
8. Giauque, A., Vadrot, A., Errante, P., Corre, C.: A priori analysis of subgrid-scale terms in compressible transcritical real gas flows. Phys. Fluids **33**(8), 085126 (2021)
9. Gori, G., Zocca, M., Guardone, A., Le Maitre, O., Congedo, P.M.: Bayesian inference of thermodynamic models from vapor flow experiments. Comput. Fluids **205**, 104550 (2020)
10. Gourdain, N., et al.: High performance parallel computing of flows in complex geometries: II. Applications. Comput. Sci. Discov. **2**(1), 015004 (2009)
11. Hoarau, J.-C., Cinnella, P., Gloerfelt, X.: Large eddy simulations of strongly non-ideal compressible flows through a transonic cascade. Energies **14**(3), 772 (2021)
12. Nicoud, F., Ducros, F.: Subgrid-scale stress modelling based on the square of the velocity gradient tensor. Flow Turbul. Combust. **62**(3), 183–200 (1999). https://doi.org/10.1023/A:1009995426001

13. Nicoud, F., Toda, H.B., Cabrit, O., Bose, S., Lee, J.: Using singular values to build a subgrid-scale model for large eddy simulations. Phys. Fluids **23**(8), 085106 (2011)
14. Novello, P., Poëtte, G., Lugato, D., Congedo, P.: Explainable hyperparameters optimization using Hilbert-Schmidt independence criterion (2021)
15. Otter, J.J., Baumgärtner, D., Wheeler, A.P.: The development of a generic working fluid approach for the determination of transonic turbine loss. In: Pini, M., De Servi, C., Spinelli, A., di Mare, F., Guardone, A. (eds.) NICFD 2020. ERCO, vol. 28, pp. 123–131. Springer, Cham (2020). https://doi.org/10.1007/978-3-030-69306-0_13
16. Saura, N., Gomez, T.: Subgrid stress tensor prediction in homogeneous isotropic turbulence using 3D-convolutional neural networks. Available at SSRN 4184202 (2022)
17. Schnerr, G., Leidner, P.: Nonclassical behavior of dense gases in axial cascades. Z. Angew. Math. Mech. **76**, 457–458 (1996)
18. Sciacovelli, L., Cinnella, P., Gloerfelt, X.: Direct numerical simulations of supersonic turbulent channel flows of dense gases. J. Fluid Mech. **821**, 153–199 (2017)
19. Vadrot, A.: Numerical simulation and modeling of compressible turbulence in dense gas flows. Ph.D. thesis, Université de Lyon (2021)
20. Vadrot, A., Giauque, A., Corre, C.: Analysis of turbulence characteristics in a temporal dense gas compressible mixing layer using direct numerical simulation. J. Fluid Mech. **893**, A10 (2020)
21. Vadrot, A., Giauque, A., Corre, C.: Direct numerical simulations of temporal compressible mixing layers in a Bethe-Zel'dovich-Thompson dense gas: influence of the convective Mach number. J. Fluid Mech. **922**, A5 (2021)

Bypass Laminar-Turbulent Transition on a Flat Plate of Organic Fluids Using DNS Method

Bijie Yang(✉), Tao Chen, and Ricardo Martinez-Botas

Imperial College London, London, UK
{b.yang15,tao.chen17,r.botas}@imperial.ac.uk

Abstract. This work aims to investigate the impact of non-ideal gas effects on turbulent boundary layers, by applying DNS to the by-pass transition for three gases: air, R1233zd(E), and MDM. Air is the baseline, and three different thermodynamic states are studied for each organic fluid. The work has investigated both the transition process and the fully developed boundary layers. For the transition process, it is found that the transition happens early in organic fluids than air. In contrast to air, more stream-wise vortices are induced by blowing-and-suction boundary conditions, and the propagation speed of Kelvin-Helmholtz instability is also higher in n organic fluids. For the fully developed boundary layer of ideal gas, it is well known that momentum and energy mixing lengths are the same as fluctuations are only driven by turbulent vorticities (SRA). However, for organic fluid cases, the acoustic mode driven by the pressure fluctuation is also found important in generating energy fluctuations.

Keywords: turbulence · transition · DNS · organic fluid

1 Introduction

Among the crucial techniques aiming at carbon neutrality, organic Rankine cycle (ORC) is the most promising technology for recovering medium- or low-grade heat. In ORC cycles, for the sake of reducing the thermal critical temperature, organic fluids are generally used. Due to the complication of molecular structures and molecular interactions, the thermodynamic properties of the organic fluids are far from ideal. However, the effects of the thermal 'non-ideality' on turbulent flows and heat transfers has been less well investigated.

Most studies of compressible turbulent boundary layers only focus on ideal gases. Morkovin [9] suggests that there is a strong similarity between compressible and in-compressible turbulent boundary layers. Van-Derist [16] proposed a density-weighted transformation to match the mean velocity profile of a compressible boundary layer to that of an equivalent in-compressible counterpart. Walz [17] derived the relation between the mean velocity profile and the mean temperature profile based on the Reynolds analogy. Recently, Direct numerical simulation (DNS) has been used as a powerful tool to carry out detailed studies on boundary layer characteristics. Huang [5] investigated energy budge of compressible flow boundary layers, and Pirozzoli [10]

M. White et al. (Eds.): NICFD 2022, ERCOFTAC 29, pp. 53–61, 2023.
https://doi.org/10.1007/978-3-031-30936-6_6

studied the statistics of the velocity fluctuation at different Reynolds numbers. They further confirmed that compressible turbulent statistics and coherent structures are analogous to that of equivalent in-compressible turbulent flows. Particularly, strong Reynolds analogy (SRA) is still valid for ideal compressible flows.

By contrast to ideal gases, research on organic fluids is limited. To the author's best knowledge, references [13] and [14] from Sciacovelli are the only studies of dense-gas effects on detailed turbulent behaviours in boundary layers. These studies focus on a particular real gas: PP11 (perfluoro-perhydrophenanthrene, $C_{14}F_{24}$), which has very high molecular weight and molecular complexity. It is found that the turbulent boundary layers of PP11 resemble 'incompressible flow', and a liquid-like behaviour of mean viscosity profile is observed. However, it is not known whether these phenomena occur at other thermodynamic states or in a less 'dense' gas. Further more, the laminar-turbulent transition process of organic fluids has not been investigated.

This work aimed to fill the gap by studying the impact of non-ideal gas effects on the by-pass laminar-turbulent transition for three gases: air, R1233zd(E), and MDM. The Reynolds number is equal to $635,000$, and the Mach number is equal to 2.25. Air is used as the baseline, and different thermodynamic states of organic fluids are studied based on the distances from their critical points. DNS has been carried out to simulate the by-pass transition. Wall frictions are applied to compare the speeds of transitions, vortex structures are used to analyse the coherent structures, and SRA is also discussed by comparing instantaneous fluctuation flow fields.

2 Methodology

The DNS solver for non-ideal fluids has been developed from the ideal-gas DNS solver (proposed by Li [7]) with an addition of the real-gas model part. Finite difference method (FDM) is used for the discretization of the Navier-Stokes equations. To achieve high numerical accuracy, the seventh-order upwind scheme is used for the convective terms, sixth-order central scheme is applied for the dissipative terms, and third-order Runge-Kutta method is adopted to advance time step.

The upwind scheme is based on the Steger-Warming splitting method [15], where fluxes are splitted into acoustic and entropy waves. Classic Steger-Warming splitting is based on the assumption that $p = \rho \cdot g(e)$. However, for organic fluids, the pressure does not follow the equation. Hence, the classic Steger-Warming splitting method has to be modified in the current work, where an correction term is added into the splitted fluxes.

Span-Wagner (SW) equation of state (EoS) are used in the current DNS. All parameters in these equations are based on the nature of the fluid in question and are determined by experimental data from various studies. For example, the parameters for the refrigerant R1233zd(E) and the siloxane MDM were determined by Mondeja [8] and Colonna [3] respectively. Viscosity and thermal conductivity of dense vapours depend not only on temperature but also on density (or pressure). Exact expressions were developed by Chung [1,2], from which the averaged absolute deviation is 4% for viscosity and 6% for thermal conductivity.

3 Cases Setup

The simulation domain for the bypass laminar-turbulent transition flow on a flat plate is shown in Fig. 1. U_∞, ρ_∞, T_∞ are inflow velocity, density and temperature respectively. The uniform incoming flows lead to a laminar boundary layer at the leading edge of the plate, and the simulation domain starts at the laminar boundary layer. A blowing and suction boundary is applied to simulate wall roughness and induce the bypass transition, and then the fully-developed turbulent boundary layer is achieved before the end of the domain.

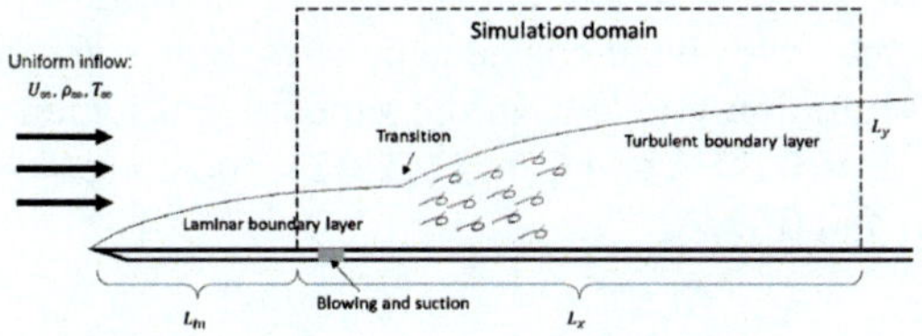

Fig. 1. Simulation domain of flat-plate flows

3.1 Air Cases

Three air cases have been carried out in the current work. The first one (named 'Case AIR') is the same as Rai [12], Pirozzoli [11], and Li [18], which is close to the experimental condition of Shutts (reported by Fernholz [4]). The characteristic length of this case is one inch ($L_{ch} = 1$ in.), the characteristic velocity and the reference thermodynamic state are based on the inflow. Parameters are normalised as:

$$\widehat{x} \equiv \frac{x}{L_{ch}}, \quad \widehat{u_i} \equiv \frac{u_i}{U_\infty}, \quad \widehat{t} \equiv \frac{tU_\infty}{L_{ch}}, \quad \widehat{T} \equiv \frac{T}{T_\infty}, \quad \widehat{p} \equiv \frac{p}{\rho_\infty U_\infty^2}, \tag{1}$$

where x is the stream-wise direction, t is time, and p is pressure. The Reynolds number ($Re_\infty \equiv \rho_\infty U_\infty L_{ch}/\mu_\infty$) is equal to 635,000, and the Mach number ($Ma_\infty \equiv U_\infty/c_\infty$) is equal to 2.25. The blowing and suction boundary condition is applied at the wall region $4.5 < \widehat{x} < 5$ to approximates the disturbance caused by wall roughness. This is achieved by introducing a randomly generated wall-normal velocity component through the region [12]. The fundamental frequency of the blowing and suction boundary condition is 2.5π.

In addition to the former case, two supplementary cases are also applied for the sake of adjusting the boundary layer thickness (named 'Case AIR S1 and AIR S2'). The differences between these air cases are shown in Fig. 2. The leading edge of the flat plate in Case AIR starts at $\widehat{x} = 0$, while it starts at $\widehat{x} = 2$ and 3 in Case AIR S1 and AIR S2, respectively. More details are given in the following paragraph.

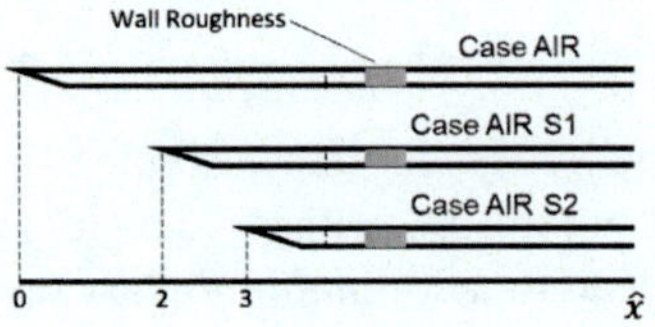

Fig. 2. Leading edge of air cases

3.2 Organic Fluids Cases

R1233zd(E) and MDM have been studied in the current work. For each fluids, three different thermodynamic states were considered. These states are located in the dilute-gas region, near the saturation line, and in the supercritical region respectively. These cases are named by Cases R1–R3 and Cases M1–M3, and corresponding T-S diagrams are shown in Fig. 3(a) and (b).

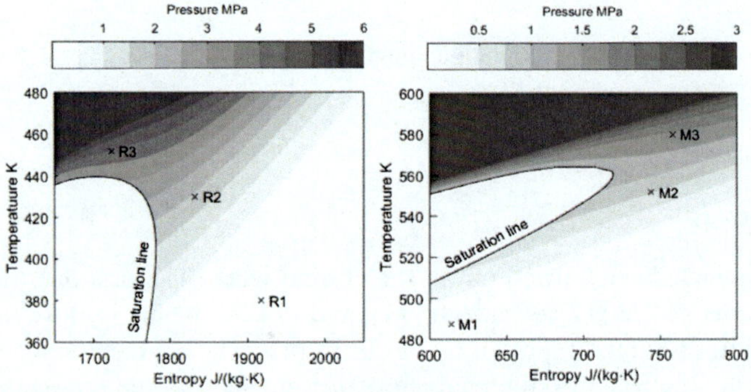

(a) *Cases R1, R2 and R3 of R1233zd(E)* (b) *Cases M1, M2, and M3 of MDM*

Fig. 3. Characteristic thermodynamic state of each organic-gas channel flow case in temperature-entropy (T-S) diagram.

Detailed information of cases has been summarised in Table 1. To get the same Re_∞, the characteristic lengths L_{ch} are different. The boundary layer thickness just upstream of the wall roughness are also listed. δ_{99} is the 99% velocity thickness, δ_1 is the displacement thickness, and θ is the momentum thickness. Seen from the table, θ of Case AIR is approximately equal to those of organic fluid cases (except Case R3), δ_{99} of Case AIR S1 is approximately equal to those of organic fluid cases, and δ_1 of Case AIR S2 is approximately equal to those of the organic fluid cases.

4 Results

4.1 Skin Friction

The skin friction coefficient is defined as $C_f \equiv 2\tau_w/\rho_\infty U_\infty^2$. The distributions of C_f along the stream-wise direction are shown in Fig. 4. Laminar boundary layers locates between

$\widehat{x} = 4$ to 4.5, where C_f is no more than 1.0. Distortion due to wall roughness is introduced at around $\widehat{x} = 4.8$. After that, bypass transition happens, with C_f significantly increase to more than 2.0. Seen from the figure, C_f of organic fluids start to increase much earlier than the air. In addition, the increase-slopes of organic fluids are slightly larger than air. This implies that organic fluids are more sensitive to boundary layer instability. Consequently, transition processes of organic fluids happen earlier and complete more quickly. Table 2 summaries the positions where the bypass transitions complete. The table suggests that MDM is slightly more unstable than R1233zd(E), which could attribute to its more significant dense-gas effect.

Table 1. Set-up information for flat-plate DNS cases.

Case	AIR	AIR S1	AIR S2	R1	R2	R3	M1	M2	M3
Gas	Air			R1233zd(E)			MDM		
T_∞/T_{cr}	–	–	–	0.86	0.98	1.03	0.86	0.98	1.03
ρ_∞/ρ_{cr}	–	–	–	0.02	0.27	1.00	0.02	0.27	1.00
$L_{ch}[mm]$	25.4	25.4	25.4	2.59	0.33	0.26	4.07	0.52	0.38
$\widehat{\delta}_{99}$	0.0180	0.0130	0.0097	0.0136	0.0133	0.0126	0.0128	0.0127	0.0126
$\widehat{\delta}_1$	0.0096	0.0069	0.0050	0.0054	0.0053	0.0047	0.0046	0.0046	0.0045
$\widehat{\theta}$	0.0016	0.0012	0.0009	0.0017	0.0016	0.0013	0.0017	0.0017	0.0016

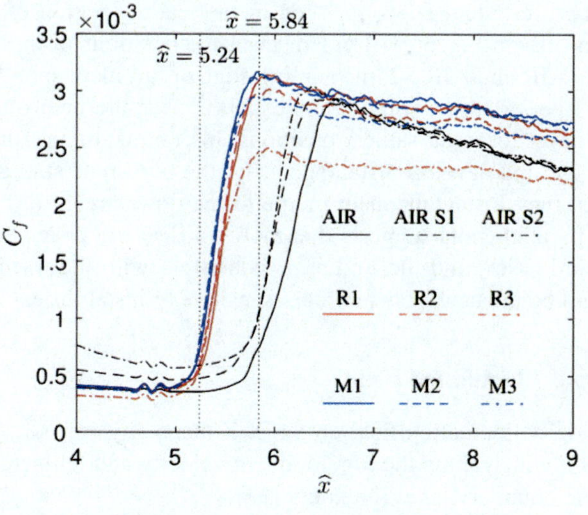

Fig. 4. Skin friction coefficient.

Table 2. The position of the end of breakdown process.

Case	AIR	AIR S1	AIR S2	R1	R2	R3	M1	M2	M3
$\widehat{x}_{br,e}$	6.62	6.41	6.39	5.92	5.95	5.94	5.80	5.80	5.79

4.2 Energy Spectra

The energy spectra between air and organic fluids are compared to illustrated the mechanisms of transition difference. Here, only the wavenumber in the X- direction, k, is analysed. Figure 5 shows the profiles of energy spectrum along the streamwise direction for wavenumbers $k = 4, 8, 16, 24, 32$ at the wall distance $y^+ = 1$. Seen from Fig. 5(a), during the transition, energy of $k = 32$ is about 10 times smaller than $K = 4$. However, for organic fluids in Fig. 5(a) and (b), the differences between $k = 4$ and $k = 32$ are much smaller. This implies that small flow structures of organic fluids have more energy, and the higher energy gives rise to early transition.

4.3 Vortex Structures

The Q criterion is used to identify vortices, and the instantaneous vortex structures of six cases are shown in Fig. 6. Here, the iso-surfaces are at $Q = 100$, and the color counter shows the wall distance $\hat{y}$. Figure 6(a) shows the vortex structures in Case AIR. The first hairpin vortex can be seen upstream of $\hat{x}_{br,s}$, where the background randomness starts to be amplified. After $\hat{x}_{br,s}$, more Ω-vortices are observed, whose head rolls upward from the wall, and significantly increasing the boundary layer thickness. As the flow moves downstream, the vortex region keep increasing until it eventually occupies the entire flow domain in the span-wise direction. At $\hat{x}_{br,e}$, the flow is totally turbulent. These phenomena are also found in all other cases shown in Fig. 6(b)–(f), although the start and end positions of the breakdown are different.

In all cases, two turbulent spots are seen in the leading region of the breakdown (labeled by 'A' and 'B' in Fig. 6(a)–(f)). However, the positions of these two spots are different. In Cases AIR and AIR S2, the leading edge of turbulent spot A is at upstream of spot B. Within a period, these two turbulent spots appear alternately and move downstream, and they never locate at same $\hat{x}$ positions. In contrast to air, for organic fluids, the two spots appear and develop instantaneously: the two spots start at similar $\hat{x}$ (see Fig. 6(c)–(f)), then they instantaneously merge to the downstream turbulent boundary layer. These two parallel spots suggests that more vortices are observed during transition in organic fluid cases, and the finding is consistent with the former that section, where organic fluid boundary layers are more sensitive to instabilities.

4.4 Instantaneous Fluctuation Field

Instantaneous velocity fluctuation field on the X-Z plane at $y = \frac{1}{2}\delta_{99,br,s}$ are compared in Fig. 7(a) and (b), which show the u-component velocity and temperature fluctuation fields. $\delta_{99,br,s}$ is the boundary layer thickness at $\hat{x}_{br,s}$.

In Case AIR, long streaks of u-component velocity perturbations are observed, corresponding to the phenomena of the bypass transition in the work of Jacob [6]. The long streaks in Case AIR are the result of stream-wise vortices, which is a 'high speed' streak caused by two tail-to-head vortices with opposite directions of rotation. The instantaneous field of temperature is significantly analogy to u-component velocity, but with opposite color. The finding solidly confirms the strong Reynold analogy (SRA), which suggests that velocity fluctuation and temperature fluctuation are strongly and negatively correlated by small turbulent vortices.

In Case M3, two 'high speed' streaks can be found. This is consistent with our finding from the analysis of the vortex structure that more stream-wise vortices are observed during transition in Case M3 than in Case AIR. By comparing velocity and temperature fluctuations, it is clear that temperature fluctuation streaks are much obscure. The blur of 'streak' structure is due to the impact of non-ideal gas effects: by contrast to ideal gas, pressure fluctuation has an obvious effect on temperature fluctuation through acoustic behaviors near the critical points, and strong Reynold analogy is challenged by the non-ideal effect.

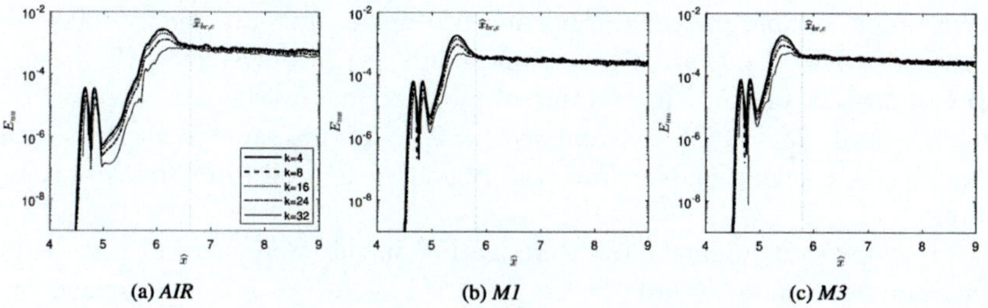

Fig. 5. Energy spectra.

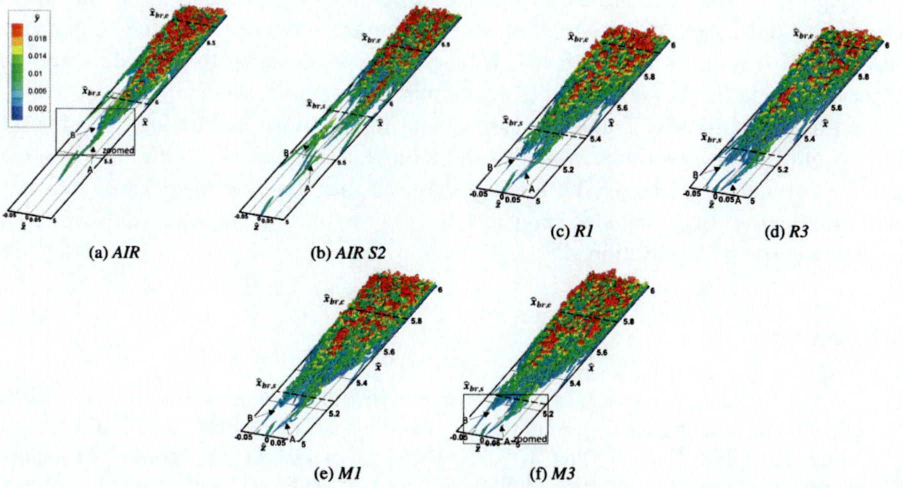

Fig. 6. Vortex structure in 3D view.

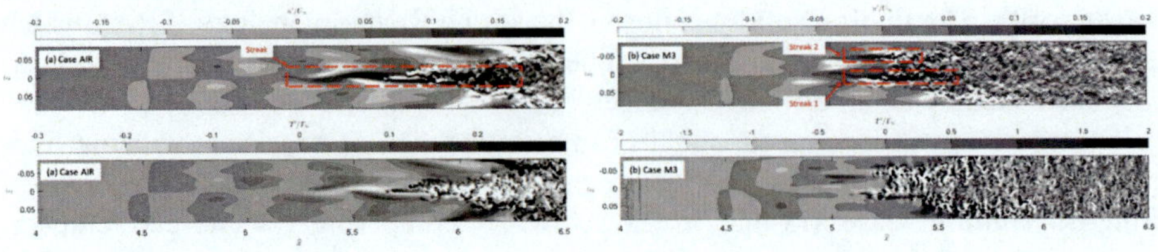

(a) *AIR* (b) *M3*

Fig. 7. Instantaneous fluctuation fields of Case AIR and Case M3.

4.5 Conclusions

The bypass-type laminar-turbulent transition of organic fluids boundary layers over a flat plate is simulated by DNS. Two organic fluids, R1233zd(E) and MDM, are studied and compared with air. Three different thermodynamic states are considered for each organic fluid. $Re_\infty = 635,000$ and $Ma_\infty = 2.25$ are the same in all DNS cases for comparability. A blowing-and-suction wall boundary is used to induce a bypass-type transition.

The breakdown of laminar flow starts earlier in the more 'dense' gas under any Reynolds number similarity criteria $(Re_x, Re_\theta, Re_{\delta 99}, Re_{\delta 1})$. All investigated organic fluid boundary layers breakdown significantly earlier than air boundary layers, but only a small difference between organic fluid cases is shown. After breakdown, for organic fluids, C_f increases faster, and it decreases more slowly after reaching the peak value.

The mechanism of transition in both air and organic fluids is due to the secondary instability of hairpin vortices. The KH instability is found at the closet point between the two stream-wise vortices, and the instability subsequently propagates along the vortex. In organic fluid cases, more stream-wise vortices are observed, and the propagation speed of KH instability is also faster. Hence the appearance of the turbulent spots is earlier in organic fluid cases, and the breakdown also occurs earlier.

For the air, similarity between velocity and temperature fluctuations implies that SRA is solidly valid. However, velocity and temperature fluctuations are less well correlated in organic gas cases. The finding suggests that, for non-ideal gases, not only small turbulent vortices, but also acoustics due to pressure fluctuations would affect the turbulent energy transportation.

References

1. Chung, T.H., Ajlan, M., Lee, L.L., Starling, K.E.: Generalized multiparameter correlation for nonpolar and polar fluid transport properties. Ind. Eng. Chem. Res. **27**(4), 671–679 (1988)
2. Chung, T.H., Lee, L.L., Starling, K.E.: Applications of kinetic gas theories and multiparameter correlation for prediction of dilute gas viscosity and thermal conductivity. Ind. Eng. Chem. Fundam. **23**(1), 8–13 (1984)
3. Colonna, P., Nannan, N., Guardone, A.: Multiparameter equations of state for selected siloxanes. Fluid Phase Equilib. **263**(2), 115–130 (2008)

4. Fernholz, H.-H., Finley, P.: A critical compilation of compressible turbulent boundary layer data. Technical report, Advisory Group for Aerospace Research and Development, Neuilly-sur-Seine, France (1977)
5. Huang, P., Coleman, G., Bradshaw, P.: Compressible turbulent channel flows: DNS results and modelling. J. Fluid Mech. **305**, 185–218 (1995)
6. Jacobs, R., Durbin, P.: Simulations of bypass transition. J. Fluid Mech. **428**, 185–212 (2001)
7. Li, X., Ma, Y., Fu, D.: DNS and scaling law analysis of compressible turbulent channel flow. Sci. China, Ser. A Math. **44**(5), 645–654 (2001)
8. Mondejar, M.E., McLinden, M.O., Lemmon, E.W.: Thermodynamic properties of trans-1-chloro-3, 3, 3-trifluoropropene (R1233zd (E)): Vapor pressure,(p, ρ, T) behavior, and speed of sound measurements, and equation of state. J. Chem. Eng. Data **60**(8), 2477–2489 (2015)
9. Morkovin, M.V.: Effects of compressibility on turbulent flows. Mécanique de la Turbulence **367**(380), 26 (1962)
10. Pirozzoli, S., Bernardini, M.: Turbulence in supersonic boundary layers at moderate Reynolds number. J. Fluid Mech. **688**, 120 (2011)
11. Pirozzoli, S., Grasso, F., Gatski, T.: Direct numerical simulation and analysis of a spatially evolving supersonic turbulent boundary layer at M = 2.25. Phys. Fluids **16**(3), 530–545 (2004)
12. Rai, M., Gatski, T., Erlebacher, G.: Direct simulation of spatially evolving compressible turbulent boundary layers. In: 33rd Aerospace Sciences Meeting and Exhibit, p. 583 (1995)
13. Sciacovelli, L., Cinnella, P., Gloerfelt, X.: Direct numerical simulations of supersonic turbulent channel flows of dense gases. J. Fluid Mech. **821**, 153–199 (2017)
14. Sciacovelli, L., Gloerfelt, X., Passiatore, D., Cinnella, P., Grasso, F.: Numerical investigation of high-speed turbulent boundary layers of dense gases. Flow Turbul. Combust. **105**(2), 555–579 (2020)
15. Steger, J.L., Warming, R.: Flux vector splitting of the inviscid gasdynamic equations with application to finite-difference methods. J. Comput. Phys. **40**(2), 263–293 (1981)
16. Van Driest, E.R.: Turbulent boundary layer in compressible fluids. J. Aeronaut. Sci. **18**(3), 145–160 (1951)
17. Walz, A.: Boundary Layers of Flow and Temperature. MIT Press, Cambridge (1969)
18. Xin-Liang, L., De-Xun, F., Yan-Wen, M., Hui, G.: Acoustic calculation for supersonic turbulent boundary layer flow. Chin. Phys. Lett. **26**(9), 094701 (2009)

High Fidelity Simulations and Modelling of Dissipation in Boundary Layers of Non-ideal Fluid Flows

Francesco Tosto[1(✉)], Andrew Wheeler[2], and Matteo Pini[1]

[1] Propulsion and Power, Delft University of Technology, Delft, Netherlands
{f.tosto,m.pini}@tudelft.nl

[2] Whittle Laboratory, University of Cambridge, Cambridge, UK
aw329@cam.ac.uk

Abstract. In this work, we investigate the sources of dissipation in adiabatic boundary layers of non-ideal compressible fluid flows. Direct numerical simulations of transitional, zero-pressure gradient boundary layers are performed with an in-house solver considering two fluids characterized by different complexity of the fluid molecules, namely air and siloxane MM. Different sets of thermodynamic free stream boundary conditions are selected to evaluate the influence of the fluid state on the frictional loss and dissipation mechanisms. The thermophysical properties of siloxane MM are obtained with a state-of-the-art equation of state. Results show that the dissipation due to both time-mean strain field and irreversible heat transfer, and the turbulent dissipation are significantly affected by both the molecular complexity of the fluid and its thermodynamic state. The dissipation coefficient calculated from the DNS is then compared against the one obtained from a reduced-order boundary layer CFD model [1] which has been extended to treat fluids modeled with arbitrary equations of state [7].

Keywords: boundary layer · dissipation coefficient · Non-Ideal Compressible Fluid Dynamics · organic Rankine cycle · Direct Numerical Simulation · turbulence

1 Introduction

The irreversible entropy generation due to viscous processes in boundary layers is one of the main loss mechanisms affecting internal flow devices such as turbomachines or heat exchangers. With regards to turbomachinery, this contribution can account for up to one-sixth of the total loss in a turbine [3]. The share increases in compressors, where the flow becomes more prone to separation due to the presence of adverse pressure gradients over the blade. Fluid flow characteristics strongly affect the viscous dissipation, which depends on the free-stream Reynolds and Mach number, the state of the boundary layer, i.e., laminar or turbulent, and the fluid parameters.

Usually, the rate of entropy generation is estimated with the so-called dissipation coefficient, C_d. Unlike the skin friction coefficient C_f, which measures the local dissipation at the wall, the dissipation coefficient shows a weak dependence on the boundary layer shape factor and is a good measure of the amount of dissipation occurring

M. White et al. (Eds.): NICFD 2022, ERCOFTAC 29, pp. 62–71, 2023.
https://doi.org/10.1007/978-3-031-30936-6_7

within the boundary layer. Nevertheless, its value strongly depends on the boundary layer regime, i.e., laminar or turbulent. As pointed out by Denton [3] and Schlichting [8], in turbulent boundary layers, the C_d shows a weak dependence on the pressure gradient. Therefore, the turbulent flat plate boundary layer flow still provides good indications of the amount of viscous dissipation.

Although analytical solutions and correlations for both C_f and C_d in boundary layer flows are well established in literature [8, 12], little knowledge has been developed on how both the complexity of the molecular structure of fluid and the thermodynamic state of the free-stream flow influence the viscous dissipation and the loss breakdown. Indeed, internal flow devices for process and energy systems such as organic Rankine cycle power system [2] operate with complex organic compounds, often in the supercritical or dense vapor state. Such fluids exhibit strong variations of thermo-physical fluid properties which sensibly affect the development of the boundary layer, the turbulence breakdown, and the dissipation. Direct numerical simulations of dense vapor boundary layer flows [9] have been performed to investigate the turbulence in non-ideal fluid dynamics. However, no simulation in thermodynamic and fluid dynamic conditions of engineering interest, using accurate models for the estimation of the thermo-physical properties of the fluid, has been performed.

In this work, we study the various sources of dissipation in adiabatic boundary layers in zero-pressure gradient non-ideal compressible fluid flows. Direct numerical simulations of transitional, zero-pressure gradient boundary layers are performed with an in-house solver. Two fluids characterized by different levels of complexity of the fluid molecules are considered, namely air and hexamethyldisiloxane (MM). Different sets of thermodynamic free stream boundary conditions are selected to evaluate the influence of the fluid state on the frictional loss and dissipation mechanisms. State-of-the-art equations implemented in a well-known fluid library [6] are used to estimate the thermo-physical properties of siloxane MM. Results are compared against those from a reduced-order model (ROM) code [7] solving the boundary layer equations in transformed coordinates for dense vapors. The final goal of the study is to develop a unified relation for the dissipation coefficient C_d as a function of the Reynolds number, the thermodynamic fluid state, the molecular complexity, and the level of flow compressibility. Such relation can then be used in engineering applications for preliminary estimations of the profile loss in internal flow devices such as turbomachinery or heat exchangers.

2 Theoretical Background

Resolving the small scales of the turbulent flow field through DNS enables the accurate investigation of the dissipation occurring in the vicinity of the wall. According to Hughes and Brighton [4], the losses can be estimated by inspecting the rise in entropy due to irreversible processes. Considering a control volume encompassing the whole boundary layer, the time-mean rate of entropy change is calculated as [11]

$$\int_S \overline{\rho s \mathbf{V}} \cdot \mathrm{d}\mathbf{S} = \int_V \overline{\theta} - \int_S \overline{\left(\frac{\mathbf{q}}{T}\right)} \cdot \mathrm{d}\mathbf{S}, \tag{1}$$

where the overline denotes time-average quantities. The two terms on the right-hand side are the time-average fluxes of entropy and reversible heat transfer across the surfaces. The dissipation is embedded in $\overline{\theta}$ and can be rewritten as

$$\overline{\theta} = \overline{\frac{\Phi}{T} + \frac{\epsilon}{T} + \frac{\zeta}{T}}. \tag{2}$$

The first two terms on the right-hand side constitute the viscous dissipation, which can be further split into a contribution from the mean velocity and strain field, $\overline{\Phi}/T$, and a contribution due to unsteady effects, i.e., the turbulent kinetic energy dissipation, $\overline{\epsilon}/T$. The third term represents the contribution due to irreversible heat transfer. $\overline{\theta}$ can be determined by calculating the instantaneous total dissipation terms ($\Phi/T + \epsilon/T$) and heat flux term (ζ/T) and averaging them over time. The contribution from the time-mean strain, Φ/T, instead, can be computed from the Favre-average flow, and this can then be subtracted from $\Phi/T + \epsilon/T$ to obtain the contribution due to turbulent dissipation. Alternatively, for a boundary layer in equilibrium, $\overline{\Phi}/T$ within the control volume can be obtained by subtracting the time-average turbulence production $\overline{Pr}$ across the boundary layer, defined as

$$\overline{Pr} = -\frac{1}{\rho u_\mathrm{e}^3}\int_0^\delta \overline{\rho u_i u_j}\frac{\partial \overline{u_i}}{\partial x_j}\mathrm{d}y, \tag{3}$$

and the reversible and irreversible contributions to heat transfer from the entropy flux, in line with Eq. 1. Indeed, for an equilibrium boundary layer, the production term $\overline{P}$ matches the turbulence dissipation, $\overline{\epsilon}$. This second approach is used in this study because the integration of the turbulence production is less mesh sensitive than that of the dissipation terms. Turbulence production is also used to compute the dissipation coefficient, which, for a compressible flow case, reads

$$C_\mathrm{d} = \overline{Pr} + \frac{1}{\rho_\mathrm{e} u_\mathrm{e}^3}\int_0^\delta \mu\left(\frac{\partial \overline{u_i}}{\partial x_j}\right)^2 \mathrm{d}y. \tag{4}$$

In this equation, the first term is the contribution to the dissipation due to the turbulent production, which matches the turbulent dissipation in equilibrium boundary layers, while the second one is the contribution due to the time-mean strain field. The dissipation coefficient here takes into account the loss of mechanical energy of the mean flow rather than that of the total mechanical energy, in agreement with the approach used to compute dissipation coefficients with conventional Reynolds-averaged Navier-Stokes solvers.

The C_d values obtained from the DNS have been compared against those obtained from a reduced-order model (ROM) solving the two-dimensional boundary layer equations in transformed coordinate and implementing the Cebeci-Smith turbulence model. This model is implemented in *BLnl*, an in-house developed Matlab code. A detailed overview of the ROM, its numerical implementation, and the turbulence model can be found in Refs. [1,7].

3 Methodology

To accurately evaluate the loss breakdown within a non-ideal boundary layer flow, direct numerical simulations of zero-pressure gradient boundary layer flows of air and siloxane MM have been performed. The simulations have been run on *3DNS*, a multi-block structured compressible Navier-Stokes solver for high-performance computations. A technical overview of the solver is provided in Refs. [11].

Three different sets of high-fidelity numerical simulations have been run. Table 1 lists the details of the test cases. All simulations have been performed using a free-stream Mach number of 0.9. The first case simulates the boundary layer flow of air modelled as an ideal gas at standard conditions, i.e., $T_\mathrm{e} = 288$ K and $p_\mathrm{e} = 1$ bar, the subscript e denoting the free-stream conditions. The iMM and niMM cases, instead, refer to simulations of siloxane MM at free-stream thermodynamic conditions close to that of a dilute gas and a dense vapor, respectively. Thermo-physical properties of siloxane MM are calculated using an entropy-based equation of state which is explicit in density and internal energy. This equation has been developed by interpolating fluid property data obtained from accurate Helmholtz energy-based equations of state [6].

Table 1. Test cases evaluated in this study. Free-stream conditions and grid resolution parameters are reported.

	fluid	$p_{\mathrm{r,e}}$	$T_{\mathrm{r,e}}$	Z_e	Ec_e	L_x [m]	H
air	air ideal gas	0.0264	2.182	1	$1.1 \cdot 10^{-3}$	0.023	1.84
iMM	siloxane MM	0.1	1	0.96	$1.5 \cdot 10^{-4}$	$5 \cdot 10^{-3}$	1.45
niMM		1.15	1.05	0.54	$6.8 \cdot 10^{-5}$	$4.42 \cdot 10^{-4}$	1.49
	fluid	x^+	$y^+_{1\mathrm{st}}$	$y^+_{10\mathrm{th}}$	z^+	$n_i \times n_j \times n_k$	
air	air ideal gas	9.06	0.88	9.85	9.06	$2500 \times 800 \times 250$	
iMM	siloxane MM	9.13	0.88	9.85	9.13	$2500 \times 800 \times 250$	
niMM		8.82	0.88	9.85	8.82	$2500 \times 800 \times 250$	

The number of grid points in each direction, as well as the corresponding values of x^+, y^+ at the first cell in the proximity of the wall, and z^+ are reported in Table 1. The plate length is defined to ensure a $Re_L = 5 \cdot 10^5$ for all cases. The corresponding channel height and width are defined as 20% and 10% of the length, respectively. The prescribed height of the domain ensures that the influence of the blockage on the flow is negligible: the measured values of Clauser's beta parameter, defined as

$$\beta_{\mathrm{Cl}} = \frac{\delta^*}{\tau_w}\frac{\mathrm{d}p}{\mathrm{d}x} \tag{5}$$

which takes into account the effect of the pressure gradient on the boundary layer, are within the $-0.15 < \beta_{\mathrm{Cl}} < 0$ range. As a consequence, the mild favourable pressure gradient does not significantly affect the boundary layer development. Stagnation temperature and pressure values are prescribed at the inlet, the static pressure value is instead prescribed at the outlet. No-slip adiabatic conditions are imposed at the wall. The upper surface is modeled as a free-slip wall, while periodic boundary conditions are imposed in the spanwise direction. To avoid non-physical leading edge effects, a Blasius laminar profile at $Re_x = 10^5$ is introduced at the inlet. The CFL number is unitary. Laminar two-dimensional simulations with the boundary conditions reported in Table 1 are first performed. Solutions from these simulations are then used to initialize the three-dimensional simulations. 3D simulations are performed in the laminar regime for the first 10000 time steps. Then, a trip is triggered at a location x such that $Re_x = 2 \cdot 10^5$.

Statistical averages are performed over the homogeneous spanwise direction and over at least two convective time units based on the free-stream velocity and the axial chord. During the simulations, time-mean flow field and turbulence quantities are calculated by means of a Favre (density-weighted) time-averaging technique.

For comparison, the same cases reported in Table 1 are simulated using the *BLnI* code. However, the free-stream Mach number is here varied to study the effect of compressibility on the overall dissipation. Both laminar, fully turbulent, and transitional regimes are numerically simulated: the transition is imposed at the same location where the trip is introduced in the DNS. The plate length matches that of the numerical simulations, and it is discretized with 200 points in the longitudinal direction. The first grid spacing in the normal-to-wall direction is set to 10^{-5} m for all cases, with a grid expansion factor of 1.02. These values have been proved to ensure grid-independent results. The wall is adiabatic.

4 Results

4.1 Analysis of Dissipation Based on DNS

Figure 1 shows the results of the loss analysis described in Sect. 2 for all the cases listed in Table 1. To avoid nonphysical results due to the flow transition induced by tripping the boundary layer, the rise in irreversible entropy production is computed by integrating over a control volume, where the inlet boundary is located right after the transition point and the outlet one at the given streamwise location. Within this control

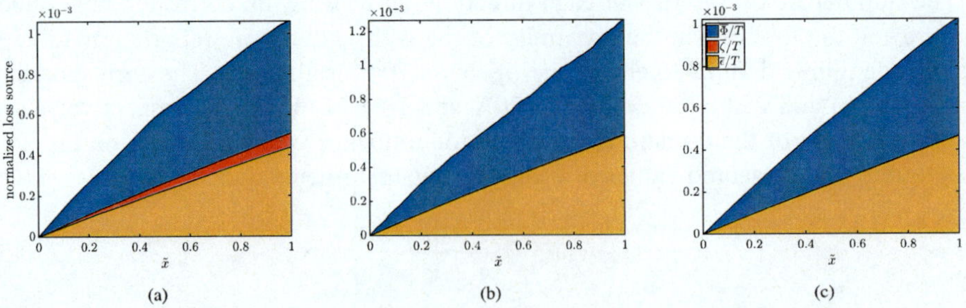

Fig. 1. Dissipation breakdown along the flat plate for the cases (a) *air*, (b) *iMM* and (c) *niMM*. The integration of each contribution is performed from $\tilde{x} = 0$ to $\tilde{x} = 1$, where $\tilde{x} = (x - x_{tr})/(L_x - x_{tr})$.

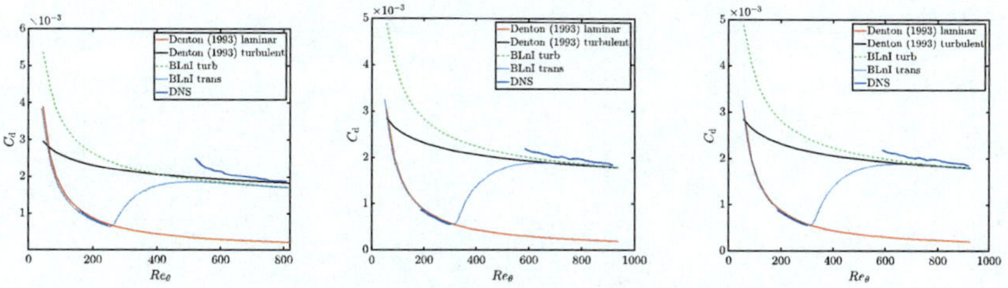

(a) (b) (c)

Fig. 2. Dissipation coefficient C_{d} *vs* Re_θ for the cases (a) *air*, (b) *iMM* and (c) *niMM*. Results from turbulent and transitional simulations conducted with *BLnl* as well as those obtained from the DNS are plotted. Correlations by Denton and Schlichting are also displayed.

volume, the boundary layer is assumed to have reached an equilibrium state. No major differences are observed between the *iMM* and *niMM* cases: a relatively higher dissipation due to the time-mean strain field is observed for the *niMM* case, arguably due to the higher value of dynamic viscosity that siloxane MM exhibits in the dense vapor state. The contribution due to irreversible heat transfer is negligible in both cases. In the case of air, instead, such contribution becomes non-negligible; however, it remains an order of magnitude lower than the other two contributions. Indeed, in perfect gases, the heat generated by the dissipation of the turbulent kinetic energy at the smallest scales induces variations of thermo-physical properties larger than those occurring in fluid characterized by high molecular complexity such as siloxane MM.

4.2 Comparison DNS-ROM

Figure 2 shows the trend of the dissipation coefficient with Re_θ computed for the cases of Table 1. The results of the fully turbulent and transitional simulations conducted with *BLnl* are also displayed. Trends of C_{d} as provided by the empirical correlations of Schlichting [3, 8] for the laminar and turbulent incompressible cases are also plotted for comparison.

In the laminar regime, the dissipation coefficient calculated from the DNS results qualitatively matches the trend predicted by means of both the empirical and the reduced-order model. The maximum deviation between the values predicted by DNS and by the correlation is of the order of 2%. In the turbulent regime, the trend of C_{d} approaches that of the incompressible Schlichting correlation for both the three cases in the proximity of the trailing edge of the plate. The $C_{\mathrm{d}} - Re_\theta$ trend in the proximity of the tripping point is not reported due to non-physical trends in that region. For the sake of clarity, the three solutions are reported in Fig. 3b. The lowest value of C_{d} in the turbulent regime is found for the case of air when $Re_\theta > 800$, suggesting that fluids made by simple molecules exhibit reduced values of dissipation at fixed M_{e} and Re_θ. The trend can also be explained by inspecting the value of the freestream Eckert number, which is inherently lower for dense vapor flows. As a consequence, in fluids made by complex molecules in both the ideal and the non-ideal thermodynamic state, the thermal and kinetic fields are decoupled [9, 10] irrespective of the freestream Mach

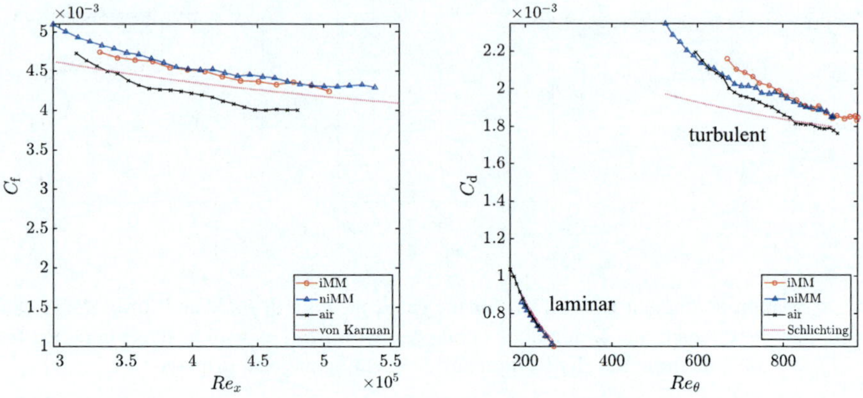

Fig. 3. (a) Skin friction coefficient *vs* Re_x and (b) dissipation coefficient *vs* Re_θ for all the cases of Table 1. Trends obtained from empirical correlations are also plotted for comparison.

number, resembling the physical behavior of an incompressible flow. This result is in line with those obtained by [7] for adiabatic laminar boundary layers of dense vapors at $M_e = 2$. Figure 3a shows the trend of the skin friction coefficient C_f for all the three investigated cases. Trends are qualitatively in agreement with those obtained for the dissipation coefficient C_d. According to Schlichting, for a zero-pressure gradient boundary layer flow in equilibrium, the relation between the dissipation and the skin friction coefficient is

$$C_d = \frac{C_f}{2}\frac{H^*}{2}, \tag{6}$$

where $H^* = \theta^*/\theta$ is the energy shape factor, being θ^* and θ the kinetic energy and momentum thicknesses, respectively. In the turbulent flow region, $H^* \sim 1.774$ for all cases, which is in line with the values computed by Jardine [5]; as a consequence, changes in the dissipation coefficient between the three cases are solely due to changes in the value of the skin friction coefficient.

Figure 4a shows the velocity profiles in wall units evaluated at a x location corresponding to $Re_\theta = 800$, while Fig. 4b depicts the compressible van-Driest velocity profile as a function of the y^+ evaluated at the same location. Results from the *BLnI* code evaluated at the same Re_θ are also reported. No major deviation from the law of the wall trend is observed for all cases in neither the viscous sublayer, the buffer zone, where the peak in turbulence production is observed (see Fig. 5a), and the logarithmic region. However, in the buffer zone and the logarithmic region, the trends obtained from the ROM do not collapse over those of the DNS. The larger deviation from the DNS trend is measured for the *air* case, where the relative difference in the u^+ value at $y^+ = 40$ is 10%. Such deviation decreases if van-Driest velocity profiles are considered: for the *air* case, at $y^+ = 40$, the relative difference is 7%. Although $M_e = 0.95$ for all the three cases, results obtained by applying the van Driest transformation do not show any major deviation from the $u^+ - y^+$ trend of Fig. 4a. The turbulence production is almost insensitive to the fluid molecular complexity and the thermodynamic state, which peaks at $y^+ \sim 10$ for all three cases.

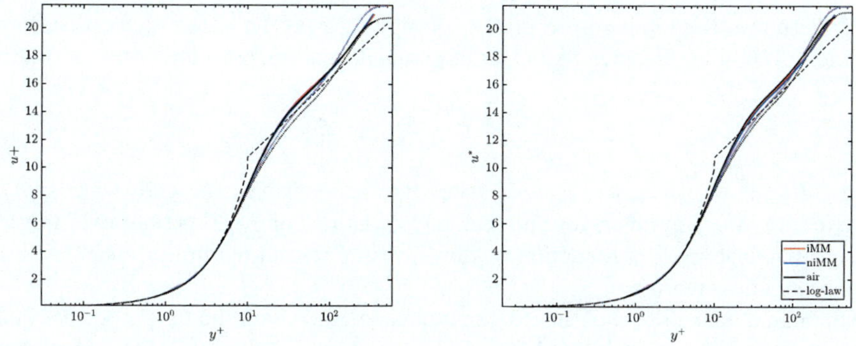

Fig. 4. Velocity profiles in (a) wall coordinates and (b) van Driest transformed coordinates. Continuous lines denote the trends obtained from the DNS, whereas dotted ones denote those obtained from the ROM. Black dashed lines denote the linear and logarithmic laws, with $k = 0.41$ and $C = 5.2$

Figure 5b displays the enthalpy profiles within the boundary layer evaluated at a x location corresponding to $Re_\theta = 800$. Results obtained from both the DNS and the ROM are plotted. It can be noted that the enthalpy increase close to the wall strongly depends on both the fluid molecular complexity and the thermodynamic fluid state. In particular, the enthalpy difference between the wall and the freestream is maximum in the case of air, while the boundary layer of siloxane MM can be assumed almost isothermal in both thermodynamic conditions. This finding is inherently related to the high heat capacity of the organic fluid, which enables to store thermal energy using the many degrees of freedom of the molecule. Among the three cases, the *niMM* case shows a nearly flat enthalpy profile, due to the increasingly high c_p in the dense vapor state.

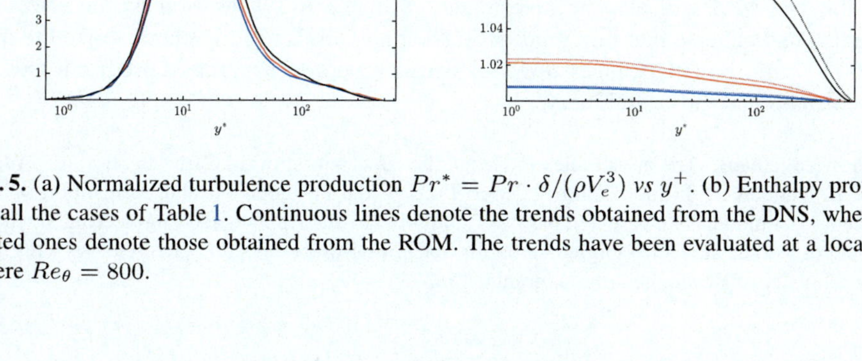

Fig. 5. (a) Normalized turbulence production $Pr^* = Pr \cdot \delta/(\rho V_e^3)$ *vs* y^+. (b) Enthalpy profiles for all the cases of Table 1. Continuous lines denote the trends obtained from the DNS, whereas dotted ones denote those obtained from the ROM. The trends have been evaluated at a location where $Re_\theta = 800$.

Furthermore, the temperature profile strongly affects the value of the dissipation coefficient. Using the Falkner-Skan variable transformation, one can write, for a turbulent flow,

$$C_{\mathrm{d}} = \frac{1}{\sqrt{Re_x}} \int_0^{\eta_{\mathrm{e}}} CR_{\mathrm{turb}} \frac{T_{\mathrm{e}}}{T} \tilde{v}^2 \mathrm{d}\eta, \tag{7}$$

where $CR_{\mathrm{turb}} = \rho(\mu + \mu_{\mathcal{T}})/(\rho_{\mathrm{e}}\mu_{\mathrm{e}})$ takes into account also the eddy viscosity $\mu_{\mathcal{T}}$. At a fixed Reynolds number, for an ideal gas, variations of T_{e}/T prevail over those of CR_{turb}, thus leading to a reduction of the C_{d} value for simple fluid molecules in the compressible flow regime.

Although a mismatch is observed, the trends obtained with the *BLnI* code are qualitatively in agreement with those obtained from the high-fidelity numerical simulations. This corroborates the hypothesis that the ROM, coupled with an algebraic turbulence model, is sufficiently accurate to preliminary estimate the dissipation rate occurring in laminar and turbulent boundary layers.

5 Conclusion

Direct numerical simulations of zero-pressure gradient boundary layer flow of air and siloxane MM operated in both the dilute gas and dense vapor state have been performed. All sources of dissipation have been discerned and analyzed. The trend of the skin friction and dissipation coefficient as a function of the Reynolds number, as well as those of the velocity and temperature profiles, have been discussed and compared against those obtained from a reduced order model implemented in an in-house code which solves the two-dimensional turbulent boundary layer equations in transformed coordinates. Based on the results obtained from the work, the following conclusions can be drawn.

1. At fixed Reynolds number, turbulent boundary layer flows of fluids made of complex molecules exhibit a higher value of both the skin friction and the dissipation coefficients than those of air. This is due to the decoupling between thermal and kinematic fields.
2. Regardless of the fluid thermodynamic state, turbulent boundary layer flows of complex molecules fluids are almost isothermal.
3. The loss contribution due to irreversible heat transfer is negligible in flows of fluids made of high molecular complexity.

Future works will investigate more in detail the structure of density and pressure fluctuations in ideal gas and dense vapor boundary layers. Other fluids other than air and siloxane MM will also be investigated with the ROM, as well as the effect of favorable and adverse pressure gradients. The final aim is to fully characterize the dissipation within boundary layer flows of dense vapors to accurately predict losses in non-conventional turbomachinery.

Acknowledgement. The authors acknowledge the contribution of dr. Carlo de Servi, dr. Adam J. Head, Dominic Dijkshoorn, and Federico Pizzi to the development and verification of the reduced-order model. This research has been supported by the Applied and Engineering Sciences Domain (TTW) of the Dutch Organization for Scientific Research (NWO), Technology Program of the Ministry of Economic Affairs, grant # 15837.

References

1. Cebeci, T.: Convective Heat Transfer. Number v. 2 in Convective Heat Transfer. Horizons Publishing (2002)
2. Colonna, P., et al.: Organic rankine cycle power systems: from the concept to current technology, applications, and an outlook to the future. J. Eng. Gas Turbines Power **137**(10), 100801 (2015)
3. Denton, J.D.: Loss mechanisms in turbomachines. In: Volume 2: Combustion and Fuels; Oil and Gas Applications; Cycle Innovations; Heat Transfer; Electric Power; Industrial and Cogeneration; Ceramics; Structures and Dynamics; Controls, Diagnostics and Instrumentation; IGTI Scholar Award, Cincinnati, Ohio, USA, p. V002T14A001. American Society of Mechanical Engineers (1993)
4. Hughues, W.F., Brighton, J.A.: Fluid Dynamics (Schaum's Outline Series), 2nd edn. McGraw-Hill, New York (1983)
5. Jardine, L.J.: The Effect of Heat Transfer on Turbine Performance, p. 185
6. Lemmon, E.W., Bell, I.H., Huber, M.L., McLinden, M.O.: NIST standard reference database 23: reference fluid thermodynamic and transport properties-REFPROP, version 10.0, National Institute of Standards and Technology (2018)
7. Pini, M., De Servi, C.: Entropy generation in laminar boundary layers of non-ideal fluid flows. In: di Mare, F., Spinelli, A., Pini, M. (eds.) NICFD 2018. LNME, pp. 104–117. Springer, Cham (2020). https://doi.org/10.1007/978-3-030-49626-5_8
8. Schlichting, H., Gersten, K.: Boundary-Layer Theory, 9th edn. Springer, Heidelberg (2016). https://doi.org/10.1007/978-3-662-52919-5
9. Sciacovelli, L., Gloerfelt, X., Passiatore, D., Cinnella, P., Grasso, F.: Numerical investigation of high-speed turbulent boundary layers of dense gases. Flow Turbulence Combust **105**(2), 555–579 (2020)
10. Tosto, F., Lettieri, C., Pini, M., Colonna, P.: Dense-vapor effects in compressible internal flows. Phys. Fluids **23**, 086110 (2021)
11. Wheeler, A.P.S., Sandberg, R.D., Sandham, N.D., Pichler, R., Michelassi, V., Laskowski, G.: Direct numerical simulations of a high-pressure turbine vane. J. Turbomach. **138**(7) (2016)
12. White, F.M., Majdalani, J.: Viscous Fluid Flow, vol. 3. McGraw-Hill, New York (2006)

Estimating Model-Form Uncertainty in RANS Turbulence Closures for NICFD Applications

Giulio Gori(✉)

Department of Aerospace Science and Technology, Politecnico di Milano, Via La Masa 34, 20156 Milano, Italy
giulio.gori@polimi.it

Abstract. The investigation of applications entailing non-ideal compressible fluid flows largely relies on Computational Fluid Dynamics (CFD) simulations. Popular CFD models rely on the Reynolds-Averaged Navier-Stokes equations (RANS). In RANS computations, turbulence models must be employed to reconstruct the Reynolds stress term arising from the time-averaged decomposition of the Navier-Stokes equations. Though literature is teeming with works supporting the development of turbulence closures for flows of fluids of common interest, little, if not just exploratory works, can be found regarding non-ideal compressible fluid flows. Moreover, a scarce amount of experimental data prevents the empirical development of turbulence models of general validity for non-ideal flows. Recently, formal estimation techniques have been developed to provide a characterization of the prediction uncertainty related to the model-form error inherent to the structure of turbulence closures. Here, we present the application of the Eigenspace Perturbation Method (EPM) from Emory et al. (2011) to a use case fundamental to non-ideal compressible fluid flows i.e., the backward facing step. The use case consists of an infinite width channel, and it is simulated both in the subsonic and supersonic regimes. Results show how predictions are affected by the model-form uncertainty inherent turbulence closures.

Keywords: Uncertainty quantification · Eigenspace Perturbation Method · Turbulence closure uncertainty · NICFD

1 Introduction

The investigation of complex non-ideal compressible fluid flows largely relies on numerical tools. In Computational Fluid Dynamics (CFD) simulations, a popular strategy to predict the flow behavior is to rely on the Reynolds-Averaged Navier-Stokes equations (RANS). In RANS computations, turbulence models must be employed, to reconstruct the Reynolds stress term arising from the time-averaged decomposition of the Navier-Stokes equations. The issue of establishing accurate turbulent closures suited to applications entailing flows of fluids of common interest (air, water, and many other)

M. White et al. (Eds.): NICFD 2022, ERCOFTAC 29, pp. 72–81, 2023.
https://doi.org/10.1007/978-3-031-30936-6_8

has been investigated since decades. In the realm of ideal compressible flows, turbulence modeling produced a large number of closures of proven accuracy and reliability that can be confidently used for investigating a wide variety of problems e.g., aeronautic applications. These often consist in empirical or semi-empirical closures that depend on a set of coefficients optimized for the specific application. On the contrary, little, if not just exploratory works, can be found regarding the modeling of turbulence in Non-Ideal Compressible Fluid Flows (NICFD). Namely, a scarce amount of experimental data prevents the empirical estimation of turbulence coefficients and the throughout validation of numerical predictions. Moreover, closures typically rely on strong inherent model-form assumptions that limit the fidelity of RANS predictions, especially for flows characterized by a pronounced streamline curvature or adverse pressure gradients [2].

To handle these aspects, engineers are prompted to resort to Uncertainty Quantification (UQ) techniques. Unfortunately, the direct assessment of the error introduced in turbulent modeling is a very demanding task [18], if not just an intractable one. In the last decade, a formal technique capable of providing an estimation of the uncertainty corresponding to turbulence modeling approximations in RANS closures, the so-called Eigenspace Perturbation Method (EPM) [4, 5, 8], was devised. Here, we present an application of the EPM to a use case fundamental to NICFD i.e., the backward facing step. The backward facing step is widely known for its application in the studies on turbulence in internal flows because it entails a sudden flow separation which results into the creation of a recirculating zone. In case a supersonic regime is attained, a shock wave is generated at the reattachment point past the recirculating zone. These phenomena are common in practical applications. Indeed, from the qualitative standpoint, the backward facing step resembles the trailing edge of a turbine blade. To lower the computational burden, the considered use case is analyzed relying on the purely two-dimensional approximation i.e., infinite width channel. In the analysis, we try to highlight how the prediction of several quantities of interest is affected by the model-form uncertainty inherent turbulence closures.

The paper is organized as follows. Section 2 exposes the fundamental concepts underlying the EPM and how it can be employed to compute the turbulence uncertainty estimates. Section 3 presents the application of the EPM to the two-dimensional backward facing step. Eventually, Sect. 4 summarizes the findings.

2 Methodology

The so-called Reynolds Stress Tensor (RST) is a byproduct of the application of the Reynolds decomposition to the Navier-Stokes equations. The RST incorporates all the effects of turbulent motions and, by definition, it is symmetric and positive semi-definite. Typically, the RST is approximated by means of a semi-empirical turbulence closure

$$\langle u_i u_j \rangle \approx \langle u_i u_j \rangle^{\mathrm{BS}}, \tag{1}$$

with $i, j = \{1, 2, 3\}$, where superscript BS stands for "baseline", u is the fluctuating component of the velocity field (according the Reynolds decomposition), and the angular brackets indicate the time-averaging operation. Despite a large body of literature dealing with the problem of formulating closures for estimating the six independent components of the RST, it is acknowledged that the strong inherent model-form assumptions limit the general credibility of RANS predictions [2].

The Eigenspace Perturbation Method [4,5] is a formal approach for estimating the uncertainty related to imposing a specific structure of RANS turbulence closure. In particular, the EPM was devised to estimate the L2 [2] uncertainty arising from the process of relating the microscopic state of a flow to macroscopic quantities. To understand the rationale underlying the EPM, we recall that the RST must be symmetric positive semi-definite. Hence, it must fulfill a set of realizability conditions that apply on its entries [14,15]

$$\langle u_i u_i \rangle \geq 0, \quad \langle u_i u_i \rangle + \langle u_j u_j \rangle \geq |2 \langle u_i u_j \rangle| , \quad \det(\langle u_i u_j \rangle) \geq 0. \tag{2}$$

The EPM applies perturbations of finite amplitude to the eigenspace of the RST, during the CFD solver iterations. Hereinafter, we will take advantage of superscript * to point out a perturbed entity. These perturbations are arbitrary within the limits defined by the conditions reported in Sys. (2). That is, the perturbed RST must anyway fulfill the realizability constraint. Not only, it has to be mentioned that recent investigations [11] show that there actually exist further conditions that must be fulfilled in order to enforce the physical plausibility (not the realizability) of perturbation. These plausibility conditions, not considered in this paper, actually restrict the range of possible perturbations to a subset of what established in Sys. (2).

To provide a clear explanation of the EPM, we refer to the decomposition of the RST into an anisotropy and a deviatoric part, bearing in mind that the realizability conditions (2) must apply also to the anisotropy tensor b_{ij}

$$\langle u_i u_j \rangle = 2k \left(b_{ij} + \frac{\delta_{ij}}{3} \right), \tag{3}$$

being k the turbulent kinetic energy and δ_{ij} the Kronecker delta. The anisotropy tensor b_{ij} can be further decomposed into its spectral form

$$b_{ij} = v_{ik} \Lambda_{kl} v_{jl}, \tag{4}$$

being v_{ik} and v_{jl} the left and the right eigenvectors and Λ_{kl} a diagonal matrix containing the eigenvalues λ_i.

The EPM perturbations target different quantities appearing in the RST definition

$$\langle u_i u_j \rangle^* = 2k^* \left(v^*_{ik} \Lambda^*_{kl} v^*_{jl} + \frac{\delta_{ij}}{3} \right). \tag{5}$$

Namely, the EPM perturbs the amount of turbulent kinetic energy (k^*), the spectral energy distribution (Λ^*_{kl}) of the anisotropy tensor, and the orientation of the anisotropy tensor basis (v^*_{ik}). In principle, the applied perturbations may be different depending on the spatial coordinate and researchers follow diverse approaches for handling the issue e.g., see [4,7,19].

In this paper, we assume uniform spatial perturbations affecting the spectral energy distribution and the orientation of the tensor basis only. That is, we avoid perturbing the value k of turbulent kinetic energy following the approach described in [8]. In particular, we consider three limiting states of turbulence componentiality that can be attained by perturbing the eigenvalues in Λ^*_{kl}. Namely, $1C$, $2C$ and $3C$, corresponding to the 1-, 2-, and 3-component turbulence. According to [8], we also modulate the production of turbulent kinetic energy $\mathcal{P}$ by perturbing the orientation of the tensor basis. In particular, we apply perturbations in order to either maximize or minimize $\mathcal{P}$. It has to be stressed that the extremal state $3C$ correspond to a isotropic condition and, therefore, in such case $\mathcal{P}$ is invariant w.r.t. the orientation of the tensor basis. In the end of the day, a total of five combinations would arise considering the three extremal states of turbulent componentiality ($1C$, $2C$ and $3C$) and two extremal states associated to the production of turbulent kinetic energy ($\mathcal{P}^{max}$ and $\mathcal{P}^{min}$). Namely, $\mathcal{P}^{max}_{1C}$ (PA), $\mathcal{P}^{max}_{2C}$ (PB), $3C$ (PC), $\mathcal{P}^{min}_{1C}$ (PD), $\mathcal{P}^{min}_{2C}$ (PE). Note that in the previous statement we also fix the labeling (PA-PE) identifying the different EPM models, recalling that BS indicates the baseline (unperturbed) turbulence closure.

The turbulent uncertainty estimates of a selected Quantity of Interest (QoI) are obtained by considering the max/min values resulting from the five EPM solutions

$$\Delta_{\mathrm{QoI}} = \max(\mathrm{QoI}^{\mathrm{PA}}, \mathrm{QoI}^{\mathrm{PB}}, \mathrm{QoI}^{\mathrm{PC}}, \mathrm{QoI}^{\mathrm{PD}}, \mathrm{QoI}^{\mathrm{PE}}) - \min(\mathrm{QoI}^{\mathrm{PA}}, \mathrm{QoI}^{\mathrm{PB}}, \mathrm{QoI}^{\mathrm{PC}}, \mathrm{QoI}^{\mathrm{PD}}, \mathrm{QoI}^{\mathrm{PE}}). \quad (6)$$

3 Results

In this section we present the test case specification and report the EPM analysis. In particular, we consider a purely two-dimensional approximation of the use case, assuming an infinite width channel. The very same test is simulated twice, assuming a different value of the static pressure imposed at the domain outlet. Namely, in test case A the back pressure is such that a subsonic flow develops in the channel, whereas for test B a supersonic stream occurs. Figure 1 reports the Mach flow field developing in each case. Note the formation of a train of reflected shock waves in test B.

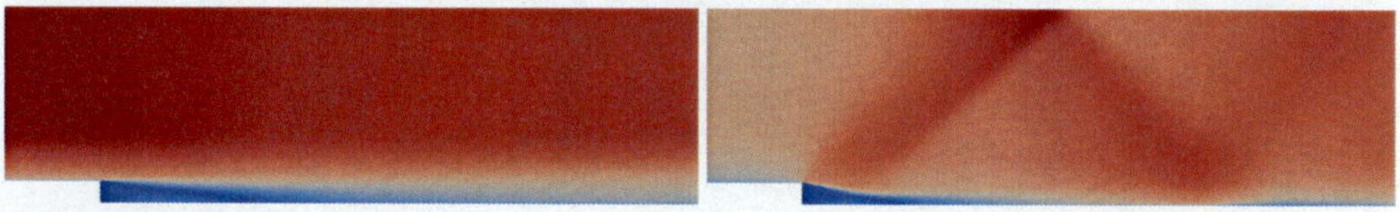

Fig. 1. Mach flow field developing in the close proximity of the step, as predicted using the baseline turbulence closure. (a) Test A. (b) Test B.

3.1 Test Case Specification

The test case consist in a subsonic flow of MDM (Octamethyltrisiloxane, $C_8H_{24}O_2Si_3$) over a backward-facing step. The geometry and the grids are taken from the NASA

Turbulence Modeling Resource (TMR) repository [1,9]. Namely, the test case labeled *2DBFS* which is a validation case for turbulence models. The reader is referred to the reference repository for finding all the details concerning the geometry of the test case. The TMR provides a large sequence of nested grids of different families that can be downloaded and used. In this work, we rely on the structured grid labeled *backstep5_3levdn.p2dfmt* for the 2D simulations. A side view of the numerical grid is reported in Fig. 2.

Fig. 2. Side view of the numerical domain employed to discretize the backward facing step channel.

The flow enters the domain on the left vertical boundary and leaves on the right. The length of the first portion of the channel ensures the full development of a turbulent boundary layer prior to the separation point. As the flow encounters the backward facing step, it separates to form a recirculating bubble. On the lower boundaries of the domain, adiabatic wall conditions apply. On the upper boundary, a symmetry condition is imposed (this is simplifying hypothesis w.r.t. the original test case). In the first section of the channel, the mesh is very coarse with a resolution increasing while approaching the step. A high resolution is employed in the region corresponding to the recirculating bubble, with elements enlarging as the flow proceeds towards the domain outlet, on the right.

The inflow conditions and the MDM fluid properties as considered in simulations are reported in Table 1. Note the different value of the static back pressure imposed at the domain outlet, to produce the subsonic test case A and the supersonic test B. The thermodynamics is modeled using the improved Peng-Robinson Stryjek-Vera (iPRSV) [16].

Table 1. Inflow and outflow conditions and MDM fluid properties

Inflow		Outflow		MDM parameter				
P^t [bar]	T^t [C°]	P^A_{outlet} [bar]	P^B_{outlet} [bar]	γ [-]	R [J/KgK]	P_{cr} [bar]	T_{cr} [C°]	ω [-]
1.0	272	0.9	0.2	1.018	35.15	14.38	292.21	0.524

In Table 1, P^t and T^t are the total pressure and temperature imposed on the inflow boundary, whereas P_{outlet} is the static pressure at the outlet. The parameter γ is the heat capacity ratio, R is the MDM gas constant, P_{cr} and T_{cr} are the MDM critical pressure and temperature, whereas ω is the fluid acentric factor.

In our computations, viscosity is assumed to be constant and equal to $\mu = 1.5281 \cdot 10^{-5}$ [Pas] and the Prandtl number is kept constant, to set the local value of the thermal conductivity. At the inlet, a turbulent to laminar viscosity ratio of 100.0 and a 5%

intensity of turbulent fluctuation are applied. The turbulence baseline closure (BS) is the Menter's Shear Stress Transport (SST) closure [10].

The CFD solver employed in this work is the open-source SU2 suite [3, 13]. SU2 is equipped with EPM capabilities [12] and can be used to evaluate several quantities of interest from field solutions. Here, CFD simulations are achieved using a standard implicit time-marching approach taking advantage of a generalized Approximate Riemann solver of Roe type with a Monotone Upstream-centered Scheme for Conservation Laws (MUSCL) [17]. No slope limiter is employed, while a single-level multi-grid approach is exploited to hasten the convergence of the numerical solution.

No mesh sensitivity analysis is presented. Indeed, the five EPM models implement a very different physics, therefore also their sensitivity to the computational grid is different. Rigorously, a grid sensitivity analysis should be devoted to each of the five EPM (and to the BS as well) solutions [6], resulting into a demanding process. Here, we are not interested in showing physically accurate solutions for each EPM model, but rather we aim at exposing the predictions' uncertainty related to the turbulence closure model-form error. This uncertainty implies also the variability due to a different sensitivity of the five EPM models to the very same computational grid. Moreover, the RANS model is known to be inaccurate in predicting separated flows. For this reason, the EPM is expected to return significantly large uncertainty estimates.

3.2 Test A

We carry out the EPM analysis of the flow field developing throughout the channel operating in the subsonic regime. Namely, we execute the BS and the five (PA-PE) EPM simulations. Figure 1(a) reports the Mach flow field resulting from the BS model. The picture shows the recirculating bubble developing past the step and the reattachment point downstream the channel. The EPM analysis targets several quantities of interest extracted along a straight line parallel to the bottom wall, originating at half of the step height and reaching the outlet boundary, on the right end side of the domain. Over this line, we define the coordinate s, being $s = 0$ the middle point of the vertical wall, at the step, and $s = 1$ a location on the outlet section.

Figure 3 reports the EPM analysis concerning, respectively from top to bottom, the x and y components of the local flow momentum vector and the scalar static pressure. For all the targeted quantities, we report the BS and the PA-PE predictions, and we highlight the corresponding Δ_{QoI} envelope using a grey shading. The EPM analysis of the flow momentum components indicates that, qualitatively, a similar flow kinematics is predicted. Nonetheless, quantitative differences are significant and lead to a maximum envelope width of $\Delta_{\mathrm{x}} = 21.99$ [Kgm/s] and $\Delta_{\mathrm{y}} = 5.61$ [Kgm/s]. Correspondingly, the magnitude of the turbulence closure uncertainty estimates define, respectively, a percentage variability of about 151.25% and -646.26% w.r.t. the BS prediction. The comparison of the local flow momentum vector components allows to appreciate a slightly different dimension of the recirculating bubble forming past the step. In particular, this can be observed as the sign variation of the x component, which is predicted to occur at different s stations in Fig. 3, in the close proximity of $s = 0.03$. Note that the line defining the s coordinate is specified without bothering about capturing the maximum extension of the bubble; rather, it just crosses the recirculating zone at a not specific height. Analogous conclusions can be drawn from the observation of the y component. Despite a different scale of the recirculating region, now it is possible to appreciate also a significantly different magnitude of the momentum associated to the ascending flow occurring in the close proximity of the step vertical wall. In particular, PA returns a value significantly larger than PC and PE predictions.

Concerning the EPM analysis of pressure, we obtain $\Delta_{\mathrm{p}} = 1847.7$ [Pa] (2.09% of the BS value). Again, the predicted trends are qualitatively similar, but the pressure rise occurs at a different location, depending on the turbulence closure. The model PA anticipates the pressure rise location, whereas PC and PE predict it to occur at a more advanced station along the s axis. Overall, the distance in between such locations is about 0.07 units.

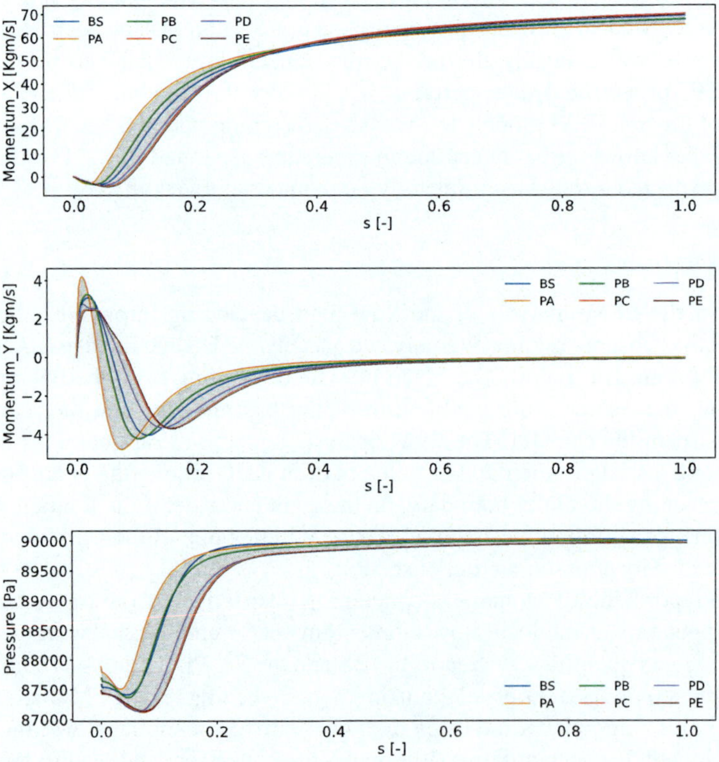

Fig. 3. Test A. EPM analysis of selected quantities in the recirculating region.

3.3 Test B

The EPM analysis is here reported for the two-dimensional channel operating in the supersonic regime. The quantities of interest subject to the EPM analysis are the same as the previous case, extracted along the very same reference segment. Figure 1(b) reports the Mach flow field developing in each case. The flow separates at the step and expands accelerating over the edge, while rotating towards the bottom wall. As the flow reaches the lower surface, it is turned back to the direction parallel to the channel axis, generating a shock wave. The shock propagates downstream, interacting with the upper wall, giving birth to a train of reflected shock waves. Note that shocks are smeared over the finite volumes in the numerical grid, indicating that a more refined grid would be necessary for this test. Nonetheless, here we are not interested in obtaining accurate and reliable predictions, but the aim is to show how the turbulence uncertainty leads to CFD models producing different predictions for the very same numerical mesh.

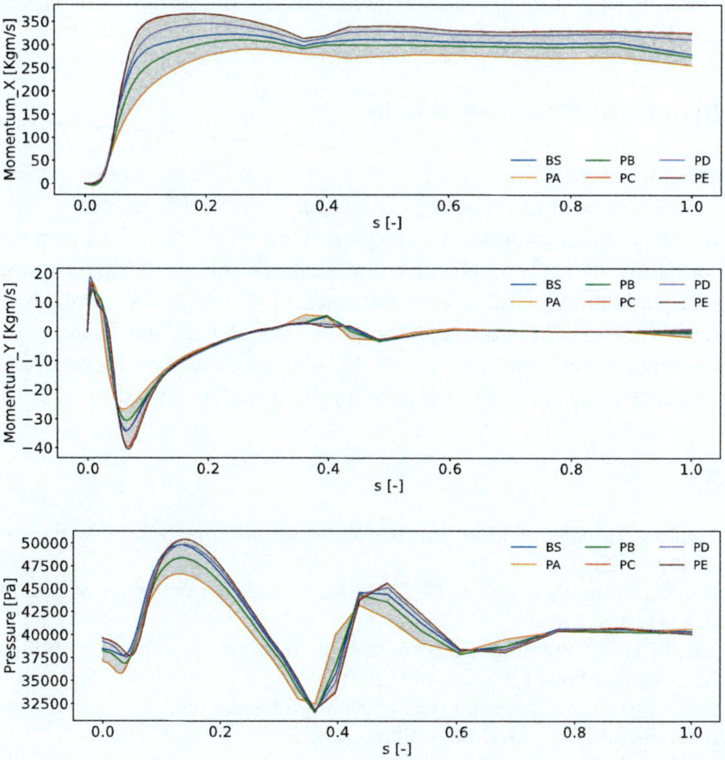

Fig. 4. Test B. EPM analysis of selected quantities in the recirculating region.

Figure 4 reports the EPM analysis concerning, respectively from top to bottom, the x and y components of the local flow momentum vector and the scalar static pressure. Again, the BS and the five EPM models predict a qualitatively similar flow kinematics. A maximum envelope width of $\Delta_x = 148.88$ [Kgm/s] and $\Delta_y = 14.34$ [Kgm/s]. Respectively, a percentage variability of about 53.96% and -42.82% w.r.t. the BS prediction. The comparison of the local flow momentum vector components, in particular the y component, show that the dimension of the recirculating region is not much affected by the turbulence uncertainty. Indeed, the y component of the momentum vector changes sign at about $s = 0.01$ for all the EPM models. This was expected since the supersonic expansion over an edge can be reasonably approximated as an inviscid phenomenon. What it is significantly affected by the turbulence closure uncertainty is the flow downstream the reattachment point, where the viscous interaction of the flow with the lower wall becomes relevant with the development of a boundary layer. Also, at $s = 0.4$ it is notable the interaction of the boundary layer with the impinging shock propagating from the first reflection point on the upper boundary. This latter interaction mechanism is of course sensitive to the modeling of the Reynolds stresses. At $s = 0.4$, it is indeed shown a localized increase of the Δ_{y} envelope. Concerning the EPM analysis of pressure, we obtain $\Delta_{\mathrm{p}} = 6327.50$ [Pa] (17.75% of the BS value). Again, the predicted trends are qualitatively similar, with quantitative differences arising among the BS and the EPM models.

4 Conclusions and Future Works

The EPM was applied to an exemplary test case in the aim of showing the relevance of structural uncertainty in turbulence closures for RANS CFD models. The relevance of such uncertainty should weight even more in view of the fact that, differently from application entailing ideal flows, NICFD applications rely on a very limited amount of data for the validation of turbulence closures. The simple but intuitive examples reported in this work should call the attention of Researchers and Engineers to implement adequate counteractions. Namely, devise and adopt design approaches robust to the epistemic uncertainty related to the form of the turbulence closure.

References

1. NASA Langley Turbulence Modeling Resource website. http://turbmodels.larc.nasa.gov. Accessed 11 Nov 2019
2. Duraisamy, K., Iaccarino, G., Xiao, H.: Turbulence modeling in the age of data. Annu. Rev. Fluid Mech. **51**, 01 (2019)
3. Economon, T., et al.: Performance optimizations for scalable implicit RANS calculations with SU2. Comput. Fluids **129**, 146–158 (2016)
4. Emory, M., Larsson, J., Iaccarino, G.: Modeling of structural uncertainties in reynolds-averaged navier-stokes closures. Phys. Fluids **25**(11), 110822 (2013)
5. Emory, M., Pecnik, R., Iaccarino, G.: Modeling structural uncertainties in reynolds-averaged computations of shock/boundary layer interactions (2011)
6. Gori, G., Le Maître, O., Congedo, P.: On the sensitivity of structural turbulence uncertainty estimates to time and space resolution. Comput. Fluids **229**, 105081 (2021)
7. Gorlé, C., Iaccarino, G.: A framework for epistemic uncertainty quantification of turbulent scalar flux models for Reynolds-averaged Navier-Stokes simulations. Phys. Fluids **25**(5), 055105 (2013)
8. Iaccarino, G., Mishra, A., Ghili, S.: Eigenspace perturbations for uncertainty estimation of single-point turbulence closures. Phys. Rev. Fluids **2**, 02 (2017)
9. Jespersen, D., Pulliam, T., Childs, M.: OVERFLOW: Turbulence Modeling Resource Validation Results. Technical report NASA-2016-01, NASA Ames Research Center, Moffett Field, CA (2010)

10. Menter, F.: Zonal Two Equation k-w Turbulence Models For Aerodynamic Flows. AIAA 1993-2906. 23rd Fluid Dynamics, Plasmadynamics, and Lasers Conference (1993)
11. Mishra, A., Iaccarino, G.: Theoretical analysis of tensor perturbations for uncertainty quantification of reynolds averaged and subgrid scale closures. Phys. Fluids **31**(7), 075101 (2019)
12. Mishra, A., Mukhopadhaya, J., Iaccarino, G., Alonso, J.: Uncertainty estimation module for turbulence model predictions in SU2. AIAA J. **57**(3), 1066–1077 (2019)
13. Palacios, F., et al.: Stanford University Unstructured (SU^2): an open-source integrated computational environment for multi-physics simulation and design. AIAA 2013-287 (2013)
14. Schumann, U.: Realizability of Reynolds-stress turbulence models. Phys. Fluids **20**(5), 721–725 (1977)
15. Simonsen, A., Krogstad, P.: Turbulent stress invariant analysis: clarification of existing terminology. Phys. Fluids **17**(8), 088103 (2005)
16. Stryjek, R., Vera, J.: PRSV - an improved Peng-Robinson equation of state for pure compounds and mixtures. Can. J. Chem. Eng. **64**, 323–333 (1986)
17. van Leer, B.: Towards the ultimate conservative difference scheme. V. A second-order sequel to Godunov's method. J. Comput. Phys. **32**(1), 101–136 (1979)
18. Xiao, H., Cinnella, P.: Quantification of model uncertainty in rans simulations: a review. Prog. Aerosp. Sci. **108**, 1–31 (2019)
19. Xiao, H., Wu, J., Wang, J., Sun, R., Roy, C.: Quantifying and reducing model-form uncertainties in Reynolds-averaged Navier-Stokes simulations: a data-driven, physics-informed Bayesian approach. J. Comput. Phys. **324**, 115–136 (2016)

Assessment of Density and Compressibility Corrections for RANS Simulations of Real Gas Flows Using SU2

D. Schuster(✉), Y. Ince, Alexis Giauque, and Christophe Corre

Univ Lyon, Ecole Centrale de Lyon, CNRS, INSA Lyon, Univ Claude Bernard Lyon 1, LMFA (Fluid Mechanics and Acoustics Laboratory), UMR 5509, 69134 Ecully, France
{dominik.schuster,alexis.giauque,christophe.corre}@ec-lyon.fr, yusuf.ince@master.ec-lyon.fr

Abstract. Reynolds-averaged Navier-Stokes (RANS) simulations are performed for turbulent perfect and real gas flows (supersonic flat plate and compressible mixing layer) with available theoretical or numerical (DNS) reference results. The model assessment is focused on versions of the SA model which include density and compressibility corrections proposed in the literature.

Keywords: RANS · compressibility · turbulence modeling · boundary layer · mixing layer

1 Introduction

Various approaches have been proposed over the past 30 years to account for compressibility effects within turbulence models, which were initially developed for incompressible flows. Dilatation-dissipation corrections have been devised by Sarkar (1989) [13] and Zeman (1990) [22] to correctly predict the growth rate of compressible mixing layers and improved by Wilcox (1992) [21] to extend the accuracy of these compressibility corrections to wall-bounded flows. In the case of the Spalart-Allmaras (SA) one-equation turbulence model, a correction has been proposed by Secundov and coworkers (1995) [16], which significantly improves the model prediction for mixing layers and is currently available in SU2 [6] along with the baseline SA model and some of its variants. Paciorri and Sabetta (2003) [12] have also developed a compressibility correction for the SA model and assessed its accuracy both for shear layer and supersonic base flow simulations. Catris and Aupoix (2000) [4] have established the need for density corrections in turbulence models to correctly predict the logarithmic region of the boundary layer in presence of large density gradients and achieve an accurate prediction of the skin friction. For the SA model, these corrections impact the turbulent diffusion term and the cross term. More recently, Pecnik et al. (2018) [11] have developed similar corrections for these terms, based on a semi-local scaling approach. The goal of the present work is to assess the influence of these density and compressibility corrections on the accurate prediction of flow features for real gas flows. To this end, some of the corrections proposed in the literature are implemented in the open-source software

M. White et al. (Eds.): NICFD 2022, ERCOFTAC 29, pp. 82–90, 2023.
https://doi.org/10.1007/978-3-031-30936-6_9

SU2, where SA as well as $k - \omega$ SST models are already available, along with real gas flow modeling. The influence of the corrections is studied in the present contribution for the SA model and two academic configurations (flat plate boundary layer and mixing layer). Note that a similar analysis of SA compressibility corrections, limited however to the computation of air flows, has been performed by Séror and Kosarev [15] using the in-house code NSHYP of the Israel Aerospace Industries. Extended results will be presented at the Conference, including the $k - \omega$ SST model with density and compressibility corrections and the simulation of the transonic turbine experimentally studied by Baumgartner, Otter and Wheeler [3] with static and total pressure measurements available for the turbulent flow of R134a.

The present analysis of conventional RANS models including density and compressibility corrections and applied to real gas flows represents a first step in the development of a reliable statistical modeling for dense gas simulations. It complements another contribution of some of the authors to NICFD 2022 devoted to the development of an ANN-based Large Eddy Simulation (LES) Subgrid-Scale (SGS) turbulence model for dense gas flows. This novel SGS model specifically dedicated to dense gas flows is currently applied, along with wall-resolved and wall-modeled conventional SGS models, to the LES simulation of the Otter *et al.* turbine. The database thus created will be used for the training of an ANN-based RANS turbulence model dedicated to dense gas flows and expected to capture some of the features specific to these flows. Section 2 reviews the baseline SA turbulence model and its density and compressibility corrections proposed in the literature. Section 3 describes the set-up of the flat plate boundary layer and the mixing layer configuration and provides some preliminary results obtained for perfect and real gas supersonic turbulent flows computed using SA and its variants.

2 Methodology

2.1 Original SA Model

The classical SA model [18] is a one-equation transport model for the modified eddy viscosity $\tilde{\nu}$, which is directly linked to the turbulent viscosity $\mu_t = \rho f_{v1}\tilde{\nu}$ for the case of compressible flows, with f_{v1} a damping function ensuring a correction of the eddy viscosity distribution in the viscous and buffer layers. The SA equation reads:

$$\frac{D\tilde{\nu}}{Dt} = \underbrace{c_{b1}\tilde{S}\tilde{\nu}}_{\text{production}} + \underbrace{\frac{1}{\sigma}\left(\nabla\cdot((\nu+\tilde{\nu})\nabla\tilde{\nu}) + c_{b2}(\nabla\tilde{\nu})^2\right)}_{\text{diffusion}} \underbrace{-c_{w1}f_w(r)\left(\frac{\tilde{\nu}}{d_w}\right)^2}_{\text{dissipation}} \qquad (1)$$

In the production term, $\tilde{S}$ is computed as $\tilde{S} = \Omega + \frac{\tilde{\nu}}{k^2 d^2} f_{v2}$ with Ω the magnitude of the rotation tensor and f_{v2} a second damping function. In the dissipation term, f_w is a non-dimensional function which adjusts the skin friction, with r a near-wall parameter involving the mixing length $\sqrt{\tilde{\nu}/\tilde{S}}$. Note that the material derivative appearing in the LHS of (1) does not lend itself naturally to a conservative finite volume discretization. Several authors (see for instance [15]) combine (1) with the mass conservation equation in order to write a model for the conservative variable $\rho\tilde{\nu}$. In the SU2 solver, as explained in [6], the conserved variable for the SA model remains $\tilde{\nu}$.

2.2 Mixing Layer Compressibility Correction

This model adaption was introduced by Sekundov et al. [9] and later referenced by Spalart [17]. The following compressibility correction term is added to Eq. (1):

$$- C_5 \frac{\tilde{\nu}^2}{a^2} \frac{\partial u_i}{\partial x_j} \frac{\partial u_i}{\partial x_j} \tag{2}$$

with a the local speed of sound and $C_5 = 3.5$ an empirical constant. This empirical model variation (here labeled SA-comp) accounts for the reduction in the spreading rate of the compressible shear layer by reducing the turbulent eddy viscosity $\tilde{\nu}$ as observed experimentally.

2.3 Edwards' Correction

Edwards and Chandra [7] developed this formulation (here labeled SA-Edwards) to improve the stability and convergence properties of the original SA model by enhancing the near-wall numerical behavior. This is achieved by modifying the calculation of $\tilde{S}$ in the production term, now expressed as:

$$\tilde{S} = \overline{S}^{1/2} \left(\frac{1}{\chi} + f_{v1} \right), \tag{3}$$

with $\overline{S}^{1/2} = \left(\frac{\partial u_i}{\partial x_j} + \frac{\partial u_j}{\partial x_i} \right) \frac{\partial u_i}{\partial x_j} - \frac{2}{3} \left(\frac{\partial u_k}{\partial x_k} \right)^2$. The near-wall parameter r is still computed from $\sqrt{\tilde{\nu}/\tilde{S}}$ but using also a different formula.

2.4 Eddy Viscosity Convection and Density Correction

Another approach, proposed by Pecnik et al. [11], is an improvement of the work by Catris and Aupoix (2000) [4]. The diffusion term of the turbulent transport equation is modified by adding density gradients, which improves the prediction of the logarithmic layer and skin friction coefficient for a variety of turbulence models. The concept of Pecnik et al. demonstrates particular improvements for flows with strong thermophysical gradients. The modified diffusion term in the original formulation is as follows:

$$Diffusion = \frac{1}{\rho} \frac{\partial}{\partial x_j} \left(\frac{\rho}{\sigma} (\nu + \tilde{\nu}) \frac{\partial \tilde{\nu}}{\partial x_j} + \frac{1}{2} \frac{\tilde{\nu}}{\sigma} (\nu + \tilde{\nu}) \frac{\partial \rho}{\partial x_j} \right) + \frac{c_{b2}}{\rho\sigma} \left(\frac{\partial \sqrt{\rho} \tilde{\nu}}{\partial x_j} \right)^2. \tag{4}$$

In order to implement (4) in SU2, the new term is expressed as the sum of a conservative term, yielding a straightforward finite volume discretization, and source terms using the gradient of the density and of the modified eddy viscosity $\tilde{\nu}$. After some algebra, the modified diffusion term is expressed as:

$$Diffusion = \frac{\partial}{\partial x_j}\left(\frac{1}{\sigma}(\nu+\tilde{\nu})\underbrace{\left(\frac{\partial \tilde{\nu}}{\partial x_j}+\frac{1}{2}\frac{\tilde{\nu}}{\rho}\frac{\partial \rho}{\partial x_j}\right)}_{B_{1d}}\right)+\underbrace{\frac{1}{\rho\sigma}(\nu+\tilde{\nu})\frac{\partial \rho}{\partial x_j}\left(\frac{\partial \tilde{\nu}}{\partial x_j}+\frac{1}{2}\frac{\tilde{\nu}}{\rho}\frac{\partial \rho}{\partial x_j}\right)}_{B_{1s}}$$
$$+\frac{c_{b2}}{\sigma}\left(\left(\frac{\partial \tilde{\nu}}{\partial x_j}\right)^2+\underbrace{\frac{\tilde{\nu}}{\rho}\frac{\partial \rho}{\partial x_j}\frac{\partial \tilde{\nu}}{\partial x_j}+\frac{1}{4}\left(\frac{\tilde{\nu}}{\rho}\right)^2\left(\frac{\partial \rho}{\partial x_j}\right)^2}_{B_2}\right) \quad (5)$$

The B_{1d} correction modifies the original conservative form of the diffusion term with a density gradient correction, the remaining part of the correction taking the form of the volume source term B_{1s}. The B_2 correction is also a volume source term, modifying the original cross-diffusion term in (1).

2.5 Martin-Hou Equation of State

To model the complex thermodynamic properties of dense gases, we use the fifth order Martin-Hou equation of state [10]. It is given by:

$$p(T,v) = \frac{RT}{v-b} + \int_{i=2}^{5} \frac{Q_i(T)}{(v-b)^{i-1}}. \quad (6)$$

where the function $Q_i(T)$ is defined as:

$$Q_i(T) = A_i + B_i T + C_i \exp(-kT/T_c), \quad (7)$$

with $k = 5.475$ and the constants A_i, B_i, C_i depend on the substance. Since the available turbine experiment described in [3], which provides reference measurements for the assessment of RANS modeling including compressibility corrections, makes use of R134a as working fluid, it is decided to compute the model flat plate boundary layer configurations using the same fluid. Furthermore, the fluid FC-70 is used for the mixing layer configuration in order to compare them to the real gas LES computations of [19]. The fluid properties of both fluids are reported in Table 1. Note also the model of Chung [5] is used to compute the transport properties (viscosity and conductivity) of R134a and FC-70. Meanwhile, air calculations are also performed for the sake of comparison both with the literature and with the real gas calculations; air is considered as an ideal gas, with viscosity modeled via Sutherland's law and a constant Prandtl number providing thermal conductivity.

Table 1. Fluid properties: molar mass M, critical temperature T_c, critical pressure P_c, critical density ρ_c, acentric factor ω, boiling temperature T_b,

	M (g/mol)	T_c (K)	P_c (Pa)	Z_c	ω	T_b (K)	$c^c_{v\infty}/R$	n
R134a	102	374.2	$4.06{\cdot}10^6$	0.260	0.3268	247.1	10.82	0.695
FC-70	821	608.2	$1.03{\cdot}10^6$	0.270	0.7584	488.1	118.7	0.493

3 Results

3.1 Numerical Setup for the Supersonic Flat Plate Boundary Layer

To examine the impact of the different variations of the SA model on the flow solution, we perform 2D simulations of the supersonic turbulent flow over a flat plate, both for air (perfect gas) and R134a (real gas). Two different operating conditions (or free-flow thermodynamic states) are considered for R134a. The first state lies in the vapour region and far from the critical point, with a thermodynamic behavior expected to be close to that of a perfect gas. The second thermodynamic state is located in the supercritical region and chosen close to the critical point where real gas effects become important. The three computed thermodynamic states are summarized in Table 2.

A no-slip, adiabatic boundary condition is applied at the plate surface, a constant pressure condition is applied at the outlet and far-field conditions are applied at the inlet. A short symmetry plane is also introduced in front of the leading edge to prevent interactions between the inlet and the plate. The free-stream Mach number is equal to $M_\infty = 3.25$. The length L of the plate is adjusted so as to ensure the Reynolds number based on the free flow conditions and the plate length is always equal to $Re = \frac{u_\infty L}{\nu_\infty} = 5 \times 10^5$. The computational mesh contains 137 points in the streamwise direction and 97 points in the spanwise direction. It was checked this grid resolution ensures grid-converged results when using the second-order Roe-MUSCL scheme available in SU2 for the mean flow equations and a second-order scalar upwind method for the eddy viscosity equation.

In the case of a supersonic flat plate turbulent boundary layer, it is expected the RANS turbulence model should be able to correctly reproduce the logarithmic law for a compressible boundary layer. Velocity profiles are reported in their dimensionless form, that is $u^+ = \frac{u}{u_\tau}$, taking also into account compressibility through the Van Driest transformation [20]:

$$u^+_{VD} = \int_0^{u^+} (\rho/\rho_w)^{1/2}\, du^+. \tag{8}$$

In the following section u^+_{VD} is plotted as a function of the non-dimensional distance to the wall $y^+ = \frac{y u_\tau}{\nu_w}$ and should match the reference theoretical profile of an incompressible turbulent boundary layer: $u+ = y+$ in the viscous layer, Spalding's law in the buffer layer and log law in the log layer ($u^+ = \frac{1}{\kappa} \ln y^+ + C$, with $\kappa = 0.41$ and $C = 5.25$). The outlet velocity profile is actually taken slightly (5%) upstream of the exit section in order to get rid of the slight influence of the outflow boundary condition.

Table 2. Freestream values of temperature T_∞ and pressure P_∞ for the 3 computed flows. The freestream velocity corresponds to $M_\infty = 3.25$ for all 3 cases.

State	Fluid	T_∞ (K)	P_∞ (Pa)	u_∞ (m/s)
1	Air	300	7039	1128
2	R134a	300	2700	535
3	R134a	400	$41 \cdot 10^5$	446

3.2 Velocity Profiles and Transport Equation Budget

The outlet velocity profiles for air and R134a with the two thermodynamic conditions under study are reported in Fig. 1. Note that all calculations have successfully converged to a steady state so that SA and its 4 variants can be compared with the theoretical reference profile. It can be first observed that differences between SA and its corrections remain overall small, especially so for the real gas. In the case of the air calculation, it seems actually the baseline SA model does already a good job at correctly predicting the logarithmic layer. This finding appears consistent with comments found in [4] where it is mentioned the effect of the density correction is smaller for SA than for some of the 2-equation model the authors experimented with (though no plot for SA is provided by the authors). Similarly, [11] reports SA results for channel flows which are only slightly modified by density corrections. In the case of R134a, differences between the baseline SA model and its variants with corrections are further reduced, especially for the supercritical thermodynamic conditions. This could be consistent with previous observations on high-Mach wall-bounded dense gases flows displaying a physical behaviour similar to incompressible flows with variable properties [14].

In order to gain some insight into the available and implemented compressibility corrections, Fig. 2 depicts the separate contribution of each term in the transport equation of $\tilde{\nu}$ as a function of the normalized distance to the wall. One noticeable feature, yet to be explained, is the significant reduction of the Pecnik *et al.* compressibility correction with respect to the other contributing terms when applied to the real gas.

3.3 Compressible Turbulent Mixing Layer

The simulation domain spans 2 m in stream-wise (x) direction and 4 m in spanwise (y) direction. A splitter plate is located in the middle of the domain and extends 1 m from the inlet. Top and bottom boundaries are slip walls and a constant static pressure is set at the outlet. The lower, subsonic stream is kept constant at $M_2 = 0.1$, whereas the upper stream is adjusted such that the desired convective $\mathrm{M}_c = \frac{u_1 - u_2}{c_1 + c_2}$ is achieved. Static pressure and static temperature of lower and upper stream equivalent. In the case of air we run at $P = 101325Pa$, $T = 293.15K$ and for FC-70 the thermodynamic operating condition is set to reduced pressure $P_r = \frac{P}{P_c} = 0.93$ and reduced specific volume $V_r = \frac{V}{V_c} = 1.85$. The energy thickness of the mixing layer is defined as $\delta(x) = y_1 - y_2$, where y_1 is the location where $u = u_2 + \sqrt{0.9}(u_1 - u_2)$ and y_2 is found at $u = u_2 + \sqrt{0.1}(u_1 - u_2)$. In the similarity region, the shear layer thickness grows linearly and thus a growth rate $\delta' = \frac{d\delta}{dx}$ can be determined. We observe velocity profiles at a streamwise distance of 0.5 m to 0.9 m from the splitter plate to ensure that the solution is self similar. As the convective Mach number increases, compressibility effects become stronger which leads to a reduction in the growth rate when the velocity ratio u_2/u_1 and density ratio $\rho_2/\rho_1 = T_1/T_2$ are kept constant. This is expressed by the compressibility factor, which is the ratio of growth rates of a compressible to an incompressible mixing layer:

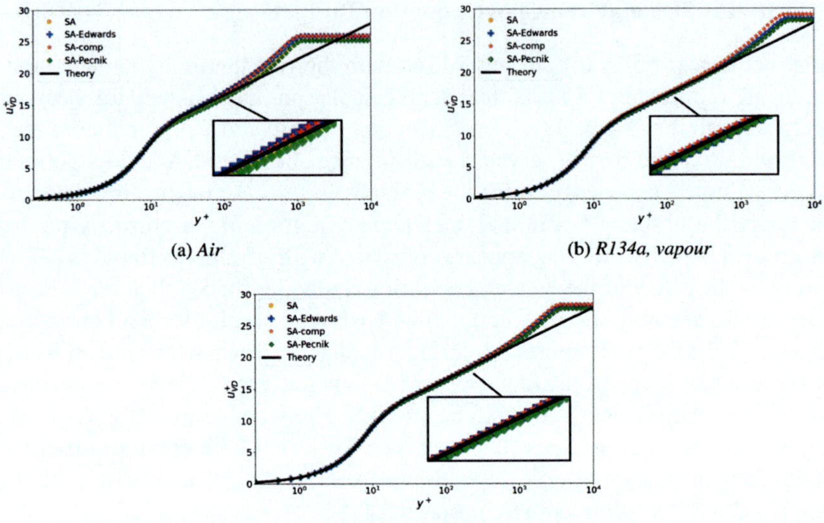

(a) *Air*

(b) *R134a, vapour*

(c) *R134a, supercritical*

Fig. 1. Computed non-dimensional velocity profile at $x = 0.95 \times L$ for the $\mathrm{M} = 3.25$ and Re = $5 \cdot 10^5$ turbulent flow over a flat plate. Comparison with the theoretical profile.

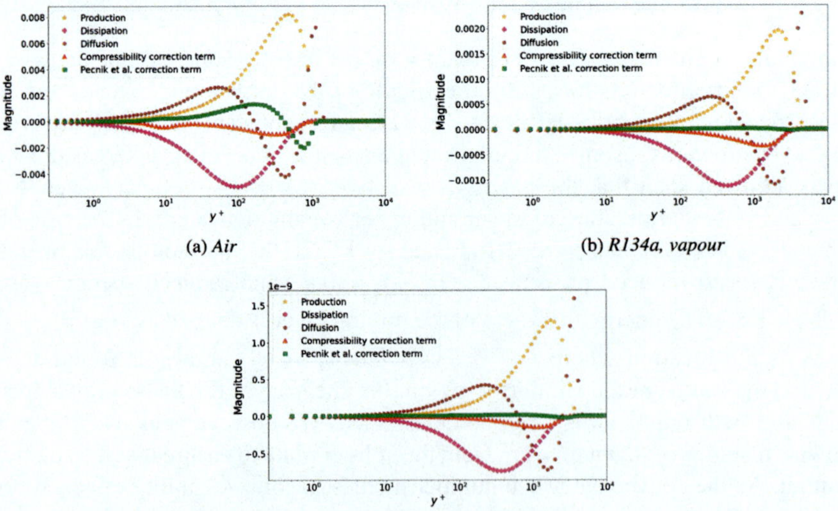

(a) *Air*

(b) *R134a, vapour*

(c) *R134a, supercritical*

Fig. 2. Computed distributions of the source terms appearing in the SA model variations: Production, destruction and diffusion terms are present in the original SA model; compressibility correction is added in SA-comp; the Pecnik et al. correction includes B_{1s}, B_{1d} and B_2 terms.

$$\Phi = \frac{\delta'_c}{\delta'_i}. \tag{9}$$

The incompressible growth rate can be obtained by using the incompressible RANS solver of SU2 at velocity and density ratios equivalent to those of the compressible case. Five different flow configurations were computed for convective Mach numbers ranging from 0.1 to 2.2. The original SA model along with compressibility and Pecnik et al.'s corrections are compared to reference results from the literature in Fig. 3. A lack of reduction in the compressible growth rate is observed for both air and FC-70 with the original SA model, therefore reaffirming the importance of compressibility corrections in mixing layers. Pecnik et al.'s correction has only a marginal influence on the compressibility factor, except for a slight improvement compared to the uncorrected SA model in the air simulation. Vadrot *et al.* observed a significant increase of the compressibility factor at high convective Mach number when using a dense gas (FC-70) instead of a perfect gas (air). The current RANS computations with the SA model using compressibility correction initially developed for perfect gas configurations show the same tendency at $M_c = 2.2$, where the reference value derived from the DNS of an equivalent temporal mixing layer (using the same Martin-Hou EoS) is almost recovered. However, for Mach number $M_c = 1.1$ the compressibility factor predicted by the RANS approach is much larger than the reference DNS value. It thus seems the compressibility correction derived for perfect gas mixing layer is not fully relevant for the real gas configuration; a specific dense gas compressibility correction should be developed.

Results obtained for the same flow configurations using $k-\omega$ SST model with density and compressibility corrections are not reported here for lack of space but will be discussed at NICFD 2022 Conference. Finally, the annular stator configuration experimentally studied by Otter *et al.*, where oblique shock-wave/ boundary layer interaction play a key role, will be also analyzed at the Conference using the baseline SA and $k-\omega$ SST model and their corrected versions. Numerical results obtained using LES with the same choice of thermodynamic description will provide reference data for the RANS simulations.

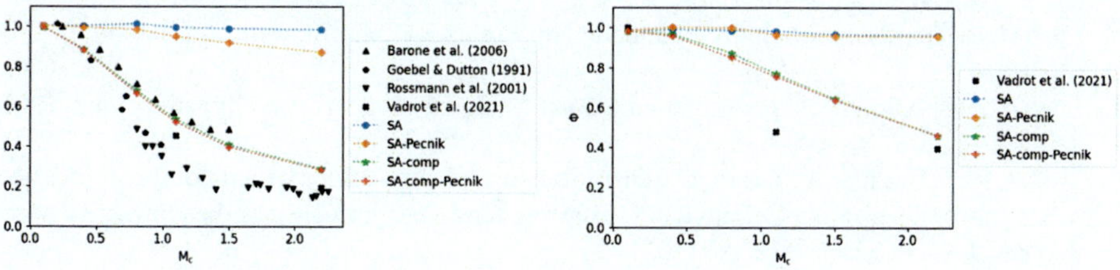

(a) *Air* (b) *FC-70*

Fig. 3. Computed compressibility factor Φ versus convective Mach number M_c compared to RANS (Barone et al. [2]), LES (Vadrot et al. [19]) and experimental data (Goebel & Dutton [8], Rossmann et al. [1]) for air and LES data (Vadrot et al.) for FC-70.

References

1. Rossmann, T., Mungal, M.G., Hanson, R.K.: Evolution and growth of large-scale structures in high compressibility mixing layers. J. Turbul. **3**, 9 (2002)
2. Barone, M.F., Oberkampf, W.L., Blottner, F.G.: Validation case study: prediction of compressible turbulent mixing layer growth rate. AIAA J. **44**(7), 1488–1497 (2006)
3. Baumgärtner, D., Otter, J.J., Wheeler, A.P.S.: The effect of isentropic exponent on transonic turbine performance. J. Turbomach. **142**(8), 1–10 (2020)
4. Catris, S., Aupoix, B.: Density corrections for turbulence models. Aerosp. Sci. Technol. **4**(1), 1–11 (2000)
5. Chung, T.-H., Ajlan, M., Lee, L.L., Starling, K.E.: Generalized multiparameter correlation for nonpolar and polar fluid transport properties. Ind. Eng. Chem. Res. **27**(4), 671–679 (1988)
6. Economon, T.D., Palacios, F., Copeland, S.R., Lukaczyk, T.W., Alonso, J.J.: SU2: an open-source suite for multiphysics simulation and design. AIAA J. **54**(3), 828–846 (2016)
7. Edwards, J.R., Chandra, S.: Comparison of eddy viscosity-transport turbulence models for three-dimensional, shock-separated flowfields. AIAA J. **34**(4), 756–763 (1996)
8. Goebel, S.G., Dutton, J.C.: Experimental study of compressible turbulent mixing layers. AIAA J. **29**(4), 538–546 (1991)
9. Gulyaev, A.N., Kozlov, V.E., Sekundov, A.N.: A universal one-equation model for turbulent viscosity. Fluid Dyn. **28**(4), 485–494 (1993)
10. Martin, J.J., Hou, Y.-C.: Development of an Equation of State for Gases. AIChE J. **2**(4), 142–151 (1955)
11. Patel, A., Diez, R., Pecnik, R.: Turbulence modelling for flows with strong variations in thermo-physical properties. Int. J. Heat Fluid Flow **73**, 114–123 (2018)
12. Paciorri, R., Sabetta, F.: Compressibility correction for the Spalart-Allmaras model in free-shear flows. J. Spacecr. Rocket. **40**(3), 326–331 (2003)
13. Sarkar, S., Erlebacher, G., Hussaini, M., Kreiss, H.: The analysis and modeling of dilatational terms in compressible turbulence. Technical report (1989)
14. Sciacovelli, L., Cinnella, P., Gloerfelt, X.: Direct numerical simulations of supersonic turbulent channel flows of dense gases. J. Fluid Mech. **821**, 153–199 (2017)
15. Séror, S., Kosarev, L.: Compressible turbulence model consistency for separated high-speed flow regimes. J. Spacecraft Rockets **54**(4), 840–862 (2017)
16. Shur, M., Strelets, M., Zaikov, L., Gulyaev, A., Koziov, V., Secundov, A.: Comparative numerical testing of one - and two-equation turbulence models for flows with separation and reattachement. In: 33rd Aerospace Sciences Meeting and Exhibit (1995)
17. Spalart, P.: Trends in turbulence treatments. In: Fluids 2000 Conference and Exhibit, p. 2306 (2000)
18. Spalart, P.R., Allmaras, S.R.: A one-equation turbulence model for aerodynamic flows. Rech. Aerosp. (1), 5–21 (1994)
19. Vadrot, A., Giauque, A., Corre, C.: Direct numerical simulations of temporal compressible mixing layers in a Bethe-Zel'dovich-Thompson dense gas: influence of the convective Mach number. J. Fluid Mech. **922**, A5 (2021)
20. Van Driest, E.R.: Turbulent boundary layer in compressible fluids. J. Spacecr. Rocket. **40**(6), 1012–1028 (2003)
21. Wilcox, D.C.: Dilatation-dissipation corrections for advanced turbulence models. AIAA J. **30**(11), 2639–2646 (1992)
22. Zeman, O.: Dilatation dissipation: the concept and application in modeling compressible mixing layers. Phys. Fluids A **2**(2), 178 (1990)

Validation of the SU2 Fluid Dynamic Solver for Isentropic Non-Ideal Compressible Flows

Blanca Fuentes-Monjas[1], Adam J. Head[1(✉)], Carlo De Servi[1,2], and Matteo Pini[1]

[1] Propulsion and Power, Delft University of Technology, Delft, The Netherlands
{a.j.head,c.m.deServi,m.pini}@tudelft.nl
[2] Energy Technology Unit, VITO, Mol, Belgium
carlo.deservi@vito.be

Abstract. This work assessed the accuracy of the SU2 flow solver in predicting the isentropic expansion of Siloxane MM through the converging-diverging nozzle test section of the Organic Rankine Cycle Hybrid Integrated Device (ORCHID) [9]. The expansion is modeled using compressible Euler equations, and assuming adiabatic flow, while the fluid thermodynamic properties are estimated using the Peng-Robinson equation of state. The boundary conditions for the experiment and simulations correspond to a stagnation temperature and pressure of $\bar{T}_0 = 253.7$ °C and $\bar{P}_0 = 18.36$ bar. At these inlet conditions the compressibility factor of the fluid is $Z_0 = 0.58$. The back pressure was equal to $\bar{P}_b = 2.21$ bar. The Mach number along the centreline, and static pressure along the nozzle surface were used as the system response quantities for the validation exercise. The studied SU2 model provides valid predictions for Mach number and static pressure. The largest deviation observed in the Mach number comparison between the simulation and experiment is in the uniform flow region of the nozzle and is equal to $E_{\text{Mach}} = 0.045$. Regarding the pressure trend, the largest discrepancy occurs in the kernel region and is equal to $E_{\text{pressure}} = 9$ kPa. At the same time, the simulated Mach number and static pressure reach a maximum absolute uncertainty of ± 0.015 and of ± 20 kPa, respectively. For both quantities, these values are reached in the region close to the throat. All the uncertainties calculated for the simulated pressure profile were larger than those of the experiments. The static pressure is particularly sensitive to the geometrical uncertainties of the nozzle profile, especially inside the kernel region. A proper characterisation of the nozzle geometry was therefore required to perform a meaningful validation of the fluid dynamic solver. The developed infrastructure can be used in the future for the validation of SU2 in different operating conditions and flow cases.

Keywords: Validation · CFD · error identification and uncertainty estimation

1 Introduction

The organic Rankine cycle (ORC) turbogenerator based on high-speed turbomachinery is a promising technology for waste heat recovery. The design optimization of the turbine is key to achieve high conversion efficiency. However, the design of these machines is complicated by the occurrence of highly non ideal supersonic flows. At the same time,

M. White et al. (Eds.): NICFD 2022, ERCOFTAC 29, pp. 91–99, 2023.
https://doi.org/10.1007/978-3-031-30936-6_10

the predictive capability of fluid dynamic solvers has not yet been rigorously assessed for these kinds of flows. As an effort in this direction, the present study focuses on the validation of the open-source CFD software SU2. Differently from the validation methodology adopted by [8], a novel mixed-uncertainty approach is employed to study the propagation of critical point properties, boundary conditions and geometric uncertainties through the computational model. The uncertainties related to the critical point properties of the fluid are considered *epistemic* uncertainties, while those associated with the boundary conditions of the experiment, as well as geometric uncertainties of the nozzle profile, constitute the *aleatory* uncertainties. These were propagated through the model using stochastic collocation. In particular, the uncertainties related to the manufacturing tolerances are modeled as in [6]. Regarding the experimental data, those pertaining to an experiment carried out in the ORCHID facility [9] and consisting of a supersonic expansion of the organic fluid hexamethyldisiloxane (MM) were considered. The validation exercise accounts also for the numerical uncertainties due to the discretization error, which were calculated using Richardson extrapolation.

The paper is structured as follows. Section 2 reports the SU2 NICFD model and the proposed uncertainty quantification and validation framework. The focus is set on the characterization and treatment of the relevant uncertainties within the validation experiment to achieve a meaningful comparison between experimental and numerical data. The results of the proposed validation workflow are presented in Sect. 4, where pressures and Mach number from the simulations are compared against the recorded experimental data. Finally, Sect. 5 reports relevant conclusions, and possible improvements that would allow for a more robust validation of the solver.

2 Model Definition and Validation Methodology

2.1 Validation Experiment

The test case considered in this work for the validation of SU2 is the expansion process of a dense vapour (siloxane MM) occurring in a planar converging-diverging nozzle. The measured quantities are the static pressure along the nozzle profile, and the Mach number along the nozzle mid-plane. The operating conditions targeted in the experiments are those corresponding to the design point of the nozzle. Notably, the total conditions at the inlet are $T_0 = 252$ °C, $P_0 = 18.4$ bar, while the back pressure is 2.1 bar. The nozzle throat is 20 mm by 7.5 mm (W x H), see [10] for more details about the nozzle geometry and experiment.

2.2 CFD Model

The flow in the nozzle is modeled as inviscid and two-dimensional to reduce the computational cost of the uncertainty quantification study. This assumption is justified also by the work of [9] and [3], where the results obtained for viscous versus inviscid, 2D versus 3D flow models of the nozzle are compared. The simulations performed in this work were carried out with the open-source SU2 solver Economon *et al.* [7] previously extended and verified for simulating NICFD flows by Pini *et al.* [14].

The calculation of the thermodynamic properties of the fluid requires the selection of a suitable *equation of state*. In this case, the polytropic (constant c_p) cubic Peng-Robinson equation of state available in SU2 was selected. The variation of c_p along the expansion process was found to be limited for the considered thermodynamic conditions. Thus, the use of a non-polytropic model was discarded to reduce the computational cost of the simulations. More specifically, the specific heat ratio was set to $\gamma = 1.0265$. Inlet/outlet Riemann boundary conditions were imposed in the simulations. For the convective fluxes the upwind Roe scheme generalized for non-ideal gases is employed together with a MUSCL reconstruction to achieve second order accuracy. The spatial gradients required by the MUSCL scheme are computed using weighted least squares. The Venkatakrishnan limiter is set to 0.3. More details regarding the solver configuration for NICFD can be found in [16].

A grid convergence study based on the methods given in [1,4] was performed to determine the optimum unstructured mesh distribution. The results of this study, which are reported in [3], showed that about forty thousands elements were required to keep a reasonable balance between numerical uncertainty and computational cost. The CFD simulations were performed on a server class AMD Opteron 6234 (2.4 Ghz, 48 cores) with 192 GB of memory.

3 Model Uncertainties

The model input uncertainties included in this study pertain to: the total inlet conditions (T_0, P_0), the critical properties used in the thermodynamic model (P_{cr}, T_{cr}) to determine the parameters of the *cubic equation of state*, and the nozzle geometry. The uncertainty associated with the critical point properties is treated as epistemic, while the other uncertainty sources are assumed aleatory.

The average and standard deviation of the normal distribution associated to the total inlet conditions are computed from the values measured inside the settling chamber of the ORCHID nozzle test section [10]. The average and the probability distribution functions of the total conditions (T_0, P_0) derived from the experiment are $\mathcal{N}(526.85, 0.3295)$ K, $\mathcal{N}(18.36, 0.01145)$ bar. The only value set as deterministic is the nozzle back pressure, fixed at a value of $\overline{P}_{\text{b}} = 2.21$ bar.

The intervals assigned to the critical point temperature and pressure are chosen based on the minimum and maximum experimental data reported in the literature. The corresponding intervals are $T_{\text{cr}} = [518.5,\ 521.6]$ K and $P_{\text{cr}} = [18.9,\ 19.27]$ bar. The lower bounds were determined by taking the minimum value of critical pressure [5] and temperature [12] and subtracting to these values the expanded uncertainty (two times the standard deviation) calculated considering the experimental data in the literature. An equivalent procedure was used to calculate the upper bound. No probability distribution is assigned to *epistemic* uncertainties.

The possible errors during the manufacturing of the nozzle profile are included in the UQ study using a modified version of the model proposed in [6]. The adapted model describes the deviation in the normal direction of the manufactured nozzle profile with respect to the nominal one, Δ, by means of a Gauss Random field. A Gauss Random field is completely defined by its average and covariance function, which is approximated by a Karhunen-Loève expansion, as

$$\Delta(s,\omega) = \bar{\Delta}(s,\omega) + C(s,x) = \bar{\Delta}(s,\omega) + \sum_{i=1}^{\infty} \sqrt{\lambda_i}\Phi_i(s)Z_i(\omega), \tag{1}$$

where s is the profile surface parametrization coordinate, being $s = 0$ at the nozzle inlet and $s = 1$ at the outlet. The terms λ_i and $\Phi_i(s)$ are the eigenvalues and eigenfunctions of the covariance function $C(s, x)$ respectively, while $Z_i(\omega)$ are the coefficients of the expansion which are defined by standard normal distributions, $\mathcal{N}(0, 1)$. The calculation of λ_i and $\Phi_i(s)$ is done by solving the Fredholm integral of the second kind using the Nÿstrom method [13].

$$\int_0^1 C(s,x)\Phi_i(x)dx = \lambda_i\Phi_i(s) \tag{2}$$

The integral in the previous equation is calculated using the algorithm described in [2] and considering a Gaussian quadrature with 500 points. The covariance in this model is defined as

$$C(s_1, s_2) = \sigma^2 \cdot \rho(s_1, s_2), \tag{3}$$

where σ is the tolerance of the manufacturing process and $\rho(s_1, s_2)$ is the correlation function that is modeled with the following squared exponential

$$\rho(s_1, s_2) = \exp\left(\frac{-|s_1 - s_2|^2}{L(s_1)\cdot L(s_2)}\right) \tag{4}$$

$$L(s) = L_0 + (L_T - L_0)\cdot \exp\left(\frac{-|s - s_{\text{throat}}|^2}{\omega^2}\right). \tag{5}$$

The parameters L_0, L_T and ω are normalized using the total nozzle surface length. Their values, that were set equal to 0.25, 0.025 and 0.25, respectively, in this study, are generally the result of a fitting procedure based on repeated measurements of the shape of the component of interest. This was, however, beyond the scope of the present work. The nozzle profiles were measured only once with a ball point probe [17]. The constants in 5 were then calibrated based on the limited data available and qualitative considerations based on the work in [6]. The tolerance, σ, was, instead, set to be 1.5×10^{-5} m based on the accuracy of the ball point probe used to measure the nozzle profiles.

The number of terms used for the truncated Karhunen-Loève expansion, Eq. 1, is chosen according to the cumulative energy, ϵ, as done in [15]. The cumulative energy is defined as $\epsilon(n) = (\sum_{i=1}^{n} \lambda_i)/(\sum_{i=1}^{N} \lambda_i)$, where N is the total number of points used for the numerical integration of Eq. 2 and λ_i sorted in descending order. A cumulative energy of 0.95 resulted in 9 eigenmodes to build the Karhunen-Loève expansion, which implies that the number of uncertainties associated to the manufacturing tolerances is nine. These uncertainties are introduced by means of the coefficients $Z_i(\omega)$ in Eq. 1, which are random variables normally distributed with mean 0 and variance 1, i.e., $\mathcal{N}(0, 1)$. Therefore, this UQ study is comprised of 11 *aleatory* uncertainties and 2 *epistemic* uncertainties.

The validation method accounts also for the numerical uncertainties due to the discretization error. These were computed using a Richardson extrapolation algorithm by resorting on the implementation in the tool ReFRESCO [11]. Numerical and model input uncertainties are combined as $\sigma_{\text{sim}} = \sqrt{\sigma_{\text{num}}^2 + \sigma_{\text{input}}^2}$ [1].

3.1 Computational Framework for CFD Solver Validation

Figure 1 shows the simulation workflow of the SU2 DakOta Quantification of Uncertainty (SU2DOQU) framework. It is a python tool built to couple the open-source suite Dakota for uncertainty quantification with the CFD solver SU2.

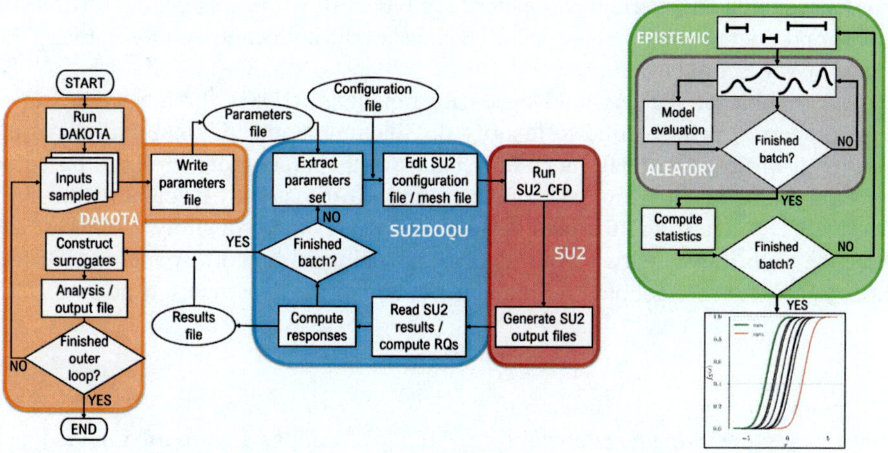

Fig. 1. SU2DOQU, Dakota and SU2 workflow (left) and flow diagram of the "Interval-valued" probability approach (right).

SU2DOQU implements two nested loops for treating both *epistemic* and *aleatory* uncertainties. The outer loop embeds an "Interval-valued" probability method for the *epistemic* uncertainties. No probability is assigned to these uncertainty sources, but a finite set of critical pressure and temperature pairs is selected from the defined intervals with a Latin hypercube sampling (LHS) method. For each pair of critical properties, the input *aleatory* uncertainties are propagated with a stochastic collocation method to calculate the cumulative probability of the chosen response quantities. The associated response functions are built by interpolation of the model predictions under different sets of sampled uncertain parameters. The interpolant used is value-based and thus the interpolation basis consist of the Lagrange interpolants. For a univariate case, where only one input uncertainty ξ is analyzed, the response functions are defined as

$$R(\xi) = \sum_{i=1}^{N} R(\xi_i) L_i(\xi) \quad (6) \quad \text{where,} \quad L_i(\xi) = \prod_{j \neq i}^{N} \frac{(\xi - \xi_j)}{(\xi_i - \xi_j)}.$$

Here, N is the number of collocation points where the model response is evaluated, i.e., the number of sampled input uncertainties combinations. These combinations are generated using a method derived from tensor product quadrature, which uses tensor product of one-dimensional quadrature rules. However, for problems with a high number of input uncertainties this method results too computationally expensive. The Smolyak sparse matrices approach allows for the reduction in the number of collocation points by selecting only certain parameters combinations while keeping the accuracy of the tensor product quadrature approach. The level of the collocation points in the sparse grid matrix was set to be 3.

The ensemble of the cumulative density functions (CDFs) for the whole set of critical properties pairs is used to identify the maximum and minimum values of the response quantities. The final uncertainty bands of the simulations are built based on the obtained CDFs distribution. The lower bound is estimated by calculating the average of the CDF identifying the lowest values for the response quantity of interest and subtracting to this average twice the standard deviation of the corresponding CDF. An analogous procedure is adopted for the upper bound of the uncertainty bands.

4 Results

The static pressure at the nozzle wall (pressure taps locations are reported in [10]) and the Mach number along the centerline of the nozzle are the response quantities of interest. For each combination of critical point properties, a total of 2979 simulations were required to build the CDFs of the response functions with the stochastic collocation method. Figure 2 shows a family of CDFs corresponding to the sampled critical properties (ten pairs) for two exemplary quantities, namely the static pressure and the Mach number at the nozzle throat location. Similar trends were observed for other response quantities characteristic of other locations in the nozzle. As an example, the uncertainty band calculated for the static pressure at the nozzle throat, see Fig. 2a, is based on the two CDFs that are on the opposite sides of the chart, depicted in orange and purple, respectively. Notice that the CDFs are independent from each other and do not intersect. An extrapolation of their trend allows then to identify the two pairs of critical point properties that yield the maximum and minimum values for all the considered response quantities. In particular, the two combinations of critical point properties are $[T_{cr}, P_{cr}]$ = [521.6 K, 18.9 bar] and $[T_{cr}, P_{cr}]$ = [518.5 K, 19.27 bar]. The first pair is constituted by the minimum of critical pressure and maximum value of critical temperature, while the second pair is the other way around. The associated CDFs ensure that the computed uncertainty bands always include the effect of any combination of critical point properties in the considered interval. For this reason, the uncertainty bands associated to the simulation results have been computed considering only these two pairs.

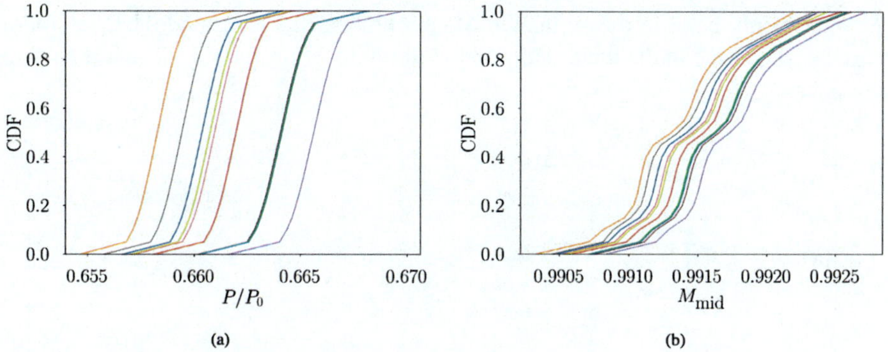

Fig. 2. CDFs corresponding to [518.72 K, 19.1 bar] —, [518.81 K, 19.33 bar] —, [519.82 K, 18.92 bar] — [520.53 K, 19.24 bar] — [521.18 K, 19.01 bar] — [521.41 K, 19.19 bar] — [519.29 K, 19.13 bar] — [519.63 K, 19.35 bar] — [518.81 K, 19.42 bar] — [520.14 K, 18.97 bar] —. **(a)** Family of CDFs for the pressure value at the nozzle wall at x = 0.0424 m from the inlet. **(b)** Family of CDFs for the centerline Mach number at x = 0.0424 m from the inlet.

Figure 3a shows the comparison between the calculated and measured static pressures along the nozzle with the corresponding uncertainty bands. The static pressure P is normalized with respect to the total pressure P_0 at the nozzle inlet, while the axial coordinate x is normalized with respect to the nozzle throat height H_{th}. The pressure distribution is well captured by the simulations. The average value of the measurements fall within the computed uncertainty bands except for a slight deviation at four locations within the kernel region of the nozzle (T09, T10, T13, T14). The maximum mismatch is observed at pressure tap T10 where the difference between the measured and simulated pressure is around 9 kPa. Possible causes of this deviation are a larger uncertainty in the throat size than that predicted by the manufacturing uncertainty model due to thermal effects, see [10], or 3D flow effects triggered by the boundary layer growth on the nozzle walls. Moreover, there is uncertainty related to the value of the nozzle back pressure, assumed fixed in the simulations. As shown also by the Mach distribution along the nozzle, in Fig. 3b, the flow is slightly over-expanded as proven by the two weak oblique shocks occurring in the final part of the channel. This effect is not quantified in the UQ study. Notice also that the uncertainty band width represents at most 1% of the average simulated pressure values, while that associated to the measurement is only 0.1%. This is the reason why the uncertainty bands of the pressure measurements are not visible in the Fig. 3a. The maximum uncertainty associated with the calculated pressure, ± 20 kPa, is found close to the throat, due to high sensitivity of throat location to manufacturing uncertainties.

Regarding the Mach distribution, the largest discrepancies are found in the straight part of the nozzle. Notably, the Mach number reduction due to the oblique shocks appears to be stronger in the experimental results. The reason thereof can be attributed again to the uncertainty associated to the nozzle back pressure. A maximum deviation of around 0.045 between the simulated and experimental Mach number is observed in correspondence of the first shock. Similarly to the uncertainty in the pressure profile, the largest uncertainty band in the simulated Mach number, that is around 0.015, is found close to the throat due to the uncertainty in the throat location caused by manufacturing inaccuracies.

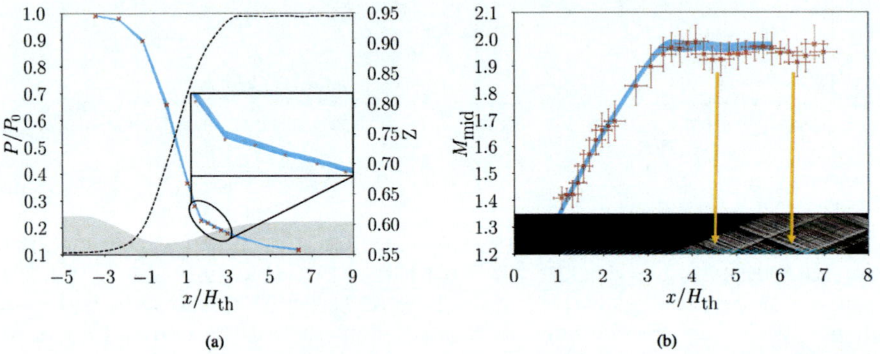

Fig. 3. Comparison between simulation (blue band) and experimental results (x). The simulation uncertainty was computed and the limits of the uncertainty band correspond to $[X_{max} + 2\sigma_{sim}, X_{min} - 2\sigma_{sim}]$, where X_{max} and X_{min} are the maximum and minimum average values of each response quantity corresponding to the different combinations of critical point properties. **(a)** Static pressure along the nozzle wall. Upper and lower wall pressure taps data is compared to the nozzle wall pressure of the simulation. The centerline compressibility factor (- -) is shown together with a zoomed-in section of the kernel region. **(b)** Centerline Mach number in the diverging part of the nozzle. A comparison with the schlieren image taken during the experiment is also shown.

5 Conclusions

In this paper, experimental pressure and Mach number data from a non-ideal isentropic supersonic expansion have been compared against the numerical results from the open-source fluid dynamic solver SU2 using an in-house uncertainty quantification framework for aleatory and epistemic uncertainties called SU2DOQU. A good match between the experimental and numerical Mach number and pressure trends was found. The maximum deviation in the pressure distribution is about 1% and occurs in the kernel region. The reason can be attributed to small differences in the area distribution and expansion ratio of the nozzle between experiment and simulations. This mismatch could be reduced by measuring the nozzle area distribution during the experiment and better characterizing the outlet static pressure of the nozzle. Similar considerations hold for the Mach number distribution, though the maximum deviation is found to occur in the straight final part of the nozzle, where two weak oblique shock waves appear. Future work will target a more comprehensive validation campaign involving experiments in other non-ideal thermodynamic conditions and the possible use of more accurate CFD models and fluid equations of state.

References

1. American Society of Mechanical Engineers (ASME) 2009 Standard for Verification and Validation in Computational Fluid Dynamics and Heat Transfer. Technical report V&V20-2009. ASME (2009)
2. Betz, W., Papaioannou, I., Strauss, D.: Numerical methods for the discretization of random fields by means of the Karhunen-Loève expansion. Comput. Methods Appl. Mech. Eng. **271**, 109–129 (2014)
3. Bills, L.: Validation of the SU2 flow solver for classical NICFD. Master's thesis, Delft University of Technology (2020)
4. Celik, I.B., Ghia, U., Roache, P.J., Freitas, C.J., Coleman, H., Raad, P.E.: Procedure for estimation and reporting of uncertainty due to discretization in CFD applications. J. Fluids Eng. **130**(7) (2008)
5. Dickinson, E., McLure, I.A.: Thermodynamics of n-alkane + dimethylsiloxane mixtures. part 3.-excess volumes. J. Chem. Soc., Faraday Trans. 1 **70**, 2328–2337 (1974)
6. Dow, E., Wang, Q.: Optimal design and tolerancing of compressor blades subject to manufacturing variability. In: 16th AIAA Non-Deterministic Approaches Conference (2014)
7. Economon, T.D., Palacios, F., Copeland, S.R., Lukaczyk, T.W., Alonso, J.J.: SU2: an open-source suite for multiphysics simulation and design. AIAA J. **54**(3), 828–846 (2015)
8. Gori, G., Zocca, M., Cammi, G., Spinelli, A., Congedo, P.M., Guardone, A.: Accuracy assessment of the non-ideal computational fluid dynamics model for siloxane MDM from the open-source SU2 suite. Eur. J. Mech. - B/Fluids **79**, 109–120 (2020)
9. Head, A.J.: Novel experiments for the investigation of non-ideal compressible fluid dynamics: The orchid and first results of optical measurements. PhD thesis, Delft University of Technology (2021)
10. Head, A.J., et al.: Mach number estimation and pressure profile measurements of expanding dense organic vapors. In: Pini, M., De Servi, C., Spinelli, A., di Mare, F., Guardone, A. (eds) Proceedings of the 4th International Seminar on Non-Ideal Compressible Fluid Dynamics for Propulsion and Power. NICFD 2022. ERCOFTAC Series, ERCO, vol. 29, p. IX, 240. Springer, Cham (2023)
11. Marin (2019). https://www.refresco.org/verificationvalidation/utilitiesvvtools/
12. McLure, I.A., Dickinson, E.: Vapour pressure of hexamethyldisiloxane near its critical point: corresponding-states principle for dimethylsiloxanes. J. Chem. Thermodyn. **8**(1), 93–95 (1976)
13. Nyström, E.J.: Über die praktische auflösung von integralgleichungen mit anwendungen auf randwertaufgaben. Acta Math. **54**, 185–204 (1930)
14. Pini, M., et al.: SU2: the open-source software for non-ideal compressible flows. J. Phys: Conf. Ser. **821**, 012013 (2017)
15. Razaaly, N., Persico, G., Congedo, P.M.: Impact of geometric, operational, and model uncertainties on the non-ideal flow through a supersonic orc turbine cascade. Energy **169**, 213–227 (2019)
16. Vitale, S., et al.: Extension of the SU2 open source CFD code to the simulation of turbulent flows of fluids modelled with complex thermophysical laws. In: Proceedings of the 22nd AIAA Computational Fluid Dynamics Conference (2015)
17. Woodward, S.D., Dury, M.R., Brown, S.B., McCarthy, M.B.: Understanding articulating arm laser line scanners for precision engineering: operator usage effects. In: Proceedings of the ASPE 2015 Annual Meeting (2015)

Multi-phase Flows

Numerical Validation of a Two-Phase Nozzle Design Tool Based on the Two-Fluid Model Applied to Wet-to-Dry Expansion of Organic Fluids

Pawel Ogrodniczak[1](✉) and Martin T. White[2]

[1] City, University of London, London, UK
pawel.ogrodniczak@city.ac.uk
[2] University of Sussex, Brighton, UK
martin.white@sussex.ac.uk

Abstract. Two-phase expansion could increase the power output of organic Rankine cycles by up to 30% in the temperature range of 150 to 250 °C compared to single-phase architectures. By employing molecularly complex fluids, it is possible to design a wet-to-dry cycle, in which the fluid transitions from a two-phase state to superheated gas during expansion. This opens the possibility of using turboexpanders, provided that the wet portion of the expansion is confined to the stator. Hence, the design of the stator becomes critical to ensure complete evaporation of the two-phase mixture. Given the rapid acceleration and large change in density across the expansion process, it may be essential to account for non-equilibrium effects when designing two-phase stators. This study aims to validate a previously developed two-phase nozzle design tool with non-equilibrium CFD simulations. The design tool assumes quasi-1D inviscid flow and employs a two-fluid model that solves separate mass, momentum and energy conservation equations for both phases. The design tool is evaluated by performing two-dimensional viscous simulations on the nozzle geometries generated from the nozzle design tool. To examine the importance of including non-equilibrium effects at the nozzle design stage, a separate nozzle was generated assuming homogeneous-equilibrium flow. The results reveal the predictions from the design tool were in relatively good agreement with the CFD simulations with the difference in streamwise variations of various flow properties typically not exceeding a few percent. However, this was not the case for expansions from low inlet pressure and low inlet vapour quality, which were found to be highly two-dimensional and could not be effectively treated by the design tool. Furthermore, a significant disparity between the CFD results and the homogeneous-equilibrium design was found, indicating that non-equilibrium effects should not be neglected when designing a nozzle for wet-to-dry cycles.

Keywords: two-phase expansion · flash boiling · two-phase nozzle · wet-to-dry cycle · organic Rankine cycle

M. White et al. (Eds.): NICFD 2022, ERCOFTAC 29, pp. 103–113, 2023.
https://doi.org/10.1007/978-3-031-30936-6_11

Nomenclature

ρ	density, kg/m^3
σ	surface tension coefficient, N/m
D	diameter, m
P	pressure, Pa
q	vapour quality
T	temperature, K
U	velocity, m/s
We	Weber number

Sub- and superscripts

cr	critical
d	droplet
L	liquid phase
mix	mixture
V	vapour phase

1 Introduction

Organic Rankine cycle (ORC) is a suitable thermodynamic cycle for power generation from low-temperature heat. However, operation at low temperatures implies low thermal efficiencies, while small temperature differences contribute to the need for larger heat exchangers [1]. These factors may lead to poor economic performance, and further improvements are needed. One of the main factors restricting further performance improvements of ORC systems is the isothermal evaporation process, which is associated with significant exergy losses [2], and may restrict the extent to which the heat source can be cooled down limiting the attainable power output [3]. Expanding from a two-phase state could potentially reduce the limitations of isothermal evaporation, allowing for a power output increase of up to 30% in the temperature range of 150 to 250 °C. However, the main challenge is the availability of an expander that can tolerate wet conditions while maintaining high efficiency. Volumetric expanders have been proposed, but the limited built-in volume ratio makes them suitable for applications with temperatures below 150 °C. Turboexpanders can achieve much higher volumetric expansion ratios, but the presence of high speed droplets in the rotor could lead to erosion issues.

To overcome these challenges, the wet-to-dry cycle has been proposed, which employs molecularly-complex fluids that could allow an expansion from a two-phase state to a superheated vapour, such that only a part of the expansion is wet. This opens the possibility of using the existing turboexpander architectures, such as the radial-inflow turbine, so long as the wet portion of the expansion is confined to the stator.

Therefore, the stator has to be designed in such a way to ensure the mixture can fully evaporate before reaching the outlet. Previous simulations of two-phase expansion in a turbine stator have shown that it is possible to design a stator that can facilitate complete mixture evaporation, delivering dry flow to the rotor [3]. However, the numerical simulations employed the homogeneous-equilibrium model. In reality, the rapid acceleration of the mixture, combined with a considerable difference in densities between the liquid and vapour phases will likely lead to the development of a large velocity slip, while a finite interphase heat-transfer rate may result in thermal non-equilibrium and delayed evaporation. A later study investigated the significance of non-equilibrium effects in a wet-to-dry expansion under operating conditions relevant to a wet-to-dry cycle [4]. Specifically, a nozzle design tool based on a quasi-1D inviscid two-fluid model was developed that solves the conservation equations for both phases. In short, the study demonstrated homogeneous-equilibrium assumption is unsuitable for simulating wet-to-dry expansion processes.

This study aims to validate the quasi-1D design tool with a CFD model capable of accounting for two-dimensional flow variations, turbulence effects and compressible flow features typical of supersonic wet-to-dry expansions. In conditions relevant to the wet-to-dry cycle, the selected working fluid, the siloxane MM, also exhibits non-ideal behaviour which must be considered. The CFD simulations are conducted and the numerical results are compared to the design predictions for validation purposes. The performance of the nozzle generated by means of the non-equilibrium design tool is also compared to the performance of a nozzle generated under the homogeneous-equilibrium assumption to establish the relative difference in nozzle performance and assess the importance of including non-equilibrium effects when designing nozzles for wet-to-dry expansion.

2 Methodology

2.1 Two-Fluid Model

The CFD simulations of the wet-to-dry expansion were performed using the two-fluid model (TFM), which is considered the most general approach to modelling flashing flows [5]. Unlike models such as the homogeneous-equilibrium, homogeneous-relaxation or algebraic slip models, it accounts for both mechanical and thermal non-equilibrium effects since the mass, momentum and energy conservation equations are solved separately for each phase. The interaction between the phases defined in terms of mass, momentum and energy exchange is accounted for via interphase models. Three interphase momentum exchange mechanisms were considered (in addition to momentum exchange resulting from interphase mass transfer), namely drag, lift and turbulent dispersion forces, for which the Ishii-Zuber [6], Tomiyama [7] and Favre Averaged Drag models were used respectively. The mass transfer was assumed to happen solely due to interphase heat transfer, which is a frequently adopted assumption in flash boiling problems characterised by relatively high pressure and temperature levels [5]. The interphase heat transfer was evaluated using the two-resistance model, which assumes that liquid and vapour exchange heat separately with an interface that is at local saturation temperature. The heat-transfer coefficient between the vapour and the interface was estimated using the Ranz-Marshall [8] correlation, while a constant Nusselt number of 6 was set on the liquid side, which is an approximation derived based on the transient heat-transfer analysis in a solid sphere. In the wet-to-dry expansion, the liquid phase has a significantly higher density than the vapour phase, and even low-quality mixtures will have a low liquid volume fraction. Hence, the dispersed phase is assumed to be in the form of spherical droplets.

2.2 Droplet Size

One of the main unknowns in the wet-to-dry expansion is the size and distribution of droplets. In the absence of experimental data, it is difficult to determine an appropriate value for the droplet size. However, one can resort to droplet breakup analysis to approximate the maximum droplet diameter that would not undergo breakup in the expansion. The parameter that has been widely used in droplet breakup studies is the Weber number, defined as:

$$We = \frac{\rho_V \left(u_V - u_L\right)^2 D_d}{\sigma} \tag{1}$$

where We is the Weber number, ρ_V the vapour density, D_d the droplet diameter, while u_V and u_L are the vapour and liquid velocities respectively. Experimental studies have demonstrated that when the ratio $\mu_L/\sqrt{\rho_L D_d \sigma}$ is less than 0.1, which was found to be true in all the investigated expansions, the critical Weber (We_{cr}) number remains constant at about 11 [9]. When the local We exceeds We_{cr} droplet breakup is expected to take place. Substituting We_{cr} into Eq. 1, one can derive an expression for the maximum stable droplet diameter that would not undergo a breakup:

$$D_{max} = \frac{We_{cr}\sigma}{\rho_V \left(u_V - u_L\right)^2} \tag{2}$$

where σ is the surface tension coefficient. Gobyzov et al. [10] studied experimentally the behaviour of water droplets in a converging-diverging nozzle traveling in a continuously accelerating air stream that reaches supersonic speeds. The authors found that breakup was initiated when We exceeded the value of 11.2, which indicates that a similar critical value of We can be expected for droplets travelling in both subsonic and supersonic conditions. It should be highlighted that the droplet size estimation technique adopted in this work makes two over-simplifications. Firstly, it presumes that the breakup occurs instantaneously when We_{cr} is exceeded, while the breakup itself reduces the diameter of the droplet just to the level needed for We to remain at the critical value. In reality, breakup occurs over a finite time, and when it happens the size of the droplet fragments may constitute only a fraction of the initial droplet size [11]. Hence, one can think of the adopted technique as a simplified way of estimating the maximum stable droplet size, rather than a model that is able to simulate the actual breakup mechanism. Despite the over-simplifications, the method is thought to be capable of estimating a realistic droplet size in the absence of empirical data.

2.3 Quasi-1D TFM Nozzle Design Tool

The wet-to-dry nozzle design tool based on the two-fluid model has been previously developed by the authors [4]. The tool assumes quasi-1D inviscid flow and uses the inverse design approach, where a pressure profile is imposed using a Bezier curve. The governing equations are then solved to converge on the imposed pressure distribution. Under the conditions relevant to the wet-to-dry cycle MM exhibits non-ideal behaviour. NIST REFPROP 10.0 [12] was employed, which uses the Helmholtz energy based equation of state that has been derived specifically for siloxane MM, with the aid of empirical and molecular simulation data [13]. It also permits calculating the fluid properties in the metastable states, which is crucial in this type of problem. Further details of the quasi-1D TFM design tool can be found in [4].

2.4 Numerical Setup

The two-fluid model has been implemented in ANSYS CFX 2021 R1. A thin nozzle slab was taken with symmetry boundary conditions applied to perform steady quasi-2D simulations. Turbulence in the continuous phase was modelled using $k - \omega$ SST model, while the dispersed phase zero equation model was applied to evaluate the eddy viscosity of the dispersed phase. Automatic wall functions were used and the mesh was refined near the wall to achieve y^+ between 30 and 300 for both phases. Look-up tables were generated using REFPROP 10.0 [12] and coupled with the CFD solver to evaluate the fluid's properties. The model has been previously validated against experimental data for flash boiling of subcooled water in which the numerical results were in good agreement with the experimental data [14] as well as other numerical studies simulating the flash-boiling experiment [15, 16]. To implement the droplet size model, the streamwise variation in droplet diameter predicted from the design tool was imposed within the CFD model using a user-defined expression. In total, six wet-to-dry expansion cases are considered with inlet pressures of 478 and 1012 kPa, and inlet vapour qualities of 0.1, 0.3 and 0.5. Small inlet velocity of 10 m/s was assumed. The mass flow rate was set arbitrarily to 0.5 kg/s, but nozzles designed for 0.05 and 5 kg/s were also investigated. No significant changes in the expansion characteristics were found for the different mass flow rates, which indicates that the results are not sensitive to the scale of application. To determine the nozzle outlet conditions, the nozzle was treated as part of a turbine designed for a wet-to-dry cycle that has a fixed condensation temperature of 40° degC. The nozzle outlet pressures were then calculated using the definition of the degree of reaction that was set to 0.5 for all cases except for the expansions from 478 kPa and low inlet qualities, where the degree of reaction had to be lowered to ensure the mixture could transition to the superheated vapour regime. The outlet Mach number for the six expansions varied between 1.6 and 1.9.

3 Results

First, the case with the inlet pressure and quality of 1012 kPa and 0.3 was considered for the mesh independence study. Three different mesh sizes were constructed with 7,000, 16,000 and 30,000 predominantly hexahedral elements (Fig. 1).

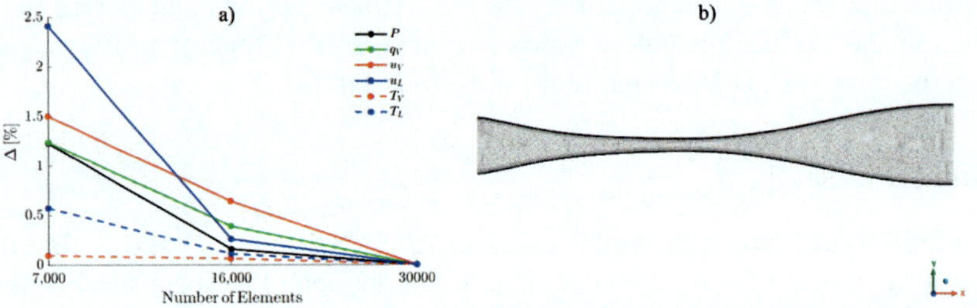

Fig. 1. a) Percentage difference in various outlet parameters for the examined mesh sizes; b) an example of the generated grid (16,000 elements mesh displayed)

Figure 1 presents the percentage difference in various outlet parameters for the examined grids. Compared to the finest mesh, the mesh with 16,000 elements gave a maximum percentage difference below 0.65%. Hence, this element size was applied to mesh all other investigated cases.

The primary objective of this work is to validate the previously developed nozzle design tool. Figure 2 reports the streamwise variations of average pressure P, mixture velocity u_{mix} (mass-fraction weighted sum of phase velocities), vapour mass fraction q_V, velocity slip u_{slip} and droplet diameter D_d for the six investigated cases, as predicted by the design tool and the CFD simulations.

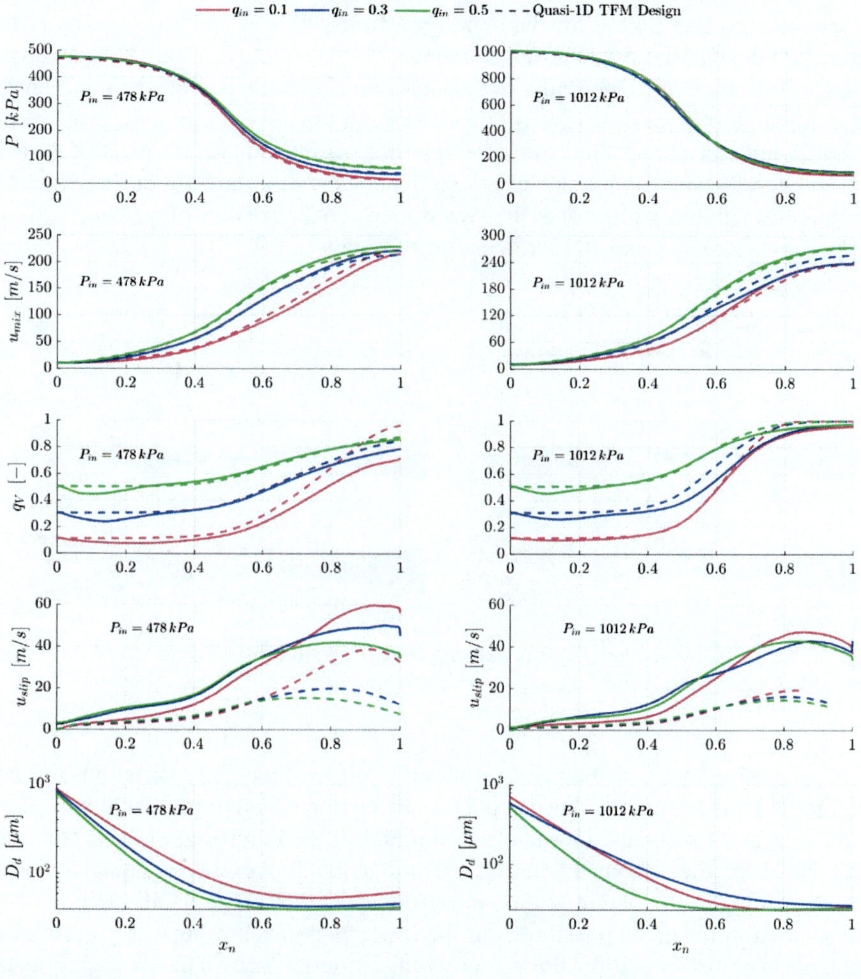

Fig. 2. Streamwise variation of average pressure P, mixture velocity u_{mix}, vapour quality q_V, slip velocity u_{slip} and droplet diameter D_d for the six expansions as predicted by the nozzle design tool (dashed) and CFD simulations (solid).

From Fig. 2 it can be noted that the design pressure profiles closely match the profiles obtained through CFD simulations. The discrepancy between the design and numerical results slightly increases when the mixture velocity is concerned. Nevertheless, the trends are conserved while the differences are relatively small, typically below 6–7% for all cases except for the expansion from 478 kPa and $q_V = 0.1$ for which the deviation is slightly higher, in the order of 9–10%. Although, a fairly good agreement in u_{mix} was found, there is a significant disparity in slip velocity, which is not completely clear and requires further investigation. The expansion was initiated with droplet diameter of about 1 mm; the droplet size quickly reduced, reaching values in the range of 30–100 μm in the diverging section. Looking at the outlet vapour quality, it can be seen that except for the expansion from 478 kPa and inlet quality of 0.1, the quasi-1D design tool predicts values that are relatively close to the those estimated by the CFD simulations. Generally, the design tool overestimates the evaporation rate when compared to numerical results. To understand the delay in mixture evaporation, one should remember that the nozzle design tool assumes quasi-1D inviscid flow. In reality, both two-dimensional and viscous effects exist and may affect the expansion characteristics. Figure 3 illustrates the two-dimensional variation of q_V and u_{mix} for the expansion from 478 and 1012 kPa and inlet quality of 0.3.

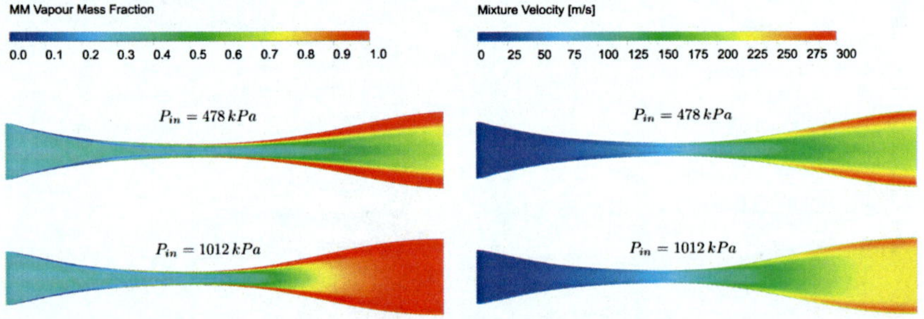

Fig. 3. Vapour mass fraction and mixture velocity for expansion from 478 and 1012 kPa and inlet quality of 0.3.

The first thing that becomes clear is that two-dimensional flow variations are non-negligible in both expansion cases as the phases tend to separate past the nozzle throat due to the larger density and inertia of the liquid droplets compared to the surrounding vapour. However, during expansion from 1012 kPa, the fluid evaporates quicker and the phase separation is not visible at the nozzle outlet. On the other hand, there is severe phase separation for the expansion from the lower pressure level that is present at the outlet. Due to phase separation, the velocity is also non-uniform in the diverging section of the nozzle. This is because, in the core of the nozzle, there is much higher liquid content which means that the overall density is higher, while the drag effects are enhanced, leading to lower velocity in the core of the nozzle. The velocity non-uniformity is also more severe in the 478 kPa case. The larger phase separation and flow non-uniformity for the lower-pressure expansion is also caused by higher density difference between liquid and vapour phases at lower pressures. The increased density difference implies a lower capability of the droplets to follow the mean flow direction. Referring back to Fig. 2, one can see that the phase change process was relatively well predicted by the design tool; however, a large discrepancy was found for the expansion from 478 kPa and inlet quality of 0.1. This is due to large density difference between liquid and vapour and excessive curvature of the nozzle geometry in the diverging section, which leads to severe phase separation and two-dimensional flow variations, which cannot be accounted for by the quasi-1D design tool.

In summary, the CFD simulations demonstrate that the nozzle design tool provides fairly accurate predictions of the wet-to-dry expansion process. However, caution should be taken when the tool is applied to low-pressure expansion when the phases tend to separate, or in cases with excessive nozzle curvature. Flow uniformity is particularly important in the context of turbomachinery, and these results indicate that CFD simulations may be necessary to ensure satisfactory flow uniformity is achieved. When it comes to the feasibility of a wet-to-dry nozzle design, it has been shown that if the inlet pressure is sufficiently high, a nozzle can be designed to expand low-quality mixtures to flows with a negligible amount of liquid.

The other question that arises is how important is it to consider non-equilibrium effects when designing a wet-to-dry nozzle, or would a nozzle geometry designed based on the homogeneous-equilibrium flow assumption provide similar performance to the nozzle generated using the non-equilibrium design tool. Figure 4 compares the performance of nozzles generated using the two above-mentioned approaches. Two expansion cases were considered, namely expansions from 478 and 1012 kPa, with the same inlet quality of 0.3. The figure shows the streamwise variation of average pressure, mixture velocity and vapour mass fraction obtained through CFD simulations and the design predictions. The comparison between CFD and design profiles indicates that the non-equilibirum model provides a much better estimation compared to the homogeneous-equilibrium model which exhibits significant deviation from the CFD results. The geometry generated with the non-equilibrium tool also provides higher outlet qualities and mixture velocities under the same boundary conditions.

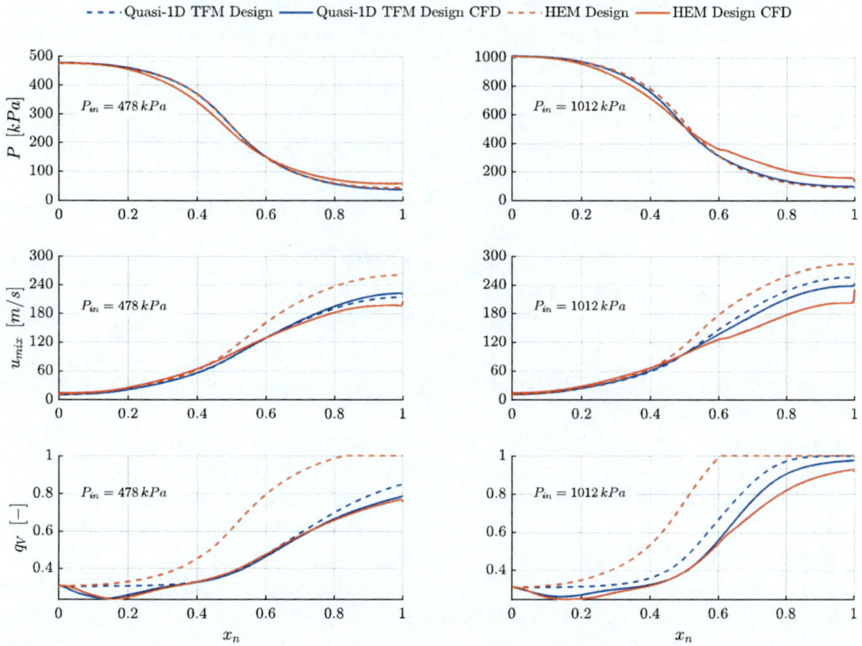

Fig. 4. Comparison between the pressure, mixture velocity and vapour mass fraction predicted by the two-fluid and homogeneous-equilibrium design tools (dashed) and the CFD simulations of the generated nozzle designs (solid).

4 Conclusions and Future Works

A non-equilibrium two-fluid CFD simulation has been carried out to investigate the wet-to-dry expansion of MM in a converging-diverging nozzle and to validate a previously developed quasi-1D non-equilibrium nozzle design tool. The results indicate that the tool can accurately predict the wet-to-dry expansion and phase-change process in most of the investigated working conditions except for expansions from low pressures and low inlet qualities which appear to be highly two-dimensional with a tendency for phase separation. Moreover, it is found that the design tool performed significantly better than a design tool based on the homogeneous-equilibrium model in generating a geometry that achieves the targeted mixture evaporation under the same working conditions. Future studies will focus on the optimisation of the converging-diverging nozzle and on the application of the design tool to design a radial turbine stator for wet-to-dry expansion.

References

1. Read, M.G., Smith, I.K., Stosic, N.: Multi-variable optimisation of wet vapour organic rankine cycles with twin-screw expanders. In: Proceedings of the 22nd International Compressor Engineering Conference, Purdue, USA, p. 7 (2014)
2. Dibazar, S.Y., Salehi, G., Davarpanah, A.: Comparison of exergy and advanced exergy analysis in three different organic Rankine cycles. Processes **8**(5), 586 (2020)
3. White, M.T.: Cycle and turbine optimisation for an ORC operating with two-phase expansion. Appl. Therm. Eng. **192**, 116852 (2021)
4. White, M.T.: Investigating the wet-to-dry expansion of organic fluids for power generation. Intl. J. Heat Mass Trans. **192**, 122921 (2022)
5. Liao, Y., Lucas, D.: Computational modelling of flash boiling flows: a literature survey. Intl. J. Heat Mass Trans. **111**, 246–265 (2017)
6. Ishii, M., Zuber, N.: Drag coefficient and relative velocity in bubbly, droplet or particulate flows. AIChE J. **25**(5), 843–855 (1979)
7. Tomiyama, A.: Struggle with computational bubble dynamics. In: Proceedings of Third International Conference on Multiphase Flow, Lyon, France, p. 6 (1998)
8. Ranz, W.E., Marshall, W.R., Jr.: Evaporation from drops: part 1. Chem. Eng. Prog. **48**(3), 141–146 (1952)
9. Guildenbecher, D.R., López-Rivera, C., Sojka, P.E.: Secondary atomization. Exp. Fluids **46**(3), 371–402 (2009)
10. Gobyzov, O.A., Ryabov, M.N., Bilsky, A.V.: Study of deformation and breakup of submillimeter droplets' spray in a supersonic nozzle flow. Appl. Sci. **10**(18), 6149 (2020)
11. Pilch, M., Erdman, C.A.: Use of breakup time data and velocity history data to predict the maximum size of stable fragments for acceleration-induced breakup of a liquid drop. Intl. J. Multiph. Flow **13**(6), 741–757 (1987)
12. Lemmon, E.W., Ian H., Bell, M.L., Huber, M., McLinden, O.: NIST standard reference database 23: reference fluid thermodynamic and transport properties-REFPROP, Version 10.0. National Institute of Standards and Technology (2018). https://www.nist.gov/srd/refprop
13. Thol, M., et al.: Fundamental equation of state correlation for hexamethyldisiloxane based on experimental and molecular simulation data. Fluid Phase Equilib. **418**, 133–151 (2016)

14. Abuaf, N., Wu, B.J., Zimmer, G.A., Saha, P.: Study of nonequilibrium flashing of water in a converging-diverging nozzle. Volume 1: Experimental. Brookhaven National Laboratory (1981)
15. Rane, S., He, L.: CFD analysis of flashing flow in two-phase geothermal turbine design. J. Comput. Design Eng. **7**(2), 238–250 (2020)
16. Liao, Y., Lucas, D.: 3D CFD simulation of flashing flows in a converging-diverging nozzle. Nucl. Eng. Des. **292**, 149–163 (2015)

Check for updates

Numerical Modelling of Cryogenic Flows Under Near-Vacuum Pressure Conditions

Theodoros Lyras(✉), Ioannis K. Karathanassis, Nikolaos Kyriazis, Phoevos Koukouvinis, and Manolis Gavaises

City, University of London, London, UK
{theodoros.lyras.2,ioannis.karathanassis,m.gavaises}@city.ac.uk

Abstract. A numerical framework for the simulation of two-phase cryogenic flows under a wide range of pressure conditions is presented in this work. Subcritical injection and near-vacuum ambient pressure conditions were assessed by numerical simulations. Two different computational approaches have been employed, namely a pressure-based solver complemented by a bubble-dynamics model, as well as a density-based solver utilising real-fluid tabulated data to describe the fluid's thermodynamic properties. The required thermodynamic-data table has been derived using the Helmholtz Equation of State (EoS) and the specific modelling approach can be applied to near-vacuum, sub-critical or even supercritical injection pressure conditions. The geometries of two single-hole injectors have been considered for investigating the flow and spray formation of liquid oxygen (LOx) and liquid Nitrogen (LN_2). Both numerical approaches were validated against available experimental data. Overall, the comparison of results to experimentally acquired data demonstrates the suitability of the employed methodologies in describing processes such cryogenic flashing-flow expansion, phase-change and flash-induced spray formation. The density-based tabulated thermodynamics approach in particular, can be considered as a complete numerical framework for treating two-phase cryogenic flows using real-fluid properties, for a wide range of conditions without the need for case-related modifications.

Keywords: cryogenic fluids · real-fluid thermodynamics · flash boiling · compressible flow

1 Introduction

Cryogenic propellants in liquid-state have been utilised since the 1960s (Harstad and Bellan 1998) and are currently the selected type of propellants for many modern upper-stage rocket engines. The combination of LOx and LH_2 has been used to fire the Ares-I vehicle (Davis and McArthur 2008), the Centaur engine of Atlas rocket (Yang 2004) and ESA's Ariane 5 upper stage engines (Coulon 2000). Depending on the rocket-stage location, the delivery of LOx to the combustion chamber can be realised in subcritical pressure conditions and due to the pressure drop in the oxidizer delivery nozzle, LOx phase-change is expected. While the initial lift-off of a space launch vehicle takes place in atmospheric conditions, the upper-stage engines are expected to ignite and operate

M. White et al. (Eds.): NICFD 2022, ERCOFTAC 29, pp. 114–124, 2023.
https://doi.org/10.1007/978-3-031-30936-6_12

in near-vacuum conditions. Those low-pressure operating conditions can lead to flash boiling of the propellants during start-up with consequent after-effects on the ignition process (Lamanna 2009), (Luo and Haidn 2016). Investigation of cryogenic flash boiling is, therefore, essential for understanding and regulating the process.

Extremely low temperature conditions and intense phase change rates are factors hindering the realisation of experimental campaigns relevant to cryogenic two-phase flows. Hendricks et al. (1976) investigated two-phase LOx and LN_2 flows through two converging-diverging nozzles for a wide range of sub- and super-critical conditions. Mayer and Tamura (1996) presented flow visualisations and measurements in a liquid rocket engine combustor. Luo and Haidn (2016), investigated a flashing LN_2 spray in a low-pressure environment. Lamanna et al. (2009) considered fully flashing LOx and ethanol sprays observed with the use of diffuse-backlight imaging. More recently, in the work of Rees et al. (2020), the characteristic morphologies of the flash boiling LN_2 sprays were determined with the use of high-speed shadowgraphy.

The scarcity of experimental data with respect to cryogenic flows has led to the incorporation of numerical simulations. Travis et al. (2012) developed a theoretical two-phase model based on the Helmholtz energy EoS used to create a hydrogen critical flow map for stagnation states. Lyras et al. (2018) utilised the volume-of-fluid method coupled with the homogeneous relaxation model (HRM) to investigate liquid nitrogen spray formation. Chen et al. (2020) proposed a two-fluid numerical model that couples an interface area density model to the HRM to investigate LN_2 spray formation. Gärtner et al. (2020) used a one-fluid approach, with tabulated thermodynamic properties and the HRM to account for phase change. Schmehl and Steelant (2009) utilized an Eulerian-Langrangian framework to simulate the pre-flow of nitrogen tetroxide (N2O4) oxidizer. Ramcke et al. (2018) used a similar numerical approach to investigate the pre-flow of LOx for satellite rocket engines.

2 Scope of Research

The lack of cryogenic-flashing experimental data in the open literature makes the need for accurate and robust numerical methods imperative, as an accurate numerical prediction of such flows can provide insight on complex phenomena relevant to compressible flow. A numerical framework for the simulation of two-phase cryogenic flows under a wide range of conditions is presented in this work. More specifically, a universal methodology based on tabulated thermodynamics applicable to a wide range of degrees superheat, R_p, is presented in a comparative manner against a phase-change model based on Knudsen's kinetic-theory. Both numerical approaches as well as the two different cylindrical injector components that have been selected in order to examine the cryogenic flow of LOx and LN_2, are presented in detail in the Methodology section. Ten different sets of varying boundary conditions have been selected for the test-cases and are presented in the Results' section along with the cryogenic flow characteristics resulting from each test-case configuration.

3 Main Research Issues

Two flow solvers, an implicit coupled pressure/velocity and an explicit density-based one, have been employed in the present investigation. The basic set of governing equations solved in both cases comprise the continuity, momentum and energy conservation equations. The complete formulations of the adopted numerical methodologies, along with the sets of equations solved can be found in detail in the works of Karathanassis et al. (2017) and Kyriazis et al. (2017) with reference to the pressure- and density-based solvers, respectively. Referring to the coupled solver, a two-phase mixture approach was implemented, including additionally Eq. 1 for the vapour transport.

$$\frac{\partial(a_v \rho_v)}{\theta t} + \nabla(a_v \rho_v \vec{u}) = \dot{R}, \tag{1}$$

where a_v, ρ_v, t and $\overrightarrow{u}$ are the vapour volume fraction, vapour density, time and velocity vector respectively. The rate of phase-change, $\dot{R}$ was calculated using the Hertz-Knudsen equation (Eq. 2) derived from the kinetic theory of gases (Fuster et al. 2010):

$$\dot{R} = \frac{\lambda A_{\text{int}}(p_{sat} - p)}{\sqrt{2\pi R_g T_{\text{int}}}}, \tag{2}$$

where λ, is an accommodation coefficient, A_{int} and T_{int} correspond to the liquid-vapour interphase area and temperature respectively, p and p_{sat} correspond to pressure and saturation pressure and R_g is the ideal gas constant. The coupled pressure-based solver has been extensively validated with reference to internally and externally flashing flows considering water as the working medium in previous authors' works, refer to (Lyras et al. 2021), (Karathanassis et al. 2017). Specifically, the model has been demonstrated to accurately capture phase-change in a converging-diverging nozzle, a (throttle) nozzle with an abrupt constriction and a rapidly depressurising duct (pipe blow-down).

In the case of the density-based solver, a single-fluid modelling approach has been formulated and the 3-D RANS equations in conservative form were considered. The fluid properties were calculated using the Helmholtz energy EoS, Eq. (3), that can be expressed as a function of dimensionless density (δ) and dimensionless temperature (τ) and were then organised into a thermodynamic table:

$$\frac{a(\rho, T)}{RT} = a(\delta, \tau) = \alpha^0(\delta, \tau) + \alpha^r(\delta, \tau) \tag{3}$$

where a is the molar Helmholtz energy, R is the specific gas constant, a^0 the dimensionless ideal gas contribution to the Helmholtz energy and finally a^r is the term of dimensionless residual Helmholtz energy. Since the Mach number is probable to obtain highly different values in the pure liquid, vapour and two-phase mixture regions, a Mach-number consistent numerical flux has been implemented based on the HLLC and the AUSM fluxes (Schmidt et al. 2008), (Liou 2006). Conservative variables at cell interfaces, required for the calculation of the fluxes, were determined using the MUSCL-Hancock reconstruction (Toro 2013), 2nd-order accurate in space. A 4th-order accurate, four-stage Runge-Kutta method has been selected for time integration. The density-based algorithm has been validated in previous works with reference to bubble- and droplet-dynamics simulations, while the accuracy of the tabulated technique based on the Helmholtz energy EoS has been verified with reference to the properties of n-dodecane (Kyriazis et al. 2017).

4 Methodology

Two geometries are represented in this study. Injector A is a single-hole cylindrical LOx injector with a length-to-diameter ratio (L/D) of 1.2 as documented in the investigation of Lamanna et al. (2015) and depicted in Fig. 1a. Injector B is a cylindrical LN_2 injecting device with L/D = 3.9, as reported in the study of Rees et al. (2020) and presented in Fig. 1b. Since both flow layouts of interest are symmetrical in the axial direction, their numerical counterparts were reduced to a 5° wedge with appropriate symmetry boundary conditions enforced on the corresponding planes. Further boundary and initial conditions are presented in Table 1.

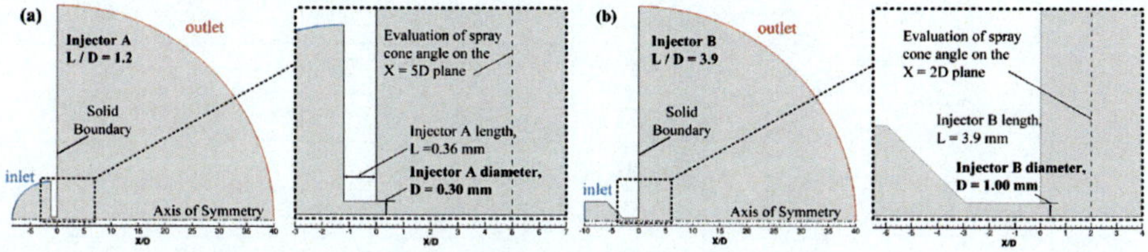

Fig. 1. Axisymmetric plane cut of the numerical domains of (a) LOx injector A and (b) LN_2 injector B.

Table 1. Summary of boundary and initial conditions imposed for the numerical simulations. Symbol e corresponds to the fluid internal energy in [J kg^{-1}], while a is the vapour volume fraction.

Boundary Conditions	Inlet		Outlet		Wall	
Pressure-based	$p = p_{in}$	$T = T_{in}$	$p = p_{out}$	$T = T_{out}$	$\vec{u} = 0$	$\partial T/\partial \boldsymbol{n} = 0$
Density-based	$p = p_{in}$	$e = e_{in}$	$p = p_{out}$	$T = T_{out}$	$\vec{u} = 0$	$\partial T/\partial \boldsymbol{n} = 0$
Initial Conditions (t = 0 s)						
$p = p_0$	$\vec{u} = 0$		$T = T_{in}$		$a_{domain} = 0$	

5 Results

The test cases examined in this study are summarised in Table 2. Specifically, cases 1 to 4 refer to LOx flow through Injector A and correspond to the experimental conditions reported by Lamanna et al. (2015), whereas cases 5 to 10 refer to Injector B and the experimental campaign of Rees et al. (2020), for which data regarding the spray cone angles are available. A transient solver has been used for all cases, yet it was confirmed that the respective flow and temperature fields reached to steady-state solutions in many of the cases. Time-averaged fields are presented for all cases. All pressure-based solver results are initially calibrated for A_{int}, based on optimal values deducted from previous investigations. On the contrary, the advantage of the tabulated thermodynamics method is that, after an initial table of fluid properties is created for a range of conditions, no further adjustment or calibration is needed, since the phase change rates are dictated by the tabulated real-gas thermodynamics.

Table 2. List of examined test-cases, including the fluid type and boundary conditions imposed for the numerical simulations

Case	Fluid	$P_{in} \cdot 10^5$ [Pa]	P_{out} [Pa]	T_{in} [K]	Superheat Degree R_p [-]
1	LOx	17	20600	113.0	33.0
2		17	14140	113.0	48.0
3		17	9100	113.0	74.0
4		17	2750	113.0	245.0
5	LN_2	8	58020	82.5	3.0
6		8	24870	82.5	7.0
7		8	6130	82.5	28.4
8		4	58020	82.5	3.0
9		4	24870	82.5	7.0
10		4	3330	82.5	52.3

5.1 Liquid Oxygen Spray Formation

Figure 2 summarises the results produced with the two approaches and evaluates the predictive capability of the two methods in terms of spray cone angle, for degrees of superheat ranging from 33 to 245. The cone angle calculation was made 5 nozzle diameters away from the injector exit. Both the pressure- and density-based solvers produce equally accurate results for high LOx superheat degree (case 4) while capturing the process with satisfactory accuracy for lower R_p.

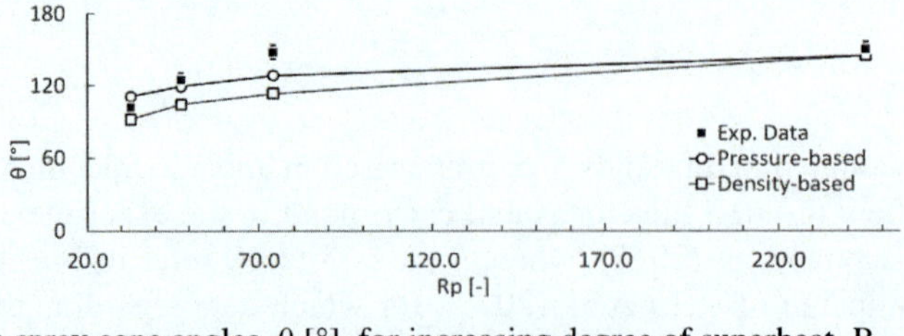

Fig. 2. Oxygen spray cone angles, θ [°], for increasing degree of superheat, R_p. Conditions correspond to cases 1–4 of Table 2. A value of $\lambda = 1.0$ and $A_{int} = 1.26 \times 10^4 m^2$ was used for the phase-change model of the pressure-based solver.

The quantity distributions shown in Fig. 3 illustrate the distinct features of flash evaporation and the differences between the two modelling approaches. The pressure and velocity distributions (Fig. 3a, Fig. 3b) predicted by the pressure- (PB) and the density-based (DB) solvers for $R_p = 245$ share the same locations of minimum and maximum values, approximately 15 nozzle diameters downstream the nozzle exit. The pressure predicted by the PB solver decreases in a marginally steeper fashion and the maximum velocity value is approximately 25% greater compared to the DB solver. This difference in maximum velocity values, stems from the variance in the density (Fig. 3c) and the temperature-field predictions (Fig. 3d). The magnitude of the assigned phase-change rate plays a key role to the topology of the two-phase flow field. In the case of the DB solver, the infinite phase-change rate that is in essence assigned, leads to states of lower temperature and higher density. Given that the mass flow rates, the spray cone angles and the pressure fields are similar for both solvers, the difference in the resulting temperature and density is the cause for lower velocity values in comparison to the results of the pressure-based solver.

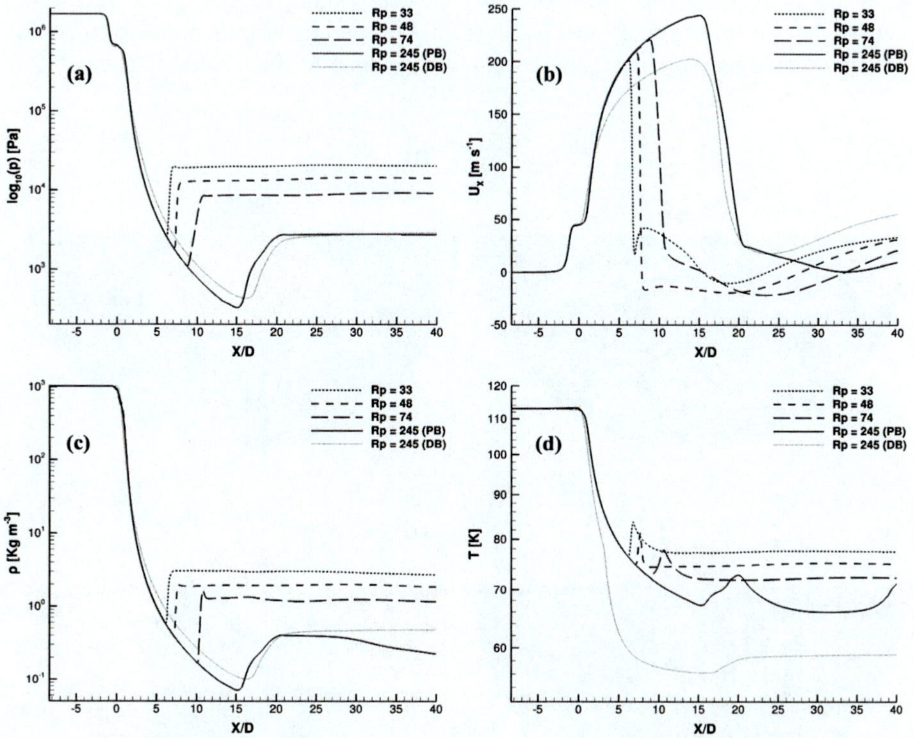

Fig. 3. Distribution of (a) Pressure, (b) Axial Velocity, (c) Density and (d) Temperature along the injector axis of symmetry as predicted for cases 1–4 of Table 2. The abbreviations PB and DB refer to the pressure- and density-based solvers, respectively.

Examination of Fig. 3 in terms of increasing R_p (R_p = 33, 48, 74 and 245(PB)) reveals the effect of the outlet (ambient) pressure on the flow characteristics. The presented results are produced by the pressure-based solver. Pressure and density fields (Fig. 3a and Fig. 3c) exhibit a steep initial decrease after the $X/D = 0$ location and an equally sharp gain that corresponds to a static shockwave and reinstates the variable's value to match an equilibrium value within the spray cone. The higher the superheat degree, the further from the nozzle exit the increase will take place and the lower the equilibrium value will be. Temperature profiles (Fig. 3d), reveal a decrease in the temperature due to the intense evaporation process and the absorption of latent heat. On the location where the flow expansion stops and the pressure increases rapidly, an amount of vapour undergoes condensation, latent heat is emitted and a brief rise in the temperature is observed. Finally, regarding the axial velocity (Fig. 3b), there is a clear forming of a recirculation area at the core of the spray cone. Negative axial velocity values are manifesting at approximately 10 to 15 nozzle diameter lengths past the location of the abrupt flow deceleration for the cases of moderate (33 to 74) values of superheat degree.

Figure 4 presents in a comparative manner the contour plots of pressure, velocity, density and temperature for case 4 of Table 2 corresponding to R_p = 245. The results show that the two solvers are in good agreement. Comparing the pressure fields (Fig. 4a), it is clear that the pressure decreases in a similar manner for both solvers. The velocity,

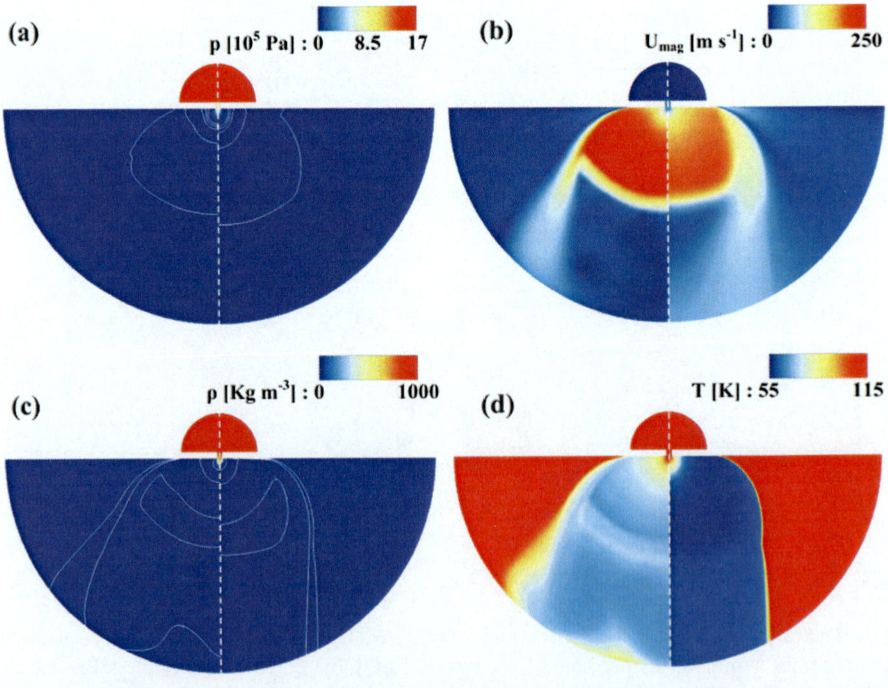

Fig. 4. Contour plots of (a) pressure, (b) velocity magnitude, (c) density and (d) temperature distributions for case 4 of Table 2. Pressure based solver results are placed on the left of each sub-plot and density-based solver results on the right.

density and temperature contour plots reveal the formation of a bell-shaped oxygen spray with an almost identical spray cone angle on the location of measurement (5 nozzle diameters downstream the nozzle outlet). The use of tabulated thermodynamics in the density-based solver imposes an infinite phase-change rate that affects the temperature ranges of the expanding flow after the nozzle exit (Fig. 4d). The density-based solver predicts, therefore, the presence of an area of lower temperature and comparable but slightly higher density (Fig. 4c) than the pressure-based solver. This change in density affects the local speed of sound and hence the magnitude of velocity acceleration of the expanding flow (Fig. 4b).

5.2 Liquid Nitrogen Spray Formation

Figure 5 summarises the results produced by the two approaches in terms of spray cone angle θ, for two different values of inlet pressure, namely 8 (Fig. 5a) and 4 bar (Fig. 5b), respectively. The cone angle was calculated 2 nozzle diameters downstream the injector exit. Both solvers produce accurate results for high values of degree of superheat. For R_p values lower than 10, numerical results are in better agreement with the experiments for cases depicted in Fig. 5a, where a higher inlet pressure is applied.

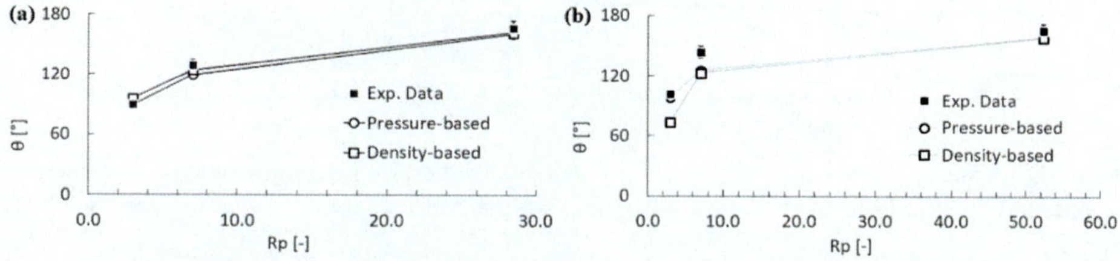

Fig. 5. Nitrogen spray cone angles for increasing degree of superheat. Conditions corresponding to (a) cases 5–7 and (b) cases 8–10 of Table 2. Spray angles are measured at a distance equivalent to 2 nozzle diameters downstream the nozzle exit.

Figure 6 presents the evolution of the field variables along the nozzle axis of symmetry for both solvers and two inlet pressure values. The results comparison reveals differences between the two methods, along with advantages and limitations. The PB solver predicts a constant pressure and density area for small positive values of the axial location X (Fig. 6a, region "1"), suggesting the presence of a liquid core exiting the nozzle. The DB solver predicts an almost instant response of the fluid to pressure change (Fig. 6b, region "1"). This response manifests as pressure spikes for the cases of higher inlet pressure and, therefore, mass flow rate and as a minor pressure drop for the cases of lower inlet pressure. Regarding flow velocity (region "2"), the pressure-based solver predicts a steeper decrease after the initial acceleration in comparison to the density-based counterpart where the axial velocity results exhibit a more subtle transition towards a constant value, for all the presented cases.

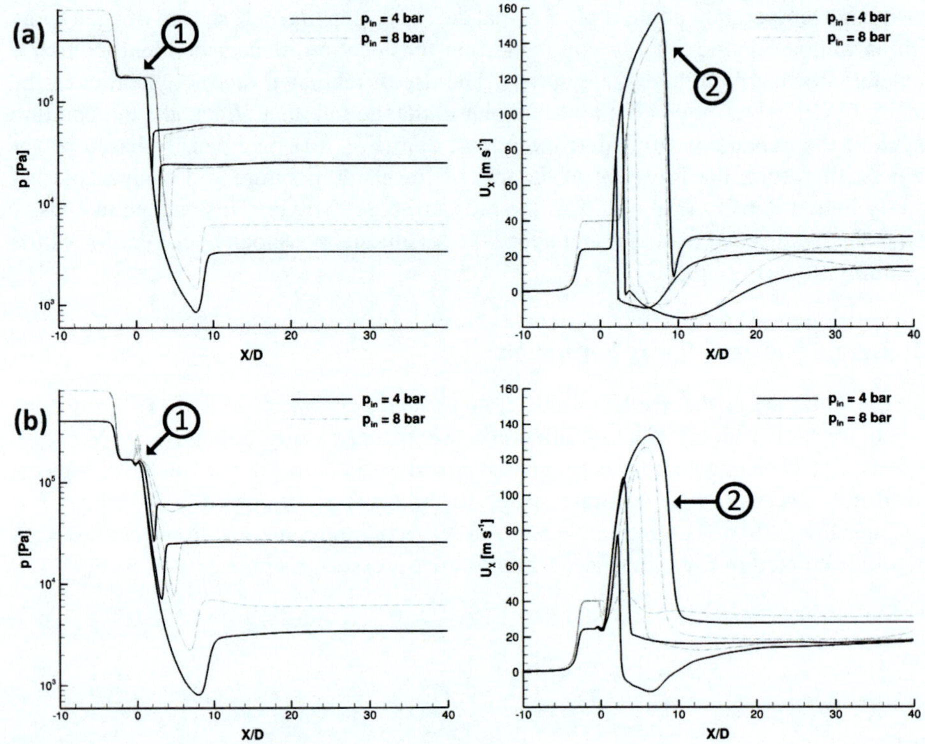

Fig. 6. Distribution of Pressure (left) and Axial Velocity (right) along the injector axis of symmetry as predicted by (a) the pressure based solver and (b) the density-based solver for cases 5–10 of Table 2.

Finally, Fig. 7 offers a visual comparison between the experimentally acquired images of Rees et al. (2020) and numerical predictions. Cases 5 to 7 of Table 2 are presented in Fig. 7 and reveal the characteristics of flashing sprays subjected to conditions that result to increasing degrees of superheat. Since the numerical results correspond to a 5° wedge of the complete domain, while the photographically acquired experimental images are the projection of the three-dimensional spray topology, the numerical images, in essence, act as a representation of a slice of the spray, revealing a hollow topology (region "1") of lower velocity and density that is formed in the core of the spray after the area of initial expansion for cases 5 and 6 that share a lower R_p.

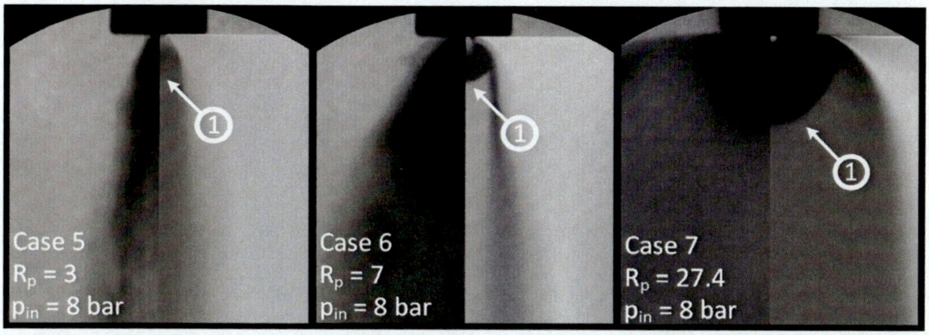

Fig. 7. Liquid Nitrogen injection, cases 5–7 of Table 2. Visual comparison between experimental images (left part of each frame) and numerically-derived results using the density based solver (right part).

6 Conclusions and Future Works

Two-phase LOx and LN_2 flows were numerically investigated for two geometrical arrangements in near-vacuum ambient pressure conditions. It has been demonstrated that the phase-change process can be accurately captured employing a bubble-dynamics model for cases of high degree of superheat that lead to flash-induced atomisation. Moreover, the density-based solver utilising real-fluid tabulated data of the fluid thermodynamic properties provides an accurate, universal modelling approach for a wide range of conditions without the need of case-dependent adjustments.

Acknowledgements. This investigation has received funding from the European Union Horizon 2020 Research and Innovation programme HAOS (Grant No. 675676). Funding from the EU Horizon-2020 Marie Skłodowska-Curie Global Fellowships AHEAD (IK, Grant no. 794831) and UNIFIED (PK, Grant no. 748784) is also acknowledged.

References

Chen, J., et al.: Numerical analysis of spray characteristics with liquid nitrogen. Cryogenics **109**, 103–113 (2020). https://doi.org/10.1016/j.cryogenics.2020.103113

Coulon, D.: Vulcain-2 cryogenic engine passes first test with new nozzle extension. ESA Bull. **102**, 123–124 (2000)

Davis, D., McArthur, J.: NASA ares i crew launch vehicle upper stage overview. In: 44th AIAA/ASME/SAE/ASEE Joint Propulsion Conference & Exhibit, Hartford (2008). https://doi.org/10.2514/6.2008-4897

Fuster, D., Hauke, G., Dopazo, C.: Influence of the accommodation coefficient on nonlinear bubble oscillations. J. Acoust. Soc. Am. **128**(1), 5 (2010). https://doi.org/10.1121/1.3436520

Gärtner, J.W., et al.: Numerical and experimental analysis of flashing cryogenic nitrogen. Int. J. Multiph. Flow **130**, 103360 (2020). https://doi.org/10.1016/j.ijmultiphaseflow.2020.103360

Harstad, K., Bellan, J.: Interactions of fluid oxygen drops in fluid hydrogen at rocket chamber pressures. Int. J. Heat Mass Transf. **41**(22), 3551–3558 (1998). https://doi.org/10.1016/S0017-9310(98)00048-9

Hendricks, R.C., Simoneau, R.J., Barrows, R.F.: Two-phase choked flow of sub-cooled oxygen and nitrogen. STIN 18376 (1976)

Karathanassis, I.K., Koukouvinis, P., Gavaises, M.: Comparative evaluation of phase-change mechanisms for the prediction of flashing flows. Int. J. Multiph. Flow **95**, 257–270 (2017)

Kyriazis, N., Koukouvinis, P., Gavaises, M.: Numerical investigation of bubble dynamics using tabulated data. Int. J. Multiph. Flow **93**, 158–177 (2017). https://doi.org/10.1016/j.ijmultiphaseflow.2017.04.004

Lamanna, G., et al.: Flashing behavior of rocket engine propellants. At. Sprays **25**(10), 837–856 (2015). https://doi.org/10.1615/AtomizSpr.2015010398

Liou, M.-S.: A sequel to AUSM, Part II: AUSM+-up for all speeds. J. Comput. Phys. **214**(1), 137–170 (2006). https://doi.org/10.1016/j.jcp.2005.09.020

Luo, M., Haidn, O.J.: Characterization of flashing phenomena with cryogenic fluid under vacuum conditions. J. Propul. Power **32**(5), 1253–1263 (2016). https://doi.org/10.2514/1.B35963

Lyras, K., et al.: Numerical simulation of sub-cooled and superheated jets under thermodynamic non-equilibrium. Int. J. Multiph. Flow **102**, 16–28 (2018). https://doi.org/10.1016/j.ijmultiphaseflow.2018.01.014

Lyras, T., et al.: Modelling of liquid oxygen nozzle flows under subcritical and supercritical pressure conditions. Int. J. Heat Mass Transf. **177**, 121559 (2021). https://doi.org/10.1016/j.ijheatmasstransfer.2021.121559

Mayer, W., Tamura, H.: Propellant injection in a liquid oxygen/gaseous hydrogen rocket engine. J. Propul. Power **12**(6), 1137–1147 (1996). https://doi.org/10.2514/3.24154

Ramcke, T., Lampmann, A., Pfitzner, M.: Simulations of injection of liquid oxygen/gaseous methane under flashing conditions. J. Propul. Power **34**(2), 395–407 (2018). https://doi.org/10.2514/1.B36412

Rees, A., et al.: About the morphology of flash boiling liquid nitrogen sprays. At. Sprays **30**(10), 713–740 (2020). https://doi.org/10.1615/AtomizSpr.2020035265

Schmehl, R., Steelant, J.: Computational analysis of the oxidizer pre-flow in an upper-stage rocket engine. J. Propul. Power **25**(3), 771–782 (2009). https://doi.org/10.2514/1.38309

Schmidt, S., et al.: Riemann techniques for the simulation of compressible liquid flows with phase-transition at all mach numbers - shock and wave dynamics in cavitating 3-D micro and macro systems. In: 46th AIAA Aerospace Sciences Meeting and Exhibit, Reno, Nevada (2008). https://doi.org/10.2514/6.2008-1238

Toro, E.F.: Riemann Solvers and Numerical Methods for Fluid Dynamics: A Practical Introduction. Springer, Heidelberg (2013). https://books.google.gr/books?id=zkLtCAAAQBAJ

Travis, J.R., Piccioni Koch, D., Breitung, W.: A homogeneous non-equilibrium two-phase critical flow model. Int. J. Hydrogen Energy **37**(22), 17373–17379 (2012). https://doi.org/10.1016/j.ijhydene.2012.07.077

Yang, V.: Liquid-Propellant Rocket Engine Injector Dynamics and Combustion Processes at Supercritical Conditions. Defense Technical Information Center, Fort Belvoir (2004). https://doi.org/10.21236/ADA428947

About the Effect of Two-Phase Flow Formulations on Shock Waves in Flash Metastable Expansions Simulations

Egoi Ortego Sampedro(✉)

Mines Paris PSL Centre d'Efficacité Energétique des Systèmes, Palaiseau, France
egoi.ortego@minesparis.psl.eu

Abstract. Two-phase expanders are promising conversion machines that could help in increasing the energy efficiency of several systems such as heat-to-power plants, liquefaction processes or chillers. Expansion of two-phase flows is complex to model since the dynamics of the liquid-gas interactions depend on several factors. This work gives a preliminary insight on the effect of various formulations for liquid-gas interaction terms on the shock waves that may appear in over-expanded flash nozzles simulations. The case study is a water thrust nozzle from literature designed for geothermal energy conversion at inlet temperatures close to 150 °C. The compared formulations differ mainly on the homogeneity assumptions regarding momentum and on the interface mass transfer term. The simulations results suggest that these assumptions have a fundamental effect on the flow behavior close to the outlet of the nozzle and in particular on the presence or not of the shock waves. This has meaningful consequences on the velocity profile in the nozzle.

Keywords: flash boiling · expansion nozzles · non-homokinetic · interfacial transfer

Nomenclature

Roman Symbols

a	Sound velocity (m/s)
C_d	Drag coefficient
C_p	Specific heat capacity at constant pressure (J/kg/K)
$\dot{m}$	Mass flow rate (kg/s)
k	Virtual mass coefficient
P	Pressure (Pa)
u	Velocity (m/s)
T	Temperature (K)

Greek Symbols

α	Volume fraction
μ	Dynamic viscosity (Pa.s)
ρ	Density (kg/m^3)

M. White et al. (Eds.): NICFD 2022, ERCOFTAC 29, pp. 125–134, 2023.
https://doi.org/10.1007/978-3-031-30936-6_13

Sub- /Super-Scripts

g	Gas or vapor
in	Inlet property
l	Liquid

Acronyms

CFD	Computational Fluid Mecanics
HRM	Homogeneous Relaxation Model
TABD	Thermal bubble-to-droplet

1 Introduction

The renewed interest in two-phase turbines in the last years can be explained by several factors. Some of them are the need for diversification of electricity production sources (by the use of waste-heat driven Organic Rankin Cycles for example), the reduction of electricity consumption of cold power production systems (especially in the context of natural refrigerants deployment), or the increase of computational capabilities for the design of two-phase components in the last 5 to 7 years that helps for the design of such machines.

However, flashing flows are highly driven by non-ideal phenomena and there is still no consensus on simulation methodology and assumptions that would lead to satisfactory results for the two-phase expanding flows modeling. This is in particular the case of the phase velocities homogeneity assumption. For example, considering the case of hot water flash expansion, Downar-Zapolski et al. (P.Downar-Zapolski, Z.Bilicki, L.Bolle and J.Franco, 1996) considered no velocity difference between liquid and vapor whereas Liao & Dirk (Liao and Dirk 2015) considered it because according to them it has great importance on the interfacial mass and heat transfer mechanisms. Besides, the velocities homogeneity assumption is very usual in the domain of two-phase expanding ejectors (Bodys et al. 2021). However, knowing that vapor has a lower density than liquid, for a given pressure gradient, the acceleration observed by the vapor is higher. This results in higher velocities for vapor than for liquid during the expansion. This was observed by various studies (Elliott 1982) (Sampedro, Breque and Nemer 2022). So velocities homogeneity assumption is likely to be incorrect. Nevertheless, the consequences of homogeneous and non-homogeneous assumptions need to be quantified.

This paper aims to analyze the effects of the velocities homogeneity assumption on the flow modeling results. In particular, the pressure and velocity profiles and the presence or not of a shock wave in simulations are analyzed. This is done on the case study of a hot water thrust nozzle tested by Ohta et al. (Ohta, Fuji, Akagawa and Takenaka 1993). The analysis of two-phase expansion nozzles presents the interest of being a relatively simple case of study that can also be considered as a real-life component used in particular as a stator in a low degree of reaction turbines. The experimental pressure profile data by Ohta et al. (Ohta, Fuji, Akagawa and Takenaka 1993) are used as a reference.

The computational fluid mechanics (CFD) model used in this paper is based on the general Euler-Euler description for two-phase flows. The interfacial mass transfer is modeled using two approaches to test the sensitivity of the simulation results to this aspect. Regarding the interface momentum coupling assumptions, two extreme situations were analyzed: the homogeneous and the non-homogeneous. The last was modeled using an interface friction formulation based on an interface drag coefficient from the literature. In addition, three supplementary values of the drag coefficient were tested to virtually observe the implications of this term in the flow behavior.

Then, the validity of the models is discussed based on the pressure profiles analysis. Besides, the paper gives a special insight into the literature models for sound velocity estimation. These models are used to analyze the CFD simulation results.

2 Methods and Equations

2.1 Models and Definitions

The general formulation of conservation equations for a two-phase flow requires a phase-per-phase description i.e. an Euler-Euler formulation. In the literature dealing with flash nozzle flows, the general multiphase description is often reduced to simpler formulations. For further insight into these aspects, the reader is invited to read the very useful review maide by Yixiang & Dirk (Yixiang and Dirk 2017) or the models' benchmark previously made by the author (Sampedro, Breque and Nemer 2022). In this paper, the general Euler-Euler formulation is adopted using the commercial software Ansys CFX. The detail of the equations is presented in a previous paper by the author (Sampedro, Breque and Nemer 2022). The CFX solver can handle, in an Euler-Euler formulation, all the assumptions to be tested on the coupling terms between phases. This is valid also for velocities homogeneity and non-homogeneity.

As a general remark for the rest of the article, since the term homogeneous is not specific enough from a semantic point of view, for clarity reasons regarding each assumption done on the coupling terms between the phases, the following definitions are adopted:

- Homokinetic: assumption of velocities equality between phases
- Homothermal: assumption of temperatures equality between phases

Regarding the mass transfer models, a substantial diversity in the interfacial transfer models exists as can be read in (Yixiang & Dirk 2017). In the present, paper a thermal phase change model explored by the author (TABD model (Sampedro, Breque and Nemer 2022)) and the so-called Homogeneous Relaxation Model (HRM (P.Downar-Zapolski, Z.Bilicki, L.Bolle and J.Franco 1996)) are used. The motivation for using two mass transfer models is to separate the **effects** of mass and momentum interfacial transfer terms on the results.

For both models, heat transfer was assumed to occur between phases. Since the test case is the same that in the present paper, the reader is invited to read the complete configuration and calibration parameters for both models used by the author in previous work (Sampedro, Breque and Nemer 2022).

Regarding the momentum transfer, the liquid/vapor interfacial drag force is included for the general case. The standard drag coefficient has a value of $Cd = 0.44$. The homokinetic case is assumed to represent an extreme case where the interfacial forces would be enough high to avoid any difference in liquid and vapor velocities in any situation. To show the effect of the interfacial force on the phasic velocities, three additional values for Cd were tested (5, 50, and 500). A homokinetic case was also tested. The HRM mass transfer based cases associated with the non-homokinetic assumption were modeled using the same interfacial area density and characteristic length as TABD.

Regarding the properties of the fluids, the liquid properties (ρ, C_p, μ) were computed as a function of the temperature computed from the enthalpy resulting from the energy balance. This means that the liquid's meta-stable condition was computed by considering it in a temperature based saturation state and not in a pressure based saturation state. The vapor was supposed to be in pressure based saturated conditions i.e. its properties were a function of the static pressure. The liquid and vapor properties of water were computed from standard IAPWS IF97 tables available in Ansys CFX. The kwSST closure was used as turbulence model.

2.2 CFD Simulation Configuration

The control volume used for the simulations is presented in Fig. 1. It represents the nozzle and an open volume at its outlet. This is a 3D domain maide by a partial revolution of 3°. The inlet conditions are static pressure, static temperature, and liquid and vapor volume fractions. The outlet is defined as an opening with static pressure, and backflow temperature and volume fractions. The dimensions and boundary conditions values are given in Sect. 3. The mesh was built by the same method and with the same characteristic sizes presented previously by the author (Sampedro, Breque and Nemer 2022). The mesh number per unit volume was similar in the nozzle than in the outlet volume. A mesh sensitivity was also presented in the cited work. The grid used here has 22500 elements.

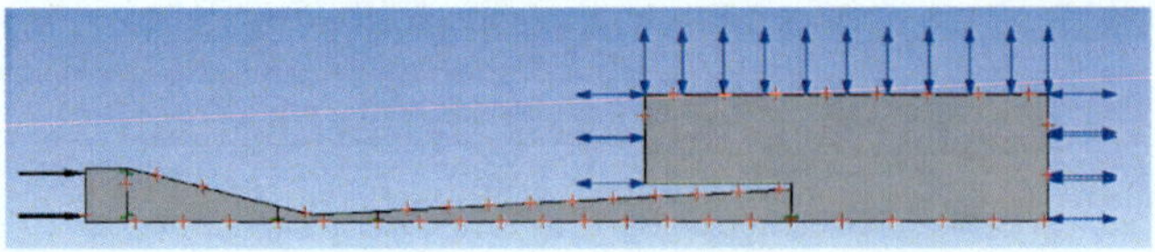

Fig. 1. computational domain; black arrows: inlet; blue arrows: outlet.

The commercial CFD code Ansys CFX 16.0 was used for CFD simulations. The CFX solver is a coupled solver using a pseudo-transient formulation; the coupled option was selected for volume fraction as well. A bounded second-order upwind scheme was selected for advection. Please refer to Ansys CFX (CFX 2020) documentation for details on numerical resolution. A steady-state simulation was done. The physical time step was set to 1.10 - 5 s. This parameter acts like a relaxation coefficient. The simulation was supposed to be converged when the mass and energy imbalance was lower than 0.5%, the inlet mass flow rate was steady and the outlet velocity was steady; all residuals were in this situation lower than 1.10–5. The total energy formulation of the energy conservation equation was selected. The flow field was initialized at 0 vapor volume fraction, at inlet temperature and pressure, and at 0m/s velocity.

2.3 Speed of Sound

The speed of sound is a parameter that helps in analyzing the flow. The two-phase flow speed of sound is not obvious to estimate since the wave propagation in a two-phase media is affected by several parameters. Among these, the following can be mentioned: interfacial mass transfer, heat and momentum transfers, and sound wave frequency (Staedtke 2006). Depending on the intensity of the interfacial coupling terms, the effective speed of sound will differ from the ideal homokinetic homothermal case.

The following lines aim to give a summary of the approaches and concepts nowadays available to handle this question.

Staedtke presents the case of the homokinetic non-homothermal case where mass exchange between phases is considered. The resulting sound velocity is the following (Staedtke 2006) (page 61):

$$a_{hke}^2 = \frac{1}{\frac{\alpha_g \rho}{\alpha_g^2 \rho_g} + \frac{\alpha_l \rho}{\alpha_l^2 \rho_l}} \tag{1}$$

This expression was also used by (White 2020) to describe the so-called "frozen homogeneous" situation. White explores as well the expression proposed by Brennen (Brennen 2005) for homokinetic homothermal situation (White 2020):

$$\frac{1}{\rho a_{Bren}^2} = \alpha_g g_g + \alpha_l g_l \tag{2}$$

The terms g_g and g_l depend on liquid and vapor phases state functions partial derivatives. Please refer to (White 2020) or (Brennen 2005) for the details.

For the non-homokinetic case, Wallis (Wallis 1969) proposed an expression of sound velocity:

$$a_{Wallis}^2 = \frac{\alpha_g \rho_l + \alpha_l \rho_g}{\frac{\alpha_g \rho_l}{\alpha_g^2} + \frac{\alpha_l \rho_g}{\alpha_l^2}} \tag{3}$$

However, according to Staedtke (Staedtke 2006), this expression leads to numerical instabilities in the computational codes using it. Staedtke presents the results of the work done by the Joint Research Center Ispra. He gives a particular insight into the virtual mass force that affects the speed of sound. This force represents the non-viscous interaction forces due to relative acceleration between phases and is associated with space and time derivatives of transfer terms. Brennen mentions them as the terms emerging from the averaging process in two-phase modeling (Brennen 2005). The formulation of this force is based on the "virtual mass" coefficient (k) in the work on Staedtke. The resulting sound velocity is the following (Staedtke 2006):

$$a^2 = \frac{\alpha_g \rho_l + \alpha_l \rho_g}{\frac{\alpha_g \rho_l}{\alpha_g^2} + \frac{\alpha_l \rho_g}{\alpha_l^2}} \frac{1 + k\frac{\alpha_g \rho_g + \alpha_l \rho_l}{\alpha_g \rho_l + \alpha_l \rho_g}}{1 + k\frac{\rho^2}{\rho_g \rho_l}} - \alpha_g \alpha_l \rho_g \rho_l (u_g - u_l)^2 \frac{(\rho_l + k\rho)(\rho_g + k\rho)}{(\rho_g \rho_l + k\rho^2)^2} \tag{4}$$

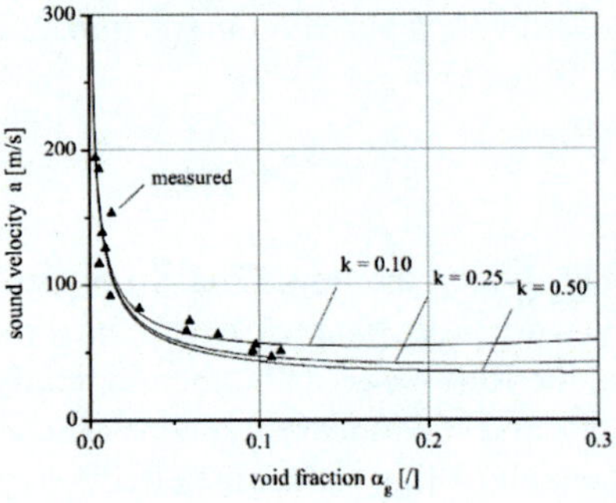

Fig. 2. Sound velocity in water/steam media as a function of the void fraction; saturated conditions at pressure p = 1 bar (Staedtke 2006)

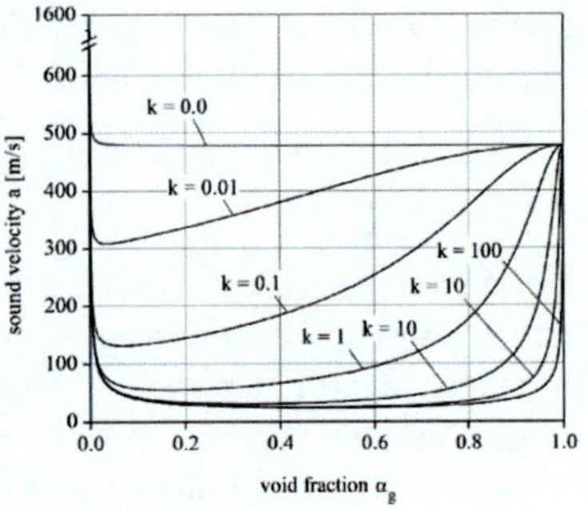

Fig. 3. Two-phase sound velocity as a function of $\boldsymbol{\alpha}$; effect of the "virtual mass" coefficient, saturated water/steam at pressure p = 1 bar, equal phase velocities. (Staedtke 2006)

The "virtual mass" coefficient of k = 0 represents a case with spatially separated phases where the momentum coupling is reduced to friction. If the friction term is small ($|u_g - u_l|/u < 0.1$), then the expression of "a" is reduced to the one of Wallis (Eq. 3). If the "virtual mass" coefficient k  ∞ then the flow is driven towards homokinetic conditions and the expression of "a" approaches "a_{hke}".

The "virtual mass" coefficient (k) was used by Staedtke as a calibration parameter for the two-phase flow computational method he presents in his book to estimate the sound velocity of water/steam flows. Staedtke mentions that the case where 0.25 ≤ k ≤ 0.5 corresponds to dispersed droplets flows and fits well with experimental observations. Figure 2 presents an example where a coefficient of k = 0.25 represents well the measurements. Figure 3 shows an extensive sensitivity analysis of the sound velocity in the entire range of void fraction (or vapor volume fraction) for the same media.

The sound velocity estimations will be used to analyze the simulation results presented below. In Ansys CFX, virtual mass modeling is not considered for two continuous phase's description (it can be for dispersed media). For that reason, the effect of parameter "k" will not be considered. Only reference cases will be used i.e. homokinetic non-homothermal (Eq. 1), homokinetic homothermal (Eq. 2), and non-homokinetic non-homothermal (Eq. 3). Virtual mass effects are important to keep in mind from a theoretical point of view.

3 Test Case

In the early '90s, a Japanese team (Ohta, Fuji, Akagawa and Takenaka 1993) worked on waste heat recovery by impulse turbines using phase change nozzles. Two nozzle geometries were tested. The first was called the B nozzle which is a fairly simple nozzle. It was extensively studied for a wide outlet pressure range. The authors measured mass flow rate, efficiency, and pressure profiles. The efficiency was obtained thanks to thrust measurements. The dimensions and the operating conditions are shown in Fig. 4 and Table 1 respectively.

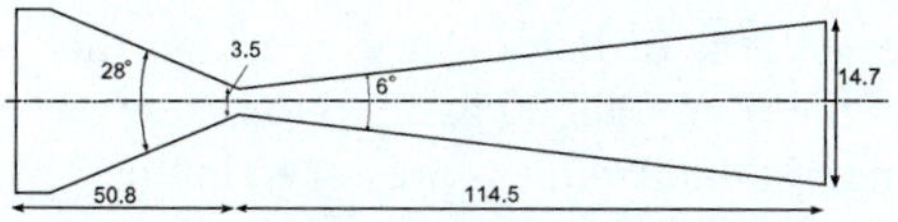

Fig. 4. Ohta B nozzle (Ohta, Fuji, Akagawa and Takenaka 1993)

Table 1. Ohta B nozzle operating points (Ohta, Fuji, Akagawa and Takenaka 1993)

$T_{in}(°C)$	$P_{in}(kPa)$	$SC(K)$	$P_{out}(kPa)$	$\dot{m}(g/s)$
148	470	1,5	18/45/73/100	122
137,5	470	12	16/43/70/100	156

Note that the mass flow in Table 1 does not depend on the outlet pressure; that means that the nozzle operates in critical conditions. The static pressure taps diameter was 0.6mm. The uncertainty on the pressure measurement was ±1%.

4 Simulation Results

4.1 Pressure Profiles

The pressure profiles will help in defining which assumption leads to the more reliable results. Pressure profiles obtained with TABD and HRM models are presented respectively in Fig. 5 and Fig. 6. Each line represents a different assumption on the interfacial momentum transfer term. The 0mm position corresponds to the throat position.

It appears that the main difference between TABD and HRM models occurs just after the throat. TABD gives pressures closer to experiments; however, the difference between models is not significant.

Regarding the assumptions on momentum transfer intensity, the difference between models appears in particular close to the nozzle outlet. The effect of this series of assumptions on the difference to experiments is also particularly visible close to the nozzle outlet. The homokinetic case presents a pressure profile characteristic of a shock wave i.e. a sudden static pressure increase. This is confirmed by velocity profiles as will be presented in Sect. 4.2 models presented. However, for the lowest value of Cd, there is no shock wave and the pressure profile is very close to the measured one.

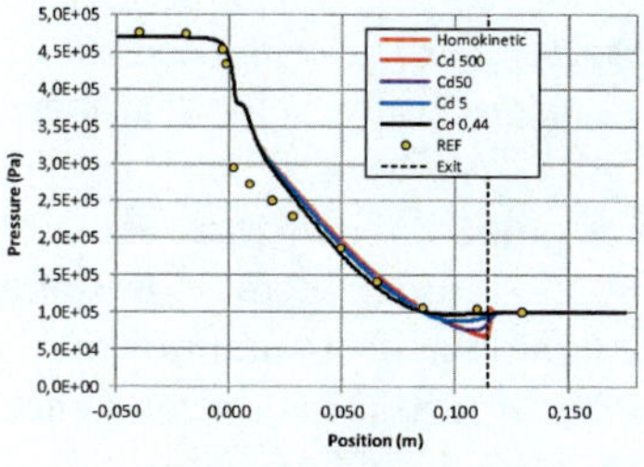

Fig. 5. pressure profiles; simulation vs experiments; TABD model.

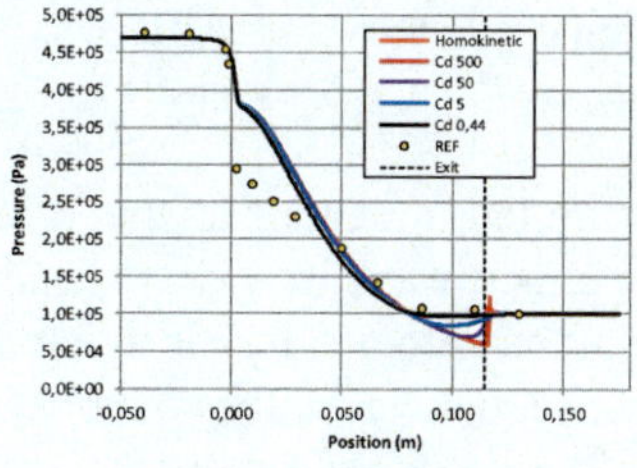

Fig. 6. pressure profiles; simulation vs experiments; HRM model.

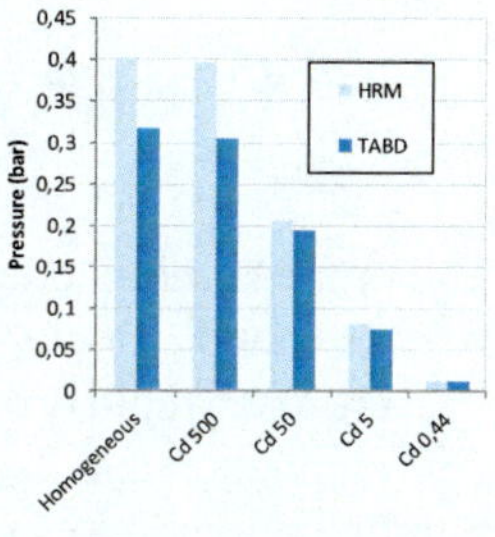

Fig. 7. pressures absolute discrepancies for last pressure tap.

The pressure tap located at 4mm before the outlet is of particular interest. At this location, the static pressure value obtained with the homokinetic assumption is lower than in reality. The absolute discrepancy to the experimental value is presented in Fig. 7. The discrepancy increases with Cd. For the nominal Cd, the discrepancy is close to 0.01bar. For the homokinetic case, it is close to 0.3bar for TABD and 0.4bar for HRM.

4.2 Velocities

Figure 8 shows the liquid and vapor velocities obtained with the TABD model. The velocity difference between vapor and liquid reduces with increasing value of Cd as expected. However, getting the same velocity would require increasing a lot the value of Cd. Also, with the increasing value of Cd, the velocities of both phases get closer to the homokinetic case.

The homokinetic case shows a shock wave close to the nozzle outlet whereas the case with Cd = 0.44 doesn't. The values computed for the sound velocity for the different models presented in Sect. 2.3 are given in Fig. 9. The figure gives also the values of the average mixture velocity before and after the nozzle outlet (respectively "av ex-4mm" and "av final"). The sound velocity models based on homokinetic assumption give low values of the speed of sound whereas the non-homokinetic model gives a high value (close to the vapor phase sound velocity).

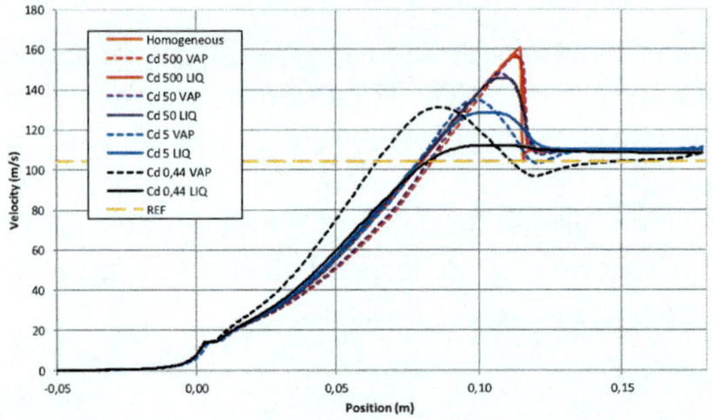

Fig. 8. vapour and liquid velocities for TABD.

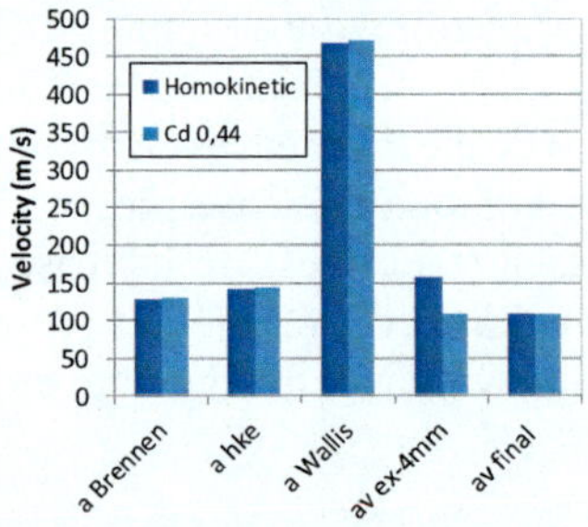

Fig. 9. sound velocities and average velocities before and after outlet; TABD.

For the homokinetic simulation, it appears that at some point, if the mixture velocity exceeds sound velocity, a shock wave may occur leading to a sudden increase in static pressure and a rapid drop of the velocity. It seems that the homokinetic assumption in the CFD simulation and its consequences on the sound velocity were modeled in coherence with literature models for homokinetic sound velocity i.e. $a_{Brennen}$ and a_{hke}.

For the case with Cd = 0.44, the mixture average velocity remains always below all the sound velocities computed. Besides, no shock wave was observed for this CFD simulation. To explain this, two hypotheses may be possible:

- The sound velocity in the non-homokinetic case is very high as suggested by the Wallis model. In this case, the momentum coupling between phases is weak and the CFD simulation represents it relatively well. This suggests that frictional and non-friction coupling terms are small and could be reduced to frictional only (at least close to the nozzle outlet in the given conditions). Please note that the Wallis model is close to the model presented by Staedtke in the case of dispersed flow (small value of "k") which corresponds to the high void fraction situation in the last third of the nozzle.

- The non-homokinetic case creates so much friction that the total enthalpy of the flow is highly degraded leading to moderate maximal velocity. This velocity is thus never higher than any sound velocity.

The last hypothesis seems to the author not probable since in the last part of the expansion, where the maximum velocities are reached, the average vapor volume fraction is very high (0.99) and the mixture density is very low (6 kg/m^3). Furthermore, given the moderate length, such energy losses seem unlikely. However, this needs to be verified in the future.

5 Conclusions and Future Works

The simulation based on the homokinetic assumption produces a two-phase mixture shock wave close to the nozzle outlet. The presence of this shock is coherent with the sound velocity models from the so-assumed homokinetic mixture theory. The non-homokinetic case doesn't show any shock wave; that seems coherent with the non-homokinetic sound velocity models.

From the comparative analysis made on the static pressure profiles, it can be concluded that whatever the mass transfer model is, the homokinetic model provides unrealistic results at high velocity. The non-homokinetic model provides results in accordance with the measurements. It can be concluded that the homokinetic assumption implies very important interface momentum coupling that provides erroneous simulated flow velocities and pressures. This has particularly remarkable effects when the homokinetic mixture velocity attains homokinetic sound velocity. It would be useful to consolidate the results on pressure profiles by measuring a larger number of pressure points close to the nozzle outlet.

Also, the interfacial momentum transfer modeling needs to make assumptions on the interfacial area density. For the present work, the parameters used to set it were defined using the TABD model. And no extended analysis of the correctness of the drag coefficient value was discussed. It would be useful to analyze it further.

Finally, according to the Wallis model, the sound velocity in the non-homokinetic case is very high, close to pure vapor sound velocity. This is coherent with the model by Staedtke. However, these are theoretical models, and experimental measurement of sound velocity in mixtures, in particular at vapor volume fractions higher than 0.7, would be extremely useful for phase-changing expansions study. Besides, it seems that the momentum coupling between phases was weak in reality as suggested by Staedtke for mist flows. For that reason, including the non-frictional momentum transfer terms was not problematic for the presented case. However, for mixtures with lower vapor/liquid density ratios, this may be no more valid and thus these terms could require further research.

References

Bodys, J., Smolka, J., Palacz, M., Haida, M., Banasiak, K., Nowak, A.J.: Experimental and numerical study on the R744 ejector with a suction nozzle bypass. Appl. Therm. Eng. **194** (2021)

Brennen, C.E.: Fundamentals of Multiphase Flows. Cambridge University Press, Cambridge (2005)

CFX, A.: V20 User's Guide (2020)

Elliott, D.G.: Theory and tests of two-phase turbines. Pasadena: Jet Propulsion Laboratory (1982)

Liao, Y., Dirk, L.: 3D CFD simulation of flashing flows in a converging-diverging nozzle. Nucl. Eng. Des. 149–163 (2015)

Ohta, J.: Performance and flow characteristics of nozzles for initially subcooled hot water (influence of turbulence and decompression rate). Int. J. Multiph. Flow **19**(1), 125–136 (1993)

Downar-Zapolski, P., Bilicki, Z., Bolle, L., Franco, J.: The non-equilibrium relaxation model for one-dimensional flashing liquid flow. Int. J. Multiph. Flow **22**, 473–483 (1996)

Sampedro, E.O., Breque, F., Nemer, M.: Two-phase nozzles performances CFD modeling for low-grade heat to power generation: mass transfer models assessment and a novel transitional formulation. Therm. Sci. Eng. Prog. (2022)

Staedtke, H.: Gasdynamic Aspects of Two-Phase Flow. WILEY-VCH, Weinheim (2006)

Wallis, G.B.: One-dimensional Two-phase Flow. McGraw-Hill, New York (1969)

White, M.: Investigating the validity of the fundamental derivative in the equilibrium and non-equilibrium two-phase expansion of MM. In: Pini, M., De Servi, C., Spinelli, A., di Mare, F., Guardone, A. (eds.) Proceedings of the 3rd International Seminar on Non-Ideal Compressible Fluid Dynamics for Propulsion and Power. NICFD 2020, Delft, vol. 28. Springer, Cham (2020). https://doi.org/10.1007/978-3-030-69306-0_1

Yixiang, L., Dirk, L.: Computational modeling of flash boiling flows, a literature survey. Int. J. Heat Mass Transfer, 246–265 (2017)

Non-equilibrium Phenomena in Two-Phase Flashing Flows of Organic Fluids

Carlotta Tammone[1](✉), Alessandro Romei[2], Giacomo Persico[2], and Fredrik Haglind[1]

[1] Technical University of Denmark, Kongens Lyngby, Denmark
{carta,frh}@mek.dtu.dk
[2] Politecnico di Milano, Milan, Italy
{alessandro.romei,giacomo.persico}@polimi.it

Abstract. Flashing flows can be found in several energy systems, but accurate and comprehensive models for the phenomenon do not exist yet. A few models have been developed for water and CO_2 flashing flows, but they are not exhaustive nor directly applicable to flashing flows of organic fluids due to the substantial differences in typical operating conditions and in thermophysical and thermodynamic properties of the fluids. These properties influence the behaviour of the two-phase flow during the flashing expansion and govern the interphase transfer phenomena. In this paper we present a comparison between results from a purposely developed 1D homogeneous equilibrium model and recent experimental data on flashing flows of R134a through a converging-diverging nozzle. The comparison between the computed mass flow rates and pressure distributions with the experimental data shows that the homogeneous equilibrium model is not able to accurately reproduce flashing flows of refrigerants and indicates a significant deviation of the system from equilibrium conditions, with underestimations of the mass flow rate up to 50% with respect to the experimental values. The difference between experimental and calculated mass flow rates is used to quantify the degree of non-equilibrium of the system and to identify the sources of non-equilibrium. The results suggest that the finite vaporization rates and the subsequent evolution of the liquid phase through metastable conditions are the primary sources of non-equilibrium in flashing flows of refrigerants and therefore, models that can relax the equilibrium assumption are needed to accurately predict the behaviour of these flows.

Keywords: two-phase flows · flashing flows · organic fluids · non-equilibrium · nozzles

Nomenclature

Roman Symbols

A	Area, m^2
c	Speed of sound, m/s
C	Friction factor, -
D	Diameter, m

M. White et al. (Eds.): NICFD 2022, ERCOFTAC 29, pp. 135–145, 2023.
https://doi.org/10.1007/978-3-031-30936-6_14

G Mass flux, kg/(s·m^2)
h Specific enthalpy, J/kg
$\dot{m}$ Mass flow rate, kg/s
p Static pressure, Pa
s Specific entropy, J/(kg·K)
v Velocity, m/s
x Quality, -
z Position along nozzle axis, m

Greek Symbols

α Void fraction, -
β Pressure ratio, -
Δ Difference, -
δ Choking margin, -
∂ Partial derivative, -
Φ Two-phase friction multiplier, -
μ Dynamic viscosity, Pa·s
ρ Density, kg/m^3
ϑ Nozzle opening angle, rad
τ Shear stress, Pa
Ψ Thermodynamic property, J/kg, J/(kg·K) or m^3/kg

Non-dimensional Numbers

Re Reynolds number
Ma Mach number

Sub-/Super-Scripts

w Wall
in Inlet property
out Outlet property
0 Stagnation property
min Minimum or corresponding to minimum
L Liquid
V Vapour
z At position z along nozzle axis
s At constant pressure
LM Lockhart-Martinelli
h Hydraulic
f Friction

1 Introduction

Two-phase flows characterized by flashing or cavitation occur in energy systems such as trilateral cycle power systems, supercritical CO_2 power systems and ejector heat pumps. The phenomenon of flashing consists in the nucleation of vapour in a liquid flow that undergoes a fast depressurization below the saturation pressure, usually through an adiabatic process. Eventually, the nucleated vapour bubbles implode if the pressure of the surrounding fluid increases back, resulting in the phenomenon known as cavitation. If not properly predicted, these phenomena can deteriorate significantly both the performance and the life of the components and therefore accurate models are crucial for the appropriate design and operation of the systems.

The fast depressurization characterizing flashing flows is coupled with a fast acceleration of the flow. Due to the high compressibility of two-phase flows, especially of organic fluids, the density, pressure and velocity fields are strongly coupled. As a results, predicting the behaviour of two-phase flashing flows is complex, due to the difficult estimation of the speed of sound and the density of two-phase mixtures. In expansion devices, like nozzles and turbines, the main issues concern the prediction of the critical mass flow rate, i.e. the maximum mass flow rate that the device is able to convey, and the corresponding outlet pressures. These are, in fact, among the most important parameters in the design phase, especially when trying to prevent the appearance of shock waves during operation. However, a complete understanding of the flashing phenomenon is still lacking and models that are both reliable and general have not been developed yet and the prediction of flashing flows in nozzles and cascades is still highly inaccurate, especially for less investigated fluids like organic fluids.

While flashing and cavitating flows of water and carbon dioxide have been investigated experimentally (Ohta et al. 1993; Nakagawa et al. 2009; Vahaji et al. 2014) and numerically (Downar-Zapolski et al. 1996; De Lorenzo et al. 2017; Angielczyk et al. 2020; Romei and Persico 2021) to a larger extent, little is known about flashing and cavitating flows of organic fluids despite their technical relevance. Just recently a first attempt to experimentally characterize flashing flows of refrigerant R134a has been made (Zhu and Elbel 2019) and a numerical investigation of flashing expansion of organic fluids in turbine nozzles has been proposed (White 2022). Investigating flashing refrigerants is particularly important because their typical operating conditions are different from those of water and CO_2 flows and because fluid-dependent transport and thermodynamic properties are expected to have a significant impact on the flow patterns and on the resulting mass transfer mechanisms. Therefore, conclusions previously drawn for water and CO_2 cannot be straightforwardly extended to other fluids without ensuring the similarity of all the concurrent mechanisms.

The simplest model used to handle two-phase flows is the homogeneous equilibrium model (HEM), which assumes that the liquid and vapour phases are in thermal, chemical and mechanical equilibrium. For flashing CO_2 the HEM has been found to predict satisfactorily the pressure distribution in converging-diverging nozzles both for subcritical inlet pressures (Angielczyk et al. 2019) and, especially, for supercritical inlet conditions (Romei and Persico 2021); conversely, previous studies on flashing water (Bartosiewicz and Seynhaeve 2013, De Lorenzo et al. 2017) suggested that the HEM fails to predict both the critical mass flow rate and the pressure distribution. In the case of flashing water, the actual vaporization rates are found to be significantly lower than the ones predicted by the HEM, especially in the region of sharp pressure gradients close to the nozzle throat. Both Bartosiewicz and Seynhaeve (2013) and De Lorenzo et al. (2017) suggested that, due to the fast depressurization, the flow experiences a delay in the vapour nucleation that forces part of the liquid phase to evolve through states of metastable equilibrium and thus modifies the thermodynamic properties of the two-phase mixture, especially the density. The same conclusion was drawn for organic fluids by White (2022), though a comparison with experimental data was not possible.

In this paper a description of the non-equilibrium phenomena affecting two-phase flashing flows is proposed, with a focus on how fluid-dependent properties influence the interphase transfer mechanisms. A comparison between the experimental data on converging-diverging nozzles running with R134a (Zhu and Elbel 2019) and results from a one dimensional model adopting the HEM is then presented in order to quantitatively assess the degree of non-equilibrium of flashing flows of R134a for typical operating conditions of motive nozzles of ejectors. In the end, conclusions are drawn on the possible sources of non-equilibrium and suggestions on how to model them are provided.

2 Sources of Non-equilibrium in Flashing Flows

When a nearly saturated fluid or a two-phase mixture experiences a depressurization, mass transfer occurs from the liquid to the vapour phase to restore a new state of thermodynamic equilibrium. However, depending on the flow conditions, mass transfer does not occur as soon as the pressure is brought below saturation, causing a delay in the nucleation of vapour which drives the flow from chemical equilibrium and brings the liquid in metastable conditions. Then, once vapour generation has begun, the two phases tend to be accelerated differently because of their different densities, promoting transfer of momentum. In addition, the finite heat capacity and conductivity of the phases lead to temperature gradients within the mixture, triggering heat transfer between the phases.

When considering thermodynamic processes in single-phase non-reacting flows, it is reasonable to assume that the flow reaches thermodynamic equilibrium conditions instantaneously, as processes happen at a molecular scale and are thus characterized by time and length scales of several orders of magnitude smaller than the typical time and length scales of the flow. In multiphase flows, on the other hand, the thermodynamic processes and the interactions among phases occur at a macroscopical level and their typical time and length scales may be comparable with the characteristic time and length scales of the flow and cannot be considered instantaneous. In general, mechanical equilibrium, i.e. pressure equilibrium across the phases interface, can be assumed to be established instantaneously upon a pressure change (Bouré et al. 1976). On the contrary, chemical and thermal equilibrium conditions between the phases are not ensured a priori in multiphase flows and the different phases may not share the same velocity field, resulting in non-homogeneous flow conditions. Therefore, an analysis of the interphase transfer mechanisms has to be carried out in order to establish whether the flow is evolving in equilibrium and homogeneous conditions or not.

Non-instantaneous heat and mass transfer phenomena give rise to local temperature gradients in multiphase flows and on average to non-equilibrium thermodynamic conditions of the mixture. In fact, the liquid phase is brought to saturation temperature conditions just in the proximity of the nucleated bubbles and evolves adiabatically elsewhere, going through metastable states and retaining a higher temperature than the saturation one. As a result, the average temperature of the mixture is well above the saturation temperature. Moreover, due to the finite vaporization rate, the mass fraction of vapour in the mixture is lower than the equilibrium value and the specific Gibbs energy, or chemical potential, of the liquid and vapour phases are not the same, leading to chemical non-equilibrium conditions. According to Bartosiewicz and Seynhaeve (2013) and De Lorenzo et al. (2017), thermal and chemical non-equilibrium phenomena are the main sources of non-equilibrium in flashing flows. However, a velocity difference between the phases, which gives rise to non-homogeneous flow conditions, cannot be excluded a priori. The tendency of the phases to have different velocities, and ultimately to separate, increases with the density ratio between liquid and vapour phases and it becomes more significant for pressures far from the critical point. The different acceleration between phases is counterbalanced by the drag force, which causes momentum transfer between the phases. In flashing flows, neither separation nor significant velocity differences between phases have been observed (Zhu and Elbel 2019) or predicted by models that allow different velocity fields for the two phases (White 2022).

3 Methods

3.1 One-Dimensional Model

A one dimensional steady-state model (1D-HEM) was developed in MATLAB to simulate converging-diverging nozzles working with two-phase flows. The model solves the discretized equations of continuity, momentum, and energy, Eqs. (1)–(2)–(3) respectively, for the two-phase mixture assuming homogeneous and equilibrium conditions. The former assumption means that the liquid and vapour phases, and consequently the two-phase mixture, share the same velocity field ($v = v_L = v_V$). The latter assumption implies that the thermodynamic properties (enthalpy, entropy and density) of the mixture are assigned based on a mass fraction average of the phase properties at saturation, as in Eq. (4). The saturation properties and the transport properties of the liquid and vapour phases were retrieved from the CoolProp library (Bell et al. 2014), which expresses the equation of state in terms of the fundamental Helmholtz free energy relation.

The solver is based on a finite volume method and it solves the equations in a conservative form. The discretization of the fluid domain is thus carried out through control volumes that are delimited by the lateral surface representing the walls of the nozzle and by two cross sections, at the generic locations z and $z + \Delta z$, where all the properties are evaluated. In this study, the nozzle was discretized in 100 control volumes with a refinement of the axial extension Δz following a geometrical progression near the throat.

$$\frac{\Delta(\rho v A)}{\Delta z} = 0 \tag{1}$$

$$\frac{\Delta(\rho v A \cdot v)}{\Delta z} + \frac{\Delta(pA)}{\Delta z} - \frac{pA_w \sin(\vartheta)}{\Delta z} + \frac{\tau_w A_w \cos(\vartheta)}{\Delta z} = 0 \tag{2}$$

$$\frac{\Delta[\rho v A(h + v^2/2)]}{\Delta z} = 0 \tag{3}$$

$$\psi = x \cdot \psi_V + (1 - x) \cdot \psi_L \quad \psi = h, s, \rho^{-1} \tag{4}$$

where τ_w is the wall stress, ϑ is the opening angle of the nozzle, A is the cross section of the nozzle and A_w is the lateral surface or wall surface that delimits the control volume between the locations z and $z + \Delta z$.

The boundary conditions at the inlet are assigned as stagnation properties, so the mass flow rate (or alternatively the mixture velocity) has to be assumed in order to have a complete set of boundary conditions on the inlet section. As in single-phase converging-diverging nozzles, a unique solution of the equations exists for each value of mass flow rate lower than the critical mass flow rate, i.e. the maximum flow rate that the nozzle can convey, and the static pressure can be assigned at the outlet section. On the contrary, when the mass flow rate reaches the critical value, the flow can have two continuous solutions, a subsonic and a supersonic solution, and the outlet pressure cannot be assigned arbitrarily without admitting shock waves within the nozzle. Values of mass flow rate higher than the critical one are not physically possible. Due to the mathematical structure of the problem, a dedicated algorithm to find the critical mass flow rate was developed, based on the approach suggested by Bouré et al. (1976) and later adopted by various authors (De Lorenzo et al. 2017; Angielczyk et al. 2020). A schematic representation of the algorithm is provided as a flow diagram in Fig. 1a. The critical mass flow rate and the corresponding subsonic solution are always computed first, as the outlet pressure given by the critical subsonic solution is the minimum pressure that can be imposed at the outlet without the presence of shock or expansion waves within or outside the nozzle. Once the critical mass flow rate is found, the adapted supersonic solution can also be computed.

To compute the flow along the nozzle, a guess value of the mass flow rate is assumed and the flow is then solved on each section i along the nozzle axis. The systems has three independent variables: pressure (P), mixture entropy (s) and velocity (v). The continuity and momentum equations are solved implicitly to retrieve the pressure and the mixture entropy respectively, whereas the energy equation is explicitly solved to retrieve the velocity. If the assigned mass flow rate is found to be below the critical value, a single solution exists in all sections of the nozzle (green curve in Fig. 1b), whereas no solution is found starting from a certain section (red curve in Fig. 1b) if the assigned mass flow rate is above the critical value. Since at each section the system of equations always has two solutions (a subsonic and a supersonic solution), a restricted range of variation has to be assigned to one of the independent variables, in this case pressure. For the subsonic solution, the upper bound for pressure corresponds to the inlet stagnation pressure $P_{0,in}$, while the lower bound is found in each section by computing the choking conditions and the corresponding value of pressure (P_{min}). This is carried out in practice by searching the pressure that minimizes the choking margin δ, defined by Eq. (5) as the distance from maximum mass flux (step 0 in Fig. 1a). The speed of sound is evaluated as a post-processing step by following its definition, Eq. (6), and by using the derivatives of density with pressure for the phases at saturation, and the derivative of void fraction.

The solution is thus independent of the speed of sound formulation adopted.

$$\delta = 1 - (\rho v A)_z / \dot{m} \tag{5}$$

$$\frac{1}{c^2} = \left(\frac{\partial \rho}{\partial p}\right)_s = \left(\frac{\partial \alpha}{\partial p}\right)_s \cdot (\rho_V - \rho_L) + \alpha \cdot \left(\frac{\partial \rho_V}{\partial p}\right)_s + (1 - \alpha) \cdot \left(\frac{\partial \rho_L}{\partial p}\right)_s \tag{6}$$

Once the range of variation of pressure is assigned, the solution is found on each section by imposing an error (δ) on mass flow rate equal to zero (step 1 in Fig. 1a). Once the solution of the flow along the nozzle is computed, the algorithm corrects the assigned mass flow rate value at each step: if the flow at current iteration is subcritical, i.e. the maximum choking margin is negative, it assigns a higher value of mass flow rate, whereas it assigns a lower value if the solver has stopped before the outlet section because it fails to find a solution.

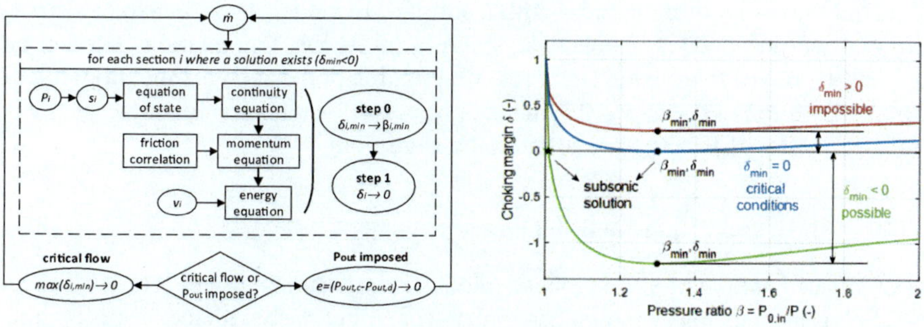

(a) (b)

Fig. 1. (a) Flow diagram representing the algorithm adopted for the solution (b) Representation of the possible solutions for different values of mass flow rates on a section of the nozzle

Once the subsonic choking conditions and the choking section are identified, the supersonic solution can be computed by imposing the critical mass flow rate and by setting a different pressure range for the sections downstream of the choking section. The supersonic pressure range has the choking pressure (P_{min}) as the upper bound and the triple point pressure as the lower bound. Differently from isentropic flow conditions, the choking section is found downstream of the throat section when viscous losses are included because the pressure gradient has a term depending on the section area gradient, $\partial A/\partial z$, and a term linked to the wall shear stress, τ_w, and it becomes zero when the two terms equalize each other. A limiting case is when the wall shear stress is so high that the pressure gradient never changes sign. In this case, the flow continues to expand in the diverging section and eventually reaches the sonic conditions on the outlet section without reaching supersonic conditions. This peculiar pressure distribution was observed in flashing CO_2 nozzles (Nakagawa et al. 2009).

The equivalent wall shear stress for the two-phase mixture, $\tau_{w,}$ is evaluated using empirical correlations. The most common formulation is the one based on the Lockhart-Martinelli parameter, Φ_{LM}, which links, as in Eq. (7), the shear stress for the two-phase mixture to the shear stress $\tau_{w,L}$ calculated by assuming that the primary phase, i.e. the liquid phase in the case of flashing, flows alone in the duct with its mass flux, G_L, as stated by Eq. (8).

$$\tau_w = \Phi_{LM} \cdot \tau_{w,L} \tag{7}$$

$$\tau_{w,L} = \frac{1}{2} C_{f,L} \frac{G_L^2}{\rho_L} \quad \text{where} \quad C_{f,L} = f\left(\mathrm{Re}_L = \frac{G_L D_h}{\mu_L}\right) \tag{8}$$

$$\Phi_{LM} = (1-\alpha)^{-1.75} \tag{9}$$

In this paper the correlation, Eq. (9), for the Lockhart-Martinelli parameter suggested by Richardson (1958) was adopted, as it was used in previous works (De Lorenzo et al. 2017; Angielczyk et al. 2020).

The developed model was verified by comparing with results obtained for water and CO_2 flashing flows by other authors with a similar 1D model (De Lorenzo et al. 2017, and Angielczyk et al. 2019, respectively) and with 2D CFD simulations (Romei and Persico 2021), both adopting the HEM for computation of thermodynamic properties. A maximum deviation of 9% and − 3% in terms of mass flow rate was found, respectively, and two-dimensional effects were found to be negligible.

3.2 Case Study and Experimental Data

A set of experimental data for an ejector motive nozzle running with R134a (Zhu and Elbel 2019) is used in this work as a benchmark for comparison with the results obtained with the 1D-HEM model. Zhu and Elbel tested flashing nozzles with both subcooled liquid and two-phase conditions at the inlet, so to highlight the role of non-equilibrium phenomena in the flashing process for different classes of incoming flows. The set of experimental data includes measurements of the mass flow rate and of the pressure distribution along the nozzle axis for different converging-diverging nozzle geometries. The authors also reported the measured arithmetical mean surface roughness, R_a = 1.6 μm, which is an important parameter in the evaluation of the wall shear stress through the friction factor, $C_{f,L}$ used in Eq. (8). In this work, experimental data from one nozzle geometry with circular cross section are used. The inlet, throat, and outlet diameters are 15 mm, 1.02 mm, and 1.37 mm, respectively. The length of the converging and diverging sections are 9.9 mm and 20.0 mm, respectively. All cases that were simulated share the same inlet stagnation pressure $P_{0,in} = 9.25 \cdot 10^5$ Pa and outlet pressure $P_{out} = 6.80 \cdot 10^5$ Pa. The authors reported an uncertainty of 0.1 g/s on mass flow rate measurements and of 2 kPa on pressure measurements.

4 Results and Discussion

Figure 2 presents a comparison between experimental and calculated pressure profiles, together with quality and Mach number distributions. The pressure profiles suggest that the flow is dominated by frictional effects, which cause the continued expansion in the

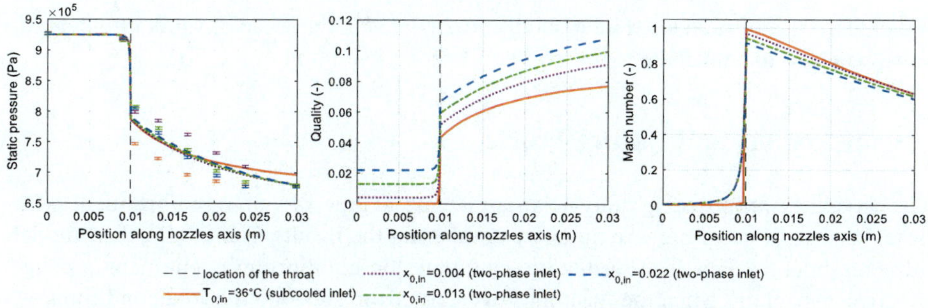

Fig. 2. Experimental (·) and 1D-HEM model pressure profiles (left), quality (center) and Mach number (right) profiles along the nozzles axis for different inlet conditions

Table 1. Experimental and 1D-HEM mass flow rates

	$T_{0,in}$ = 36 °C	$x_{0,in}$ = 0.004	$x_{0,in}$ = 0.013	$x_{0,in}$ = 0.022
Experimental mass flow rate (g/s)	15.7	11.9	10.6	9.8
1D-HEM mass flow rate (g/s)	8.0	7.8	7.5	7.3
Relative error (%)	49.0%	34.5%	29.2%	25.5%

diverging section despite never reaching supersonic conditions, as shown by the Mach numbers profiles. When the flow is already in two-phase conditions at the inlet, a subsonic solution is found for mass flow rates below the critical value for the imposed pressure at the nozzle outlet. On the contrary, for the case of subcooled liquid inlet conditions the critical outlet pressure is found to be higher than the imposed pressure and therefore a lower pressure at the outlet cannot be imposed without the formation of shock waves within the diverging section of the nozzle. Since the experimental visualizations do not show the formation of a shock wave, the critical subsonic solution was assumed to be the closest one to the real conditions of the flow for this case. Overall, the HEM is not able to predict the pressure profile accurately for subcooled liquid and very low quality ($x_{0,in}$ = 0.004) conditions at the inlet, but it improves for higher quality ($x_{0,in}$ = 0.013 and $x_{0,in}$ = 0.022) inlet conditions. However, as highlighted in a previous work (Angielczyk et al. 2019), the pressure profile in the diverging section is highly influenced by the adopted correlation for the friction losses.

A comparison in terms of mass flow rate indicates more clearly whether the adopted model is suitable to predict the evolution of the flow. Mass flow rates are found to be largely under-predicted by the HEM, as shown in Table 1. This has already been observed for water flashing flows (De Lorenzo et al. 2017), but it could not be evaluated for CO_2 flashing flows due to the lack of mass flow rate measurements. The causes of such discrepancy between model results and experimental data are the finite vaporization rates and the consequent evolution of part of the liquid phase in adiabatic metastable conditions, which increases the mass flow rates passing through the nozzle for the same pressure levels. Moreover, as suggested by the fact that the mass flow rate is worst predicted in case of subcooled inlet conditions, a delay in the onset of vapour nucleation is also believed to contribute to the non-equilibrium evolution of the flow.

5 Conclusions and Future Work

This paper has proposed a study to evaluate the degree of non-equilibrium in two-phase flashing flows of organic fluids by comparing the results from a 1D-HEM model with experimental data. The results indicate that non-equilibrium phenomena are significative, especially when the flashing flow starts from subcooled liquid conditions. A first physical interpretation suggests that thermal non-equilibrium phenomena are more likely to be responsible for the discrepancy between the results from the 1D-HEM and the experimental data. This indicates a need for more complex two-phase flow models that are able to accommodate non-equilibrium features. An example is represented by the delayed equilibrium model (DEM), which has been formulated and successfully adopted for water. A critical aspect of these non-equilibrium models is that they are based on semi-empirical correlations for the vaporization rate and to what extent they can be applied to organic fluids is still unclear and it will therefore be the focus of our future research.

References

Angielczyk, W., Seynhaeve, J.M., Gagan, J., Bartosiewicz, Y., Butrymowicz, D.: Prediction of critical mass rate of flashing carbon dioxide flow in convergent-divergent nozzle. Chem. Eng. Process. **143**, 107599 (2019)

Angielczyk, W., Bartosiewicz, Y., Butrymowicz, D.: Development of delayed equilibrium model for CO_2 convergent-divergent nozzle transonic flashing flow. Int. J. Multiph. Flow **131**, 103351 (2020)

Bartosiewicz, Y., Seynhaeve, J.M.: Delayed equilibrium model (DEM) of flashing choked flows relevant to LOCA. Multiph. Sci. Technol. **25**(2–4), 117–131 (2013)

Bell, I.H., Wronski, J., Quoilin, S., Lemort, V.: Pure and pseudo-pure fluid thermophysical property evaluation and the open-source thermophysical property library CoolProp. Ind. Eng. Chem. Res. **53**(6), 2498–2508 (2014)

Bouré, J.A., Fritte, A.A., Giot, M.M., Réocreux, M.L.: Highlights of two-phase critical flow: on the links between maximum flow rates, sonic velocities, propagation and transfer phenomena in single and two-phase flows. Int. J. Multiph. Flow **3**(1), 1–22 (1976)

De Lorenzo, M., Lafon, P., Seynhaeve, J., Bartosiewicz, Y.: Benchmark of delayed equilibrium model (DEM) and classic two-phase critical flow models against experimental data. Int. J. Multiph. Flow **92**, 112–130 (2017)

Downar-Zapolski, P., Bilicki, Z., Bolle, L., Franco, J.: The non-equilibrium relaxation model for one-dimensional flashing liquid flow. Int. J. Multiph. Flow **22**(3), 473–483 (1996)

Ohta, J., Fujii, T., Akagawa, K., Takenaka, N.: Performance and flow characteristics of nozzles for initially subcooled hot water (influence of turbulence and decompression rate). Int. J. Multiph. Flow **19**(1), 125–136 (1993)

Richardson, B.L.L.: Some Problems in Horizontal Two-Phase Two-Component Flow. Argonne, IL (United States) (1958)

Romei, A., Persico, G.: Computational fluid-dynamic modelling of two-phase compressible flows of carbon dioxide in supercritical conditions. Appl. Therm. Eng. **190**, 116816 (2021)

Vahaji, S., Akbarzadeh, A.A., Date, A., Cheung, S.C.P., Tu, J.Y.: Efficiency of a two-phase nozzle for geothermal power generation. Appl. Therm. Eng. **73**(1), 229–237 (2014)
White, M.T.: Investigating the wet-to-dry expansion of organic fluids for power generation. Int. J. Heat Mass Transf. **192**, 122921 (2022)
Zhu, J., Elbel, S.: Measurement of static pressure profiles of vortex flashing R134a flow expanded through convergent–divergent nozzles. Int. J. Refrig **108**, 258–270 (2019)

A Pressure-Based Model for Two-Phase Flows Under Generic Equations of State

Barbara Re[1](✉), Giuseppe Sirianni[1], and Rémi Abgrall[2]

[1] Department of Aerospace, Science and Technology, Politecnico di Milano, Milano, Italy
{barbara.re,giuseppe.sirianni}@polimi.it
[2] Institute of Mathematics, University of Zurich, Zurich, Switzerland
remi.abgrall@math.uzh.ch

Abstract. We present a diffuse interface method for a pressure-based Baer-Nunziato type model for compressible two-phase flows, which allows the use of generic equations of state to describe each phase. The model is made dimensionless by means of a special pressure scaling that recovers the correct scaling of the discrete governing equations in the zero Mach limit, and overcomes the difficulties related to the lack of a clear notion of reference speed of sound in non-equilibrium two-phase flows. The model is equipped with pressure and velocity relaxation terms to impose the mechanical equilibrium between phases after their independent evolution. Two different finite volume schemes are presented. First, a 1D semi-implicit staggered scheme is introduced to show the capability of the model to work with the Peng-Robinson EOS when each phase evolves close to the saturation curve. Then, a preliminary 2D explicit scheme, which does not include the relaxation terms, is presented as a first step toward the development of an unstructured 2D scheme for compressible two-phase flows at all Mach numbers. The validity of the preliminary 2D monolithic implementation of the hyperbolic operator is illustrated through the simulation of a shock-bubble interaction with air and helium.

Keywords: compressible two-phase flows · Baer-Nunziato model · low Mach · finite volume scheme

1 Introduction

Compressible two-phase flows occur in NICFD regimes, especially in applications involving organic fluids and CO_2 flows. An example is represented by Organic Rankine cycles (ORCs) operating with the so-called wet-to-dry expansion allowed by the high molecular complexity of the working fluids, which may result in a positive slope of the vapor-liquid equilibrium (VLE) curve in the temperature–entropy (T–s) plane. The two-phase expansion could considerably increase the power performance of waste heat recovery systems [15]. Another

M. White et al. (Eds.): NICFD 2022, ERCOFTAC 29, pp. 146–155, 2023.
https://doi.org/10.1007/978-3-031-30936-6_15

example is the transportation of dense CO_2 through pressurized pipelines along the Carbon Capture and Storage (CCS) chain, where a harmful pipeline failure likely results in two-phase flows [4]. To investigate the flow behavior in these and similar conditions, simulation tools able to model two-phase flows in NICFD regimes are mandatory.

Diffuse-interface methods (DIMs) are efficient strategies to model unsteady compressible multi-component or multi-phase flows, consisting of two or more immiscible fluids or components, that exhibit some level of separation at a scale above the molecular level [12]. DIMs do not describe a local instantaneous realization of the flow, but its behavior on average, so they are well suited to dispersed flows, e.g., droplets dispersed in a continuous carrier phase, but they are flexible enough to describe almost pure fluids as well. These methods rely on augmented systems of governing equations that specifically model the behavior of the continuum close to the interfaces, whereas they recover the pure fluid behavior in the bulk. The dynamic interfaces separating different fluids or components are not explicitly tracked, but they are reconstructed from an indicator function.

The cornerstone of the DIM class is the Baer and Nunziato (BN) model, which assumes that each component evolves with its own pressure, temperature, and density. Hence, each component is treated as a separate continuum with its own thermodynamic model. This last feature is of paramount importance to be able to select the proper equation of state (EOS) to describe the thermodynamic behavior in the NICFD regime.

From the full non-equilibrium models, reduced models are derived assuming the equilibrium between one or more variables, e.g. the five-equation model by Kapila assumes mechanical equilibrium among phases, which evolve with the same pressure and velocity. This assumption simplifies the model, reducing the number of governing equations, but introduces serious computational challenges, such as the non-conservative term in the equation for the volume fraction that depends on the divergence of the velocity, and the non-monotone behavior of the sound speed to the volume fraction [13]. Furthermore, modeling important phenomena such phase slip and non-equilibrium liquid/vapor transition is important, especially in two-phase flow applications involving dense gases that undergo mass transfer [3].

We consider a recent pressure-based BN-type model [9] able to deal with generic EOSs and equipped with relaxation parameters, which can be used to control whether the phasic pressure and/or velocity are driven toward the equilibrium. In this work, we show its capability to work with different EOSs, and we present the preliminary results in 2D.

2 The Model

The pressure-based BN-type model developed by Re and Abgrall [9] reads

$$\frac{\partial \alpha_i}{\partial t} = -\mathbf{u}_{\mathrm{IN}} \cdot \boldsymbol{\nabla}\alpha_i + \mu\Delta_i P \tag{1a}$$

$$\frac{\partial(\alpha_i\rho_i)}{\partial t} + \boldsymbol{\nabla}\cdot(\alpha_i\rho_i\mathbf{u}_i) = 0 \tag{1b}$$

$$\frac{\partial(\alpha_i\rho_i\mathbf{u}_i)}{\partial t} + \boldsymbol{\nabla}\cdot(\alpha_i\rho_i\mathbf{u}_i\otimes\mathbf{u}_i + \alpha_i P_i\mathbb{I}) = P_{\mathrm{IN}}\boldsymbol{\nabla}\alpha_i - \lambda\Delta_i\mathbf{u} \tag{1c}$$

$$\begin{aligned} M_{\mathrm{ref}}^2\left[\alpha_i\frac{\partial P_i}{\partial t} + \alpha_i\mathbf{u}_i\cdot(\boldsymbol{\nabla}P_i) + (\alpha_i\rho_i c_i^2)\boldsymbol{\nabla}\cdot\mathbf{u}_i\right] &= \kappa_i\left[\mathbf{u}_{\mathrm{IN}}\cdot\boldsymbol{\nabla}\alpha_i - \mu\Delta_i P - \boldsymbol{\nabla}\cdot(\alpha_i\mathbf{u}_i)\right] \\ &\quad + M_{\mathrm{ref}}^2\Big[\rho_i c_{i,\mathrm{IN}}^2\left[(\mathbf{u}_{\mathrm{IN}}-\mathbf{u}_i)\cdot\boldsymbol{\nabla}\alpha_i - \mu\Delta_i P\right] \\ &\quad - \kappa_i\lambda(\mathbf{u}_{\mathrm{IN}}-\mathbf{u}_i)\cdot\Delta_i\mathbf{u}\Big] \end{aligned} \tag{1d}$$

where the subscript $i = \{1, 2\}$ indicates the phase, α the volume fraction, ρ the density, $\mathbf{u}$ the velocity, P the pressure, c the speed of sound, μ and λ are relaxation parameters which can be used to reproduce different multi-phase flow topologies. For instance, λ can be defined in terms of the specific interfacial area and the acoustic impedances of the fluids [12], whereas μ is an homogenization parameter whose physical meaning has been justified with considerations based on the second law of thermodynamics by Baer and Nunziato [2]. The operator Δ_i expresses the signed jump in a variable between phases, e.g., $\Delta_1\mathbf{u} = \mathbf{u}_1 - \mathbf{u}_2$. The subscript IN indicates interfacial quantities: $\mathbf{u}_{\mathrm{IN}}$ and P_{IN} are closure terms and they are defined as weighted averages

$$\mathbf{u}_{\mathrm{IN}} = \frac{\alpha_1\rho_1\mathbf{u}_1 + \alpha_2\rho_2\mathbf{u}_2}{\alpha_1\rho_1 + \alpha_1\rho_1} \quad \text{and} \quad P_{\mathrm{IN}} = \alpha_1 P_1 + \alpha_2 P_2\,, \tag{2}$$

whereas $c_{i,\mathrm{IN}}$ is the intefacial speed of sound, which does not have any thermodynamic meaning but it is defined by analogy with the speed of sound:

$$c_i^2 = \chi_i + \kappa_i\frac{P_i + e_i}{\rho_i} \quad \text{and} \quad c_{i,\mathrm{IN}}^2 = \chi_i + \kappa_i\frac{P_{\mathrm{IN}} + e_i}{\rho_i}, \quad \text{with} \quad \kappa = \left(\frac{\partial P}{\partial e}\right)_\rho \quad \text{and} \quad \chi = \left(\frac{\partial P}{\partial \rho}\right)_e, \tag{3}$$

where e is the internal energy per unit of volume. Equations (1b)–(1d) are repeated for both phases, while Eq. (1a) is solved only for phase 1, as the constraint $\alpha_1 + \alpha_2 = 1$ holds. Thanks to the symmetry of the model, which phase is solved as 1 does not affect the results.

2.1 The Low-Mach Scaling

The system of Eqs. (1) is dimensionless and it includes a special pressure scaling, according to which the dimensionless pressure P is defined as $P = \left(\tilde{P} - \tilde{P}_{\mathrm{ref}}\right)/\tilde{\rho}_{\mathrm{ref}}\tilde{u}_{\mathrm{ref}}^2$, where $\sim$ indicates a dimensional quantity, and the subscript "ref" a reference quantity. In addition, a reference Mach number M_{ref} is defined as $M_{\mathrm{ref}}^2 = \tilde{\rho}_{\mathrm{ref}}\tilde{u}_{\mathrm{ref}}^2/\tilde{P}_{\mathrm{ref}}$, which expresses the global level of compressibility of the flow field. This peculiar scaling overcomes the singularity in the limit $M \to 0$ that affects standard compressible schemes that assume a non-dimensional scaling based on a single reference velocity, for instance a reference speed of sound, which results in the term $1/M^2$ in front of the pressure gradient in the momentum equation. In this context, we remark that for non-equilibrium two-phase flows, there is not a clear notion of mixture speed of sound, so the choice of a reference speed of sound is ambiguous, if not impossible. Moreover, the adopted scaling allows the recovery of the correct limit of the governing equations in the incompressible regime [9].

2.2 Thermodynamic Model

The model of Eqs. (1) assumes a generic description of the thermodynamics of each fluid, with the only requirement to express the pressure EOS as $e = e(\rho, P)$. Most of the EOSs used for academic and industrial purposes, especially cubic ones, can easily meet this requirement, so they can model the thermodynamic behavior of each component. Moreover, different EOSs can be adopted for different phases.

Particular attention should be paid while scaling the thermodynamic definition involving the pressure, as they can involve the additional term $\tilde{P}_{\text{ref}}$. For instance, the relation between the dimensional and dimensionless speed of sound reads

$$\tilde{c}^2 = \tilde{\chi} + \tilde{\kappa}\frac{\tilde{P} + \tilde{e}}{\tilde{\rho}} = \left[\chi + \kappa\frac{P+e}{\rho}\right]\tilde{u}_{\text{ref}}^2 + \frac{\kappa}{\rho}\frac{\tilde{P}_{\text{ref}}}{\tilde{\rho}_{\text{ref}}} = \left[c^2 + \frac{\kappa}{\rho M_{\text{ref}}^2}\right]\tilde{u}_{\text{ref}}^2\,. \tag{4}$$

In this work, we consider two different EOSs: the stiffened gas model and the Peng-Robinson EOS. Considering the former, it is possible to directly define the dimensionless definitions according to the special low Mach scaling described in the previous subsection. The pressure EOS is defined as

$$P(e,\rho) = (\gamma - 1)e - \gamma P_\infty - (\gamma - 1)\rho q - 1/M_{\text{ref}}^2\,, \tag{5}$$

where γ, P_∞, and q are substance-specific parameters [7]. The compatible caloric EOS, derived under the polytropic assumption following [10] is

$$e(T,\rho) = c_v\rho T + \rho q + P_\infty \qquad \text{or} \qquad P(T,\rho) = (\gamma - 1)c_v\rho T - P_\infty - \frac{1}{M_{\text{ref}}^2}\,, \tag{6}$$

where T is the temperature, and c_v is the isochoric specific heat.

Concerning the Peng-Robinson EOS, we use an external thermodynamic library developed at SINTEF (Norway), which exploits the concept of the corresponding states to enhance the accuracy of specific properties, such as the density and the speed of sound, computed through the Peng-Robinson EOS for the liquid phase [16]. Then, the results of the library are made dimensionless, according to the special scaling described above, e.g. in Sec. 2.1 and Eq. (4).

3 Numerical Method

The solution strategy for the system of Eqs. (1) is here briefly summarized. A detailed description for the 1D can be found in [9]. The Strang splitting approach is used to solve the system of partial differential equations (PDEs): at each time step, two operators are consecutively applied. First, the hyperbolic operator solves the PDEs composed without the relaxation terms, then the relaxation operator solves the system of ordinary differential equations (ODEs) composed only by the temporal derivatives and the relaxation terms.

3.1 Hyperbolic Operator in 1D: A Semi-implicit Staggered Scheme

A semi-implicit time integration scheme is used to reach an acceptable computational efficiency without running into a too severe restriction on the time step. The acoustic terms, generally linear, are integrated implicitly in time, while the convective terms are integrated explicitly. In this way, the stability constraint on the time step is based only on the convective velocity, but the non-linearity of the system is weaker than the full implicit time integration.

The pressure-based formulation of the BN-type model is solved through a segregate solution approach, according to which, at each time step, the momenta (or the velocities) are first predicted considering the pressures at the previous time step, then, the new pressures are computed, and, finally, the momenta are updated for the pressure correction.

A finite volume scheme is used to spatially discretize the governing equations. A staggered grid is used to combine different discretizations. For simplicity, the convective terms are approximated using first-order Rusanov fluxes, which circumvent the need to solve a Riemann problem at cell interfaces. A central finite difference approximation is adopted for the gradient of the pressure in Eq. (1c) and the divergence of the velocity in Eq. (1d). A robust discretization of the non-conservative terms involving the gradient of the volume fraction is derived following the non-disturbance conditions, which states that a flow uniform in pressure and velocity should remain uniform in these variables at the next time steps [1].

3.2 Source Operator

Velocity and pressure relaxation give rise to two ODE problems, which include only time derivatives and the right-hand side terms of Eqs. (1) involving λ and μ. These systems are solved sequentially: first, the solution computed by the hyperbolic operator (labeled $\mathrm{L}_{\mathrm{hyp}}$) is evolved towards the velocity equilibrium by the operator $\mathrm{R}_{\mathrm{vel}}$; then, this result is used as initial condition for $\mathrm{R}_{\mathrm{pres}}$ to relax the pressures. The solution strategy at each time step can be summarized as:

$$\mathbf{U}^{n+1} = \mathrm{R}_{\mathrm{pres}}\left(\mathbf{U}^{\diamond}\right) \longleftarrow \mathbf{U}^{\diamond} = \mathrm{R}_{\mathrm{vel}}\left(\mathbf{U}^{\triangleright}\right) \longleftarrow \mathbf{U}^{\triangleright} = \mathrm{L}_{\mathrm{hyp}}\left(\mathbf{U}^{n}\right), \tag{7}$$

where the subscripts $\triangleright$ and $\diamond$ indicate intermediate solution between the time step t^n and t^{n+1}.

We impose instantaneous relaxation, e.g., we consider λ and μ as infinite. As explained in [8], the results of $\mathrm{R}_{\mathrm{vel}}$ are

$$\alpha_i^\diamond = \alpha_i^\triangleright, \qquad \rho_i^\diamond = \rho_i^\triangleright, \qquad \mathbf{u}^\diamond = \mathbf{u}_1^\diamond = \mathbf{u}_2^\diamond = \frac{\alpha\rho\mathbf{u}_1^\triangleright + \alpha\rho\mathbf{u}_2^\triangleright}{\alpha\rho_1^\triangleright + \alpha\rho_2^\triangleright}, \qquad \text{and}$$
$$P_i^\diamond = P_i^\triangleright - \rho_i \frac{\kappa_i}{2}\left(\mathbf{u}_{\mathrm{IN}}^\triangleright - \mathbf{u}_i^\triangleright\right)\cdot\left(\mathbf{u}^\diamond - \mathbf{u}_i^\triangleright\right).$$

The ODE problem for the pressure relaxation includes an equation for α_1, deriving from Eq. (1a), and one equation for each pressure, coming from Eq. (1d). We integrate it using a backward Euler scheme, but we introduce in the definition of the interface speed of sound $c_{\mathrm{IN},i}^2$ suitable approximations $\overline{P_{\mathrm{IN},i}}$ and $\overline{e_i}$ (defined in Algorithm 1), which are kept constant during time integration to mitigate its non-linearity [11]. The discrete equations for the pressures read

$$f_P(P_i, \alpha_i) = M_{\mathrm{r}}^2\left(P_i - P_i^\diamond\right) - M_{\mathrm{r}}^2 \chi_i (\alpha\rho_i)^\diamond \left(\frac{1}{\alpha_i} - \frac{1}{\alpha_i^\diamond}\right) + \left[M_{\mathrm{r}}^2 \kappa_i\left(\overline{P_{\mathrm{IN},i}} + \overline{e}_i\right) + \kappa_i\right] \log\left(\frac{\alpha_i}{\alpha_i^\diamond}\right) = 0. \tag{8}$$

Combining these expressions and imposing pressure equilibrium, we obtain the following non-linear function

$$\begin{aligned} f_\alpha(\alpha_1) = P_1^{n+1} - P_2^{n+1} = M_{\mathrm{r}}^2 &\left[\left(P_1^\diamond - P_2^\diamond\right) + \chi_1\left(\frac{(\alpha\rho_1)^\diamond}{\alpha_1} - \frac{(\alpha\rho_1)^\diamond}{\alpha_1^\diamond}\right) - \chi_2\left(\frac{(\alpha\rho_2)^\diamond}{1-\alpha_1} - \frac{(\alpha\rho_2)^\diamond}{\alpha_2^\diamond}\right)\right] \\ &- \left[M_{\mathrm{r}}^2\kappa_1\left(\overline{P_{\mathrm{IN},1}} + \overline{e}_1\right) + \kappa_1\right]\log\left(\frac{\alpha_1}{\alpha_1^\diamond}\right) + \left[M_{\mathrm{r}}^2\kappa_2\left(\overline{P_{\mathrm{IN},2}} + \overline{e}_2\right) + \kappa_2\right] \\ &\times \log\left(\frac{1-\alpha_1}{\alpha_2^\diamond}\right) = 0, \end{aligned} \tag{9}$$

where the only unknown is α_1. We solve this equation iteratively, performing at each iteration k, one iteration of Newton's method, but with a convergence criterion based on the pressure difference, as illustrated in Algorithm 1.

3.3 Hyperbolic Operator in 2D: A Fully-Explicit Co-Located Scheme

We also present a preliminary 2D hyperbolic operator based on a fully-explicit co-located finite volume scheme. A monolithic solver is developed in this framework, considering, as done in 1D, the Rusanov fluxes for the spatial discretization of the convective terms and imposing the non-disturbance condition [1] for the non-conservative terms. On the other hand, for the divergence of the velocity in Eq. (1d), we use the Rhie and Chow interpolation of the velocity at cell interfaces to circumvent checker-boarding problems due to the low pressure-velocity coupling at low Mach number [14].

Algorithm 1 . Pseudo-code to solve the non-linear system resulting from the implicit time integration of the ODE system for pressure relaxation.

$\alpha_i^k \leftarrow \alpha_i^\diamond$, $\overline{P_{\mathrm{IN},i}} \leftarrow P_{\mathrm{IN}}^\diamond$, $\overline{e_i} \leftarrow P_i^\diamond$ ▷ Initialization
for $k = 0$ to $k_{\max}$ **do**
  $P_i^{k+1} \leftarrow$ solution of $f_P(P_i^{k+1}, \alpha_i^k)$ ▷ Solve Eq. (8)
  if $|P_1^{k+1} - P_2^{k+1}| < \epsilon$ **then**
    Exit loop ▷ $\mathbf{U}^{n+1} \leftarrow \mathbf{U}^{k+1}$
  else
    $\alpha_1^{k+1} \leftarrow \alpha_1^k - f(\alpha_1^k)/f'(\alpha_1^k)$ ▷ Solve Eq. (9)
    $\alpha_2^{k+1} \leftarrow 1 - \alpha_1^{k+1}$
    $\overline{e_i} \leftarrow \frac{1}{2}\left(e_i(\rho_i^k, P_i^{k+1}) + e_i^\diamond\right)$
    $\overline{P_{\mathrm{IN},i}} \leftarrow \frac{1}{2}\left(P_i^{k+1} + P_{\mathrm{IN}}^\diamond\right)$
  end if
end for

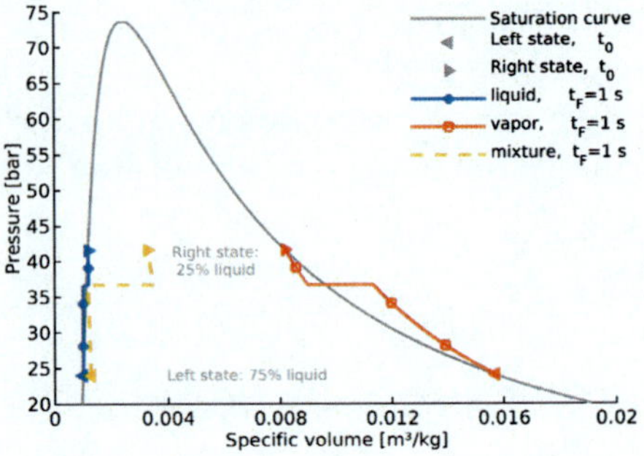

Fig. 1. Results of the CO_2 shock-tube test with instantaneous relaxation. The blue and the red lines show the states of the liquid and the vapor, respectively, within the whole domain. The yellow, dashed line shows the behavior of the mixture.

4 Results

The first test we present is a 1D shock-tube with instantaneous pressure and velocity relaxation, which means that mechanical equilibrium is imposed at each time step. The tube is filled with a liquid-gas mixture of CO_2 at saturated conditions. In the left chamber, the liquid volume fraction is $\alpha_{\mathrm{liq,L}} = 0.75$ and the temperature of both phase is $T_{\mathrm{L}} = 260$ K. In the right chamber, there is $\alpha_{\mathrm{liq,R}} = 0.25$ and $T_{\mathrm{R}} = 280$ K. The domain is 60 m long, and it is discretized with 600 uniform cells. The final time is $t_{\mathrm{f}} = 1$ s, reached in 2000 time steps. The results are shown in the $P - v$ plane in Fig. 1: each phase evolves close to "its side" of the VLE curve, governed by the Peng-Robinson EOS. On the other hand, if we look at the behavior of the mixture, whose density is defined as $\overline{\rho} = \alpha\rho_1 + \alpha\rho_2$, it evolves in the two-phase region, as expected. But, thanks to the modeling choice of having two pressures and two velocities, each phase can be described by its own EOS, without the need for an EOS for the mixture.

The second test is a preliminary assessment of the hyperbolic operator in 2D, and we do not consider any relaxation, so λ and μ are zero, and pressure and velocity disequilibrium can occur between phases. We present the simulation of the interaction of a shock-wave in air with a cylindrical bubble of helium. The domain is a rectangle ($[0, 300] \times [0, 47.5]$ mm) representing only half of the problem, thanks to the symmetry about the center-line, and it is discretized using an average grid spacing $\Delta x = 2$ mm and time step $\Delta t = 0.1$ μs. The initial flow field is separated into three regions: the cylindrical bubble, with a diameter of 5 mm, centered at $x_{\mathrm{B}}^0 = 196$ mm and characterized by $\alpha_{\mathrm{He}} = 0.93$, the left and the right regions separated by the shock-wave located at $x_{\mathrm{s}}^0 = 235$ mm with $\alpha_{\mathrm{air}} = 0.93$. The left and the bubble regions are initialized with the pre-shock pressure $P_{\mathrm{L}}^0 = 10^5$ Pa and density $\rho_{\mathrm{L,air}}^0 = 1.4$ kg/m^3; the right region is initialized with the post-shock state corresponding to a shock Mach number in air of $M_s = 1.22$. For helium, a uniform initial density $\rho^0_{He} = 0.2546$ kg/m^3 is imposed everywhere. The final simulation time is $t_f = 900\mu$s, reached in 9000 time steps.

Figure 2 shows the flow field at different times, before, during, and after the shock-bubble interaction. The bubble deformation is qualitatively in agreement with the numerical data by Daude et al. [5] and experimental results by Hass and Sturtevant [6]. This can be better appreciated in Fig. 3, where the bubble interface obtained with the proposed method is superimposed on the experimental results. More specifically, the background picture is the shadow photograph taken from Fig. 7(h) in [6], 427 μs and 674 μs after the shock-bubble interaction, and the violet lines represent the contour lines $\alpha_{He} = 0.3$ and $\alpha_{He} = 0.7$ at the same time. A good agreement shows the validity of the proposed approach.

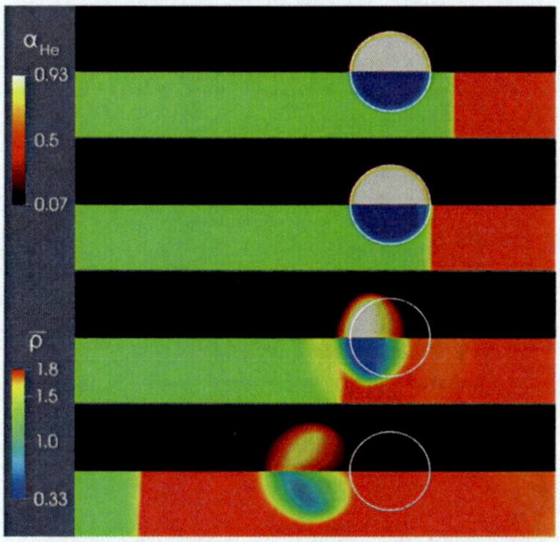

Fig. 2. Shock and air/He bubble interaction. Contour plots of helium volume fraction α_{He} (upper part) and mixture density $\overline{\rho}$ (bottom part), at different times. The white circle displays the initial position of the bubble.

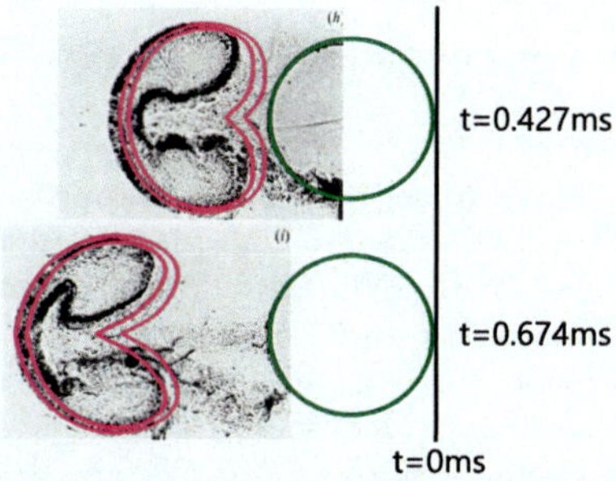

Fig. 3. Shock and air/He bubble interaction. Comparison of the position and shape of the bubble between experiment data from [6] (background picture) and the numerical data (violet isolines at $\alpha_{He} = \{0.3, 0.7\}$, 427 μs and 674 μs after the shock-bubble interaction, i.e., at the simulation times $t = 467$ μs and $t = 714$ μs. The green circles display the initial positions of the bubble.

5 Conclusions and Future Works

We presented a solution strategy for a pressure-based Baer-Nunziato-type model for compressible two-phase flows. The rejection of the assumption of thermo-mechanical equilibrium allows the description of each phase with its own thermodynamic model, which could consist of any generic pressure EOS in the form

$e = e(P, \rho)$ and the compatible caloric EOS. Two numerical methods were tested: a semi-implicit finite-volume scheme over staggered grids in 1D, and a co-located explicit finite-volume scheme in 2D. The former one was equipped with instantaneous pressure and velocity relaxation, and it showed the possibility to simulate flow problems where the two-phase mixture evolves within the VLE curve, without the need to define a mixture EOS. The latter has been the first step toward an all-Mach monolithic 2D numerical method for the simulation of two-phase flows over unstructured grids. The preliminary implementation of the explicit hyperbolic solver has been validated through the simulation of the interaction between a shock wave in air and a bubble of helium.

The current and future developments follow two main paths: the BN-type model will be equipped with new relaxation terms to include mass and heat transfer, and the implementations of the numerical scheme will be improved by including higher order discretization and the source operator also in 2D.

Acknowledgements. R.A. is partially funded by SNF Grant 200020_204917. G. S. was partialy funded by R.A's UZH Einrichtungkredit.

References

1. Abgrall, R.: How to prevent pressure oscillations in multicomponent flow calculations: a quasi conservative approach. J. Comput. Phys. **125**(1), 150–160 (1996). https://doi.org/10.1006/JCPH.1996.0085
2. Baer, M.R., Nunziato, J.W.: A two-phase mixture theory for the deflagration-to-detonation transition (DDT) in reactive granular materials. Int. J. Multiphase Flow **6**, 861–889 (1986). https://doi.org/10.1016/0301-9322(86)90033-9
3. Brown, S., Martynov, S., Mahgerefteh, H., Chen, S., Zhang, Y.: Modelling the non-equilibrium two-phase flow during depressurisation of CO_2 pipelines. Int. J. Greenhouse Gas Control. **30**, 9–18 (2014). https://doi.org/10.1016/j.ijggc.2014.08.013
4. Brown, S., Martynov, S., Mahgerefteh, H., Proust, C.: A homogeneous relaxation flow model for the full bore rupture of dense phase CO_2 pipelines. Int. J. Greenhouse Gas Control. **17**, 349–356 (2013). https://doi.org/10.1016/j.ijggc.2013.05.020
5. Daude, F., Galon, P.: On the computation of the Baer-Nunziato model using ALE formulation with HLL- and HLLC-type solvers towards fluid-structure interactions. J. Comput. Phys. **304**, 189–230 (2016). https://doi.org/10.1016/j.jcp.2015.09.056
6. Haas, J.-F., Sturtevant, B.: Interaction of weak shock waves with cylindrical and spherical gas inhomogeneities. J. Fluid Mech. **181**, 41 (1987). https://doi.org/10.1017/S0022112087002003
7. Métayer, O.L., Saurel, R.: The Noble-Abel Stiffened-Gas equation of state. Phys. Fluids. **28**, 046102 (2016). https://doi.org/10.1063/1.4945981
8. Re, B., Abgrall, R.: Numerical simulation of weakly compressible multiphase flows with a Baer-Nunziato type model. World Congr. Comput. Mech. ECCOMAS Congr. **600**, 1–12 (2021). https://doi.org/10.23967/wccm-eccomas.2020.247
9. Re, B., Abgrall, R.: A pressure-based method for weakly compressible two-phase flows under a Baer-Nunziato type model with generic equations of state and pressure and velocity disequilibrium. Int. J. Numer. Meth. Fluids **94**(8), 1183–1232 (2022). https://doi.org/10.1002/fld.5087

10. Re, B., Guardone, A.: An adaptive ALE scheme for non-ideal compressible fluid dynamics over dynamic unstructured meshes. Shock Waves **29**, 73–99 (2019). https://doi.org/10.1007/s00193-018-0840-2
11. Saurel, R., Abgrall, R.: A multiphase Godunov method for compressible multifluid and multiphase flows. J. Comput. Phys. **150**(2), 425–467 (1999). https://doi.org/10.1006/JCPH.1999.6187
12. Saurel, R., Pantano, C.: Diffuse-interface capturing methods for compressible two-phase flows. Annu. Rev. Fluid Mech. **50**, 105–130 (2018). https://doi.org/10.1146/annurev-fluid-122316-10.1146/annurev-fluid-122316-050109
13. Saurel, R., Petitpas, F., Berry, R.A.: Simple and efficient relaxation methods for interfaces separating compressible fluids, cavitating flows and shocks in multiphase mixtures. J. Comput. Phys. **228**, 1678–1712 (2009). https://doi.org/10.1016/j.jcp.2008.11.002
14. Sirianni, G., Re, B., Abgrall, R., Guardone, A.: Momentum weighted interpolation for unsteady weakly compressible two-phase flows on unstructured meshes. J. Comput. Appl. Math. **428**, 115209 (2023). https://doi.org/10.1016/j.cam.2023.115209
15. White, M.T.: Cycle and turbine optimisation for an ORC operating with two-phase expansion. Appl. Thermal Eng. **192**, 116852 (2021). https://doi.org/10.1016/j.applthermaleng.2021.116852
16. Wilhelmsen, Ø., et al.: Thermodynamic modeling with equations of state: present challenges with established methods. Indus. Eng. Chem. Res. **56**, 3503–3515 (2017). https://doi.org/10.1021/acs.iecr.7b00317

Validation and Application of HEM for Non-ideal Compressible Fluid Dynamic

Liyi Chen(✉), Michael Deligant, Mathieu Specklin, and Sofiane Khelladi

Arts et Metiers Institute of Technology, CNAM, LIFSE, HESAM University, 75013 Paris, France
{alienor.chenliyi,michael.deligant,mathieu.specklin, sofiane.khelladi}@ensam.eu

Abstract. In industrial applications such as the Organic Rankine Cycle (ORC), accurate predictions of the system performance rely on simulations of the flow properties of the working fluid. During operation, fluids may exhibit large property changes and non-ideal gas behavior due to phase changes. Especially in recent years, a large number of applications of supercritical fluids have been developed, and the simulation of fluid flow in trans/supercritical conditions has become important due to the drastic changes in its thermophysical properties near the critical point. As a result, the general analytical equations of state (EOS) such as perfect gas EOS and stiffened-gas EOS cannot accurately model the flow of various fluids in organic cycles. In this work, we used the finite volume method coupled with CoolProp, a fluid thermophysical property library, to consider the properties of the real gas. Homogeneous equilibrium model (HEM) is used to deal with the two-phase flows. Rusanov and SLAU approximate Riemann solver are used for the flux computation.

The present solver has been used and assessed on a variety of test-cases, including compressible flow in single phase, two-phase and supercritical shock tubes. The accuracy of the simulation results is satisfactory compared to other numerical and analytical solutions. Simulations of supersonic flow in a converging-diverging nozzle were also carried out, and the results were reasonable compared with the experimental results. The solver can reflect the real gas behavior and can be further used in the future for the development of other fluid applications in industry.

Keywords: real gas effect · compressible flow · two-phase flow · supercritical fluid · volume finite method

1 Introduction

The study of fluid heat transfer in the energy field is of great importance. In fact, in various energy systems, fluids generally undergo drastic state changes, such as ORC cycles, two-phase nozzle applications, supercritical nozzles, etc., which may change from supercritical to single-phase or two-phase flow in a very limited space and time [1,2]. Therefore, in order to study and develop the heat transfer potential of fluids, efficient and accurate CFD simulations are necessary. In particular, the accuracy of the

M. White et al. (Eds.): NICFD 2022, ERCOFTAC 29, pp. 156–165, 2023.
https://doi.org/10.1007/978-3-031-30936-6_16

thermodynamic properties' calculation should be one of the most important characteristics to be dealt with. Equation of state (EOS) is mainly used to determine the importance of fluid properties. In the last decade, simple EOS such as ideal Gas EOS, Stiffened Gas (SG) EOS have been adopted in the calculation of air, water and other complex fluids [3]. Even for same fluid, the choice of simple EOS varies for different states. For example, for the calculation of water, the Ideal Gas EOS (for steam) coupled with SG EOS (for liquid water) is used [4]. For CO_2 at low pressure conditions, the ideal gas EOS can ensure a certain accuracy; however, as the temperature or the pressure increases, in the near or trans-critical region, the effect of the real gas becomes apparent and a more accurate equation of state, such as the Peng & Robinson or Span-Wagner model, needs to be used [5]. In summary, simple EOS allows fast properties' calculations, but is not considered globally-valid and sometimes even provides unrealistic results. In this paper we use CoolProp thermodynamic parameter library, where all the thermodynamic properties are determined based on the exact Helmholtz EOS [6]. CoolProp library contains properties of dozens of fluids and the development of corresponding fluid solvers helps to explore the applications for different fluids.

The work in this paper includes the development of a real gas solver coupled to the CoolProp parameter library and tested with relevant examples as well as applications. The results are compared with existing simulation results or experimental results and are in good agreement. It is shown that this solver is globally valid and can accurately simulate the flow in different phase states, including two-phase mixture as well as trans-critical flow.

2 Governing Equation

Based on the conservation equations of momentum and total energy, the conservation equations for one-dimensional, inviscid, compressible flows in tubes with variable cross-section area can be written as:

$$\begin{aligned} \partial_t (\rho A) + \partial_x (\rho u A) &= 0 \\ \partial_t (\rho u A) + \partial_x (\rho u^2 A + pA) - p\partial_x A &= 0 \\ \partial_t (\rho \varepsilon A) + \partial_x (\rho \varepsilon u A + puA) &= 0 \end{aligned} \tag{1}$$

The specific internal energy $\varepsilon = e - (1/2)u^2$. For the two-phase conditions, we use the Homogeneous Equilibrium Model (HEM) where the slip between phases is neglected and instantaneous thermal, mechanical and chemical equilibria are assumed. As a consequence, in HEM, the liquid-gas mixture is supposed to be at its saturation point, and the two phases share the same velocity, the same pressure, the same temperature. And the closure laws are:

$$\begin{aligned} e &= xe_v(p) + (1-x)e_l(p) \\ v &= xv_v(p) + (1-x)v_l(p) \end{aligned} \tag{2}$$

where the specific volume is $v = 1/\rho$ and the quality is $x = \frac{mass\ of\ vapor}{total\ mass}$. With the constraints expressed in Eq. (2), the internal energies and the specific volumes of vapor and liquid are those of the saturation curve. That is:

$$e_v = e_v(p_{\text{sat}}), e_l = e_l(p_{\text{sat}}), v_v = v_v(p_{\text{sat}}), v_l = v_l(p_{\text{sat}}) \tag{3}$$

System (1) can be also expressed in vertical form as follows:

$$\frac{\partial}{\partial t}U + \nabla \cdot \mathbf{F}(U) = \mathbf{S}(U) \tag{4}$$

where

$$U = \begin{pmatrix} \rho A \\ \rho u A \\ \rho \varepsilon A \end{pmatrix}, \mathbf{F}(U) = \begin{pmatrix} \rho u A \\ \rho u^2 A + pA \\ \rho \varepsilon u A + puA \end{pmatrix} \text{ and } \mathbf{S}(U) = \begin{pmatrix} 0 \\ p \partial_x A \\ 0 \end{pmatrix}$$

U is the set of conservative variables, $\mathbf{F}$ is the flux function, and $\mathbf{S}$ is the source term. The time evolution of U evolves as the divergence of the flux and the gain from source. Using the Finite-Volume method [7], the solution can be updated from timestep n to the next timestep $n+1$ as:

$$V_i U_i^{n+1} = V_i U_i^n - \Delta t^n \mathbf{F}(U_i^n) + \Delta t^n p_i^n R_i^n \tag{5}$$

where the volume V_i is the control volume and

$$R_i^n = \begin{pmatrix} 0 \\ A_{i+1/2} - A_{i-1/2} \\ 0 \end{pmatrix}$$

The numerical procedure is based on an approximate Riemann solver, following the Rusanov scheme that needs to know an averaged state between the states $U_{i-1/2}$ and $U_{i+1/2}$ of the Riemann problem at the grid interfaces. The Rusanov flux $\mathscr{F}(U^n)_i$ is given by

$$\mathscr{F}(U^n)_i = \mathscr{F}\left(U_{i-1/2}^n, U_{i+1/2}^n\right) = \frac{1}{2}\left[\mathbf{F}\left(U_{i-1/2}^n\right) + \mathbf{F}\left(U_{i+1/2}^n\right)\right] - \lambda\left(U_{i+1/2}^n - U_{i-1/2}^n\right) \tag{6}$$

where $\lambda = \max\left(c_{i-1/2} + |u_{i-1/2}|, c_{i+1/2} + |u_{i+1/2}|\right)$ with $c_{i-1/2}$ (resp. $c_{i+1/2}$) the sound velocity associated to state $U_{i-1/2}^n$ (resp. $U_{i+1/2}^n$). According to the results of De Lorenzo [3], the speed of sound of the HEM has two discontinuities for the void fraction $\alpha_v = 0$ and $\alpha_v = 1$. For a more convenient and reasonable computational purpose, we define the sound speed in two-phase conditions using the expression given by Wood [4,8]:

$$\frac{1}{\rho c^2} = \frac{\alpha_l}{\rho_l c_l^2} + \frac{\alpha_v}{\rho_v c_v^2} \tag{7}$$

An EOS is needed to close the systems (1). In the present work, we use CoolProp Library, which is based on the Helmholtz Energy equations of state, to evaluate the thermal properties:

$$\begin{aligned} p &= p(e, \rho) = p^{CoolProp}(e, \rho) \\ T &= T(e, \rho) = T^{CoolProp}(e, \rho) \end{aligned} \tag{8}$$

3 Numerical Tests

3.1 Shock-Tubes Problem in a Duct with Constant Cross-Section

The ability of this method to predict pressure wave propagation was first evaluated in a constant cross-section pipeline. The length of the shock tube used is L, the initial discontinuity occurs at the pipeline coordinate $x = 0.5L$, and the initial conditions in the pipeline are as follows:

$$(\rho, u, p, T, \alpha_v) = \begin{cases} \rho_L, u_L, p_L, T_L, \alpha_{vL}, & x < 0.5L; \\ \rho_R, u_R, p_R, T_R, \alpha_{vR}, & x \geq 0.5L \end{cases}$$

The simulation results are performed in comparison with the numerical solutions obtained by Daude [9] and Terashima [10] or the available analytical solutions, respectively.

3.1.1 Test 1: Single-Phase Gas Shock-Tube

The first test-case considered in this paper consists of a $L = 1$ m long tube filled by air at rest. The pressure discontinuity is initially located at $x = 0.5$ m. The tube is composed of a high-pressure chamber with $p_L = 1$ bar, $\rho_L = 1\,\text{kg/m}^3$ and a low-pressure chamber with $p_R = 0.1$ bar, $\rho_R = 0.125\,\text{kg/m}^3$. The corresponding initial conditions was originally proposed by Sod [11]. The numerical solutions obtained with 200, 500 and 1000 cells in conjunction with the Courant number $C = 0.5$ are compared with the analytical solution of this Riemann problem which is composed of a left rarefaction wave, a contact discontinuity and a right shock-wave as shown in Fig. 1. Good agreement is obtained as all of the intermediate states are obtained in the numerical solution with a satisfactory manner.

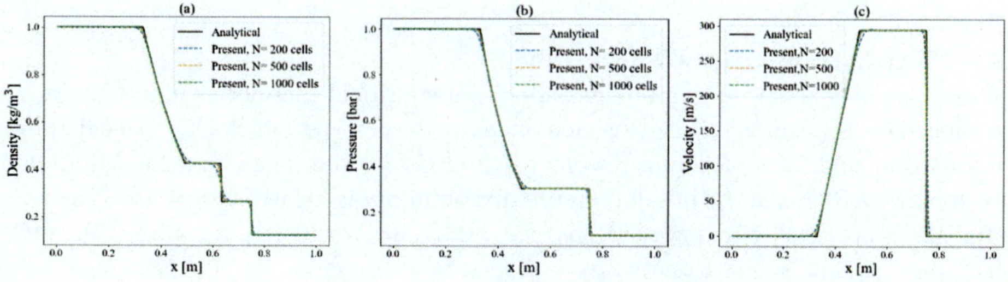

Fig. 1. Numerical solutions obtained on the single-phase gas shock-tube at $t = 4.5125$ ms with 200,500 and 1000 cells in conjunction with $C = 0.5$; comparison with the analytical solution: (a) pressure, (b) density and (c) velocity.

3.1.2 Test 2: Two-Phase Shock-Tube

Test 2 corresponds to a water-steam two-phase shock-tube, tube long $L = 100$ m and the initial pressure discontinuity is located at $x = 50$ m. The two initial states are $p_L = 150$ bar, $p_R = 100$ bar, $\alpha_{vL} = 0.1$, $\alpha_{vR} = 0.5$, $u_L = u_R = 0.0$ m/s corresponding to a pressure and temperature jumps in a two-phase flow at the saturation point. The numerical results obtained using 200, 500 cells with a Courant number $C = 0.8$ at $t = 81$ ms are plotted in Fig. 2.

The present results compared well with the numerical solutions previously obtained with the result [9] using 5000 nodes and a CFL (Courant-Friedrichs-Lewy) condition of 0.8. Daude considers the steam-water tables based on the 1984 NBS/NRC as EOS [12]. Three discontinuities (the rarefaction wave located at $x \approx 40$ m, the contact discontinuity at $x \approx 50$ m and the shock wave at $x \approx 60$ m) are clearly visible on the void fraction profile. As a consequence, it allows for the comparison with the present numerical solutions showing that the three discontinuities are well captured even using 500 cells.

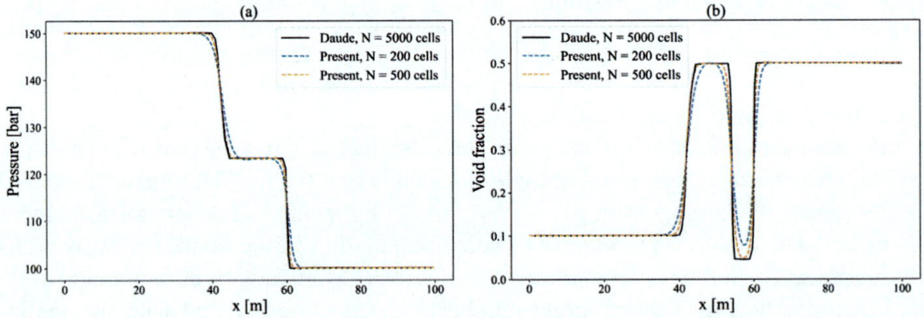

Fig. 2. Numerical solutions obtained on the two-phase liquid-steam shock-tube at $t = 81$ ms with 200 and 500 cells with $C = 0.8$; comparison with the results issued from Daude [9]: (a) pressure and (b) void fraction.

3.1.3 Test 3: Trans-Critical Shock Tube

Simulations of a shock tube problem under a supercritical thermodynamic condition, introduced by Terashima [10], are carried out as the test case in which CO_2 is used as the working fluid and Soave-Redlich-Kwong (SRK) EOS is used. The initial conditions are tube long $L = 10$ m and the initial pressure discontinuity is located at $x = 5$ m. The two initial states are $p_L = 10p_0$, $p_R = 0.1p_0$, $\rho_L = 0.01\rho_0$, $\rho_R = 0.1\rho_0$, $u_L = u_R = 0.0$ m/s, where $\rho_0 = \rho_{cr,CO_2} = 348.8\,\text{kg/m}^3$, $p_0 = p_{cr,CO_2} = 73.7$ bar.

The numerical results obtained using 200, 500 and 1000 cells at $t = 2.745$ ms are plotted in Fig. 3. The present results have good agreement with the numerical solutions previously obtained with the results of Terashima [10] using 8001 nodes. It can be seen from the close-up view that in the existing results, the speed of the shock wave is slightly larger than that of the SRK model, and satisfactory accuracy can be obtained when the grid number is 1000. The result obtained here demonstrates the applicability of the present method to supercritical flows with shock wave.

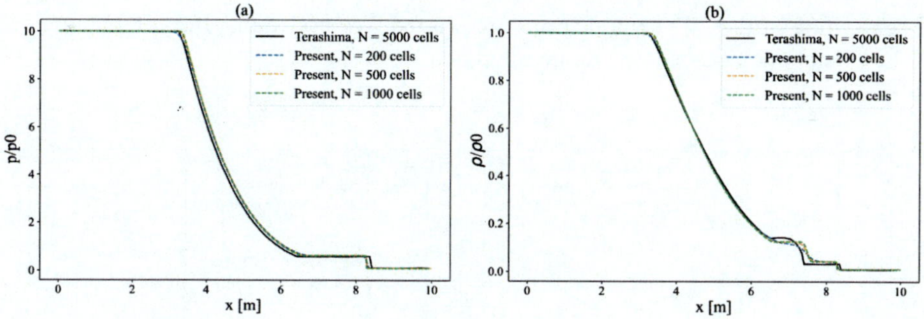

Fig. 3. Numerical solutions obtained on trans-critical shock-tube at $t = 2.745$ ms with 200,500,1000 cells with $C = 0.8$; comparison with the results issued from Terashima [10]: (a) p/p_0 and (b) ρ/ρ_0.

3.2 Shock-Tubes Problem with an Abrupt Change of Area

Single-phase liquid water shock-tube problem involving a sudden change of duct cross-sections is considered in this section. The present test-case was initially proposed in Tiselj [13]. The computational domain is 5 m long. A shock-tube is initially located at $x = 2$ m whereas an abrupt expansion of area with a cross-section ratio of 20 is located at $x = 3$ m. The initial conditions are given in Table 1.

Table 1. Initial condition for the single-phase liquid shock-tube

Position	p(bar)	T(K)	α_v	u(m/s)	A(m^2)
$x \in [0.0; 2.0)$	100	413.43	0	0	0.02
$x \in [2.0; 3.0)$	50	412.94	0	0	0.02
$x \in [3.0; 5.0]$	50	412.94	0	0	0.4

The 1-D numerical results obtained using 1000 cells at $t = 1$ ms are plotted in Fig. 4, comparing with the numerical solutions previously obtained by the results of Daude [9] using 5000 nodes and steam-water tables EOS based on the 1984 NBS/NRC [12]. On the pressure and velocity profiles the rarefaction wave travels on the left side and is visible at $x \approx 0.5$ m. On the contrary to the results of Daude, there is a discontinuity at the expansion ($x = 3$ m). Then after the expansion, the transmitted shock wave is visible at $x \approx 3.5$ m whereas the reflected rarefaction wave due to the expansion is located at $x \approx 2.5$ m. Daude couples the 1D model with 3D equation at the cells of abrupt change of area. But in the present work, the 1D model didn't consider the contribution coming from the corner and the wall, so the results are different from those in the references, mainly from the discontinuity at the abrupt change of the cross section. Due to the difference between the 1-D numerical results and reference, a 2-D computation with N = 5284 cells is also considered. The numerical profiles of the pressure and velocity at the center-line of the 2-D computations are plotted. We notice that the computation in 2-D has the better agreement with the reference.

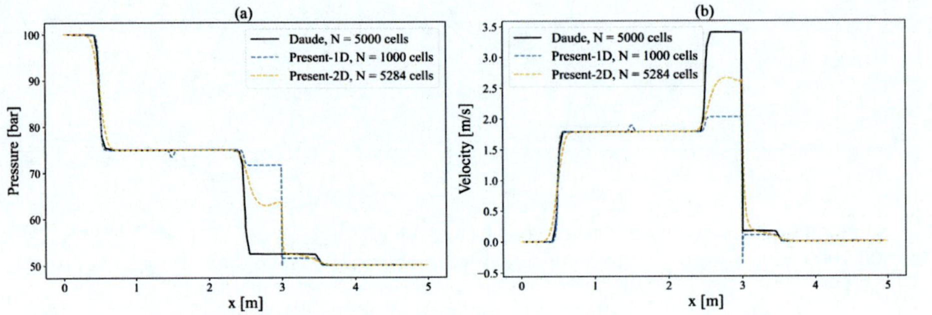

Fig. 4. Numerical solutions obtained on the liquid shock-tube interaction with an abrupt expansion at $t = 1$ ms with 1D-1000cells, 2D-5284cells; comparison with the results issued from Daude [9]: (a) pressure and (b) velocity.

3.3 Modified Shu-Osher Problem

The Shu-Osher problem has been widely used to assess the dissipation characteristics of numerical schemes. Here, the original Shu-Osher problem is extended to a supercritical thermodynamic condition presented by Terashima [10]. The EOS used in reference is SRK model. The working fluid is nitrogen, and the critical values are stated in the advection problem. The initial conditions are set as

$$(\rho, u, p) = \begin{cases} 3.8570\rho_0, M_{ref}c_{s,L}, 10.333p_0, & -5 \leq x \leq -4m; \\ \rho_0(1.0 + 0.2sin5x), 0.0, p_0, & -4 \leq x \leq 5m \end{cases}$$

where $\rho_0 = 50.0\,\text{kg/m}^3$, $p_0 = 40$ bar, $M_{ref} = 2.629, c_{s,L} = c^{EOS}(\rho_0, p_0)$.

The initial condition means a moving shock wave with Mach number M_{ref} intersects a density sinuous wave. 200 and 2000 grid points are used and 2001 grid points data from Terashima are plotted as the reference solution. Figure 5 shows the profiles of density and pressure. Firstly, the speed of sound in the reference is estimated using the ideal gas EOS while in the present work a real gas approach is adopted, which is more realistic. The shock wave propagates thus faster in the present case than the reference result. Secondly, the difference between peaks is mainly due to the dissipative character of the chosen solver. The present results using 2000 grid points are in good agreement with the reference solutions, showing that the present method can resolve the high-frequency waves behind the shock wave with reasonable grid points.

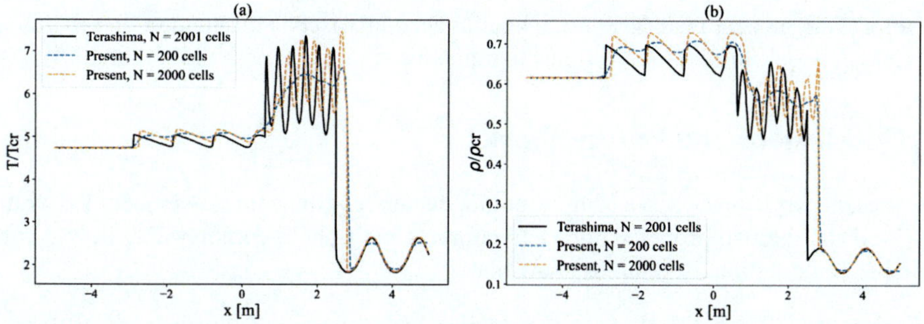

Fig. 5. Numerical solutions obtained of modified Shu-Osher problem at t = 5.903 ms with 200 and 2000 cells; comparison with the results issued from Terashima [10]: (a) T/T_{cr} and (b) ρ/ρ_{cr}.

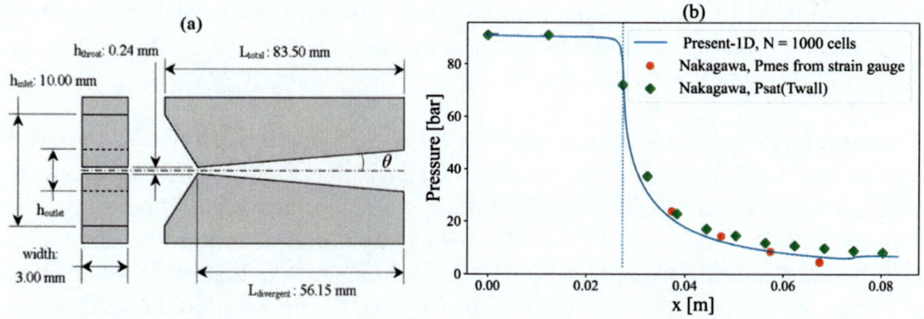

Fig. 6. (a) Schematic view of the converging-diverging nozzle ($\theta = 0.612°$) and (b) Pressure profile along the center-line of the nozzle compared with the pressure measurements by Nakagawa [14].

3.4 Converge and Diverge Nozzle

Convergence-divergence nozzles are an important component in two-phase CO_2 injectors, converting pressure energy into kinetic energy so that secondary flows can be entrained.

Here, 1D convergent-divergent nozzles with operating conditions similar to those of a two-phase CO_2 injector application are investigated. An experimental study of a simulated nozzle was carried out by Nakagawa [14]. The inlet conditions are $p = 91$ bar and $T = 309.65$K, while at the outlet, the pressure is fixed at 5 bar. The CO_2 is supercritical at the inlet and two-phase at the outlet. The total nozzle length was 83.50 mm and the divergence angle was 0.612°, as shown in Fig. 6 (a). The distribution of pressure along the center-line is reported in Fig. 6 (b) and compared with the direct pressure measurements from strain gauge (red circles) and the saturation pressure (green diamonds) corresponding from the temperature measurements of Nakagawa [14]. A small difference can be found after the throat. It may come from physical modelling, as the two-phase model we used in this article (here HEM) assumes that the two phases are in thermodynamic and mechanical equilibrium, without body forces and viscous effects.

Therefore, the present results do not take into account viscous dissipation and thermodynamic nonequilibrium, those effects that may be significant near the throat.

4 Conclusions and Future Works

The present work proposes a simple homogeneous equilibrium solver coupled with the CoolProp thermal property library to compute multiphase compressible flows. The available computational examples show that:

1. This method is capable of capturing changes in thermodynamic properties, allowing finer solutions to be obtained by refining the mesh and reducing dissipation;
2. Actual solver can be used to perform direct simulations of different fluids without the need to use the corresponding analytical EOS;
3. It has been initially demonstrated that the solver can obtain more realistic results for fluids in single-phase - two-phase - trans-critical - super-critical all flow states, and can be considered globally-valid for fluids.

The current solver has been extended to 2D and 3D models. However, the different call to the CoolProp library is computationally expense, and increases drastically with the number of cells in 2D and 3D. Taking the single-phase gas shock-tube (Test 1) as an example, for the case of 500 cells, the calculation time required in the present work using the CoolProp library is about 110 times longer than that using the ideal gas EOS. For more complex cases such as two-phase flow, the time required is even longer. Considering the computational economy and accuracy, the next step will be to develop a Look-up Table method based on the CoolProp library, taking into account the turbulence model and the viscous and heat transfer effects of the flow process. In addition, for the simulation of two-phase flow, favorable agreement is obtained compared to the reference results that equally apply the HEM. In the next step, we consider extending the solver to compressible non-equilibrium two-phase flows, which is more reasonable for the real case.

References

1. Qinghua Deng, Yu., Jiang, Z.H., Li, J., Feng, Z.: Condensation and expansion characteristics of water steam and carbon dioxide in a Laval nozzle. Energy **175**, 694–703 (2019)
2. Haida, M., et al.: System model derivation of the CO2 two-phase ejector based on the CFD-based reduced-order model. Energy **144**, 941–956 (2018)
3. De Lorenzo, M., Lafon, P., Di Matteo, M., Pelanti, M., Seynhaeve, J.-M., Bartosiewicz, Y.: Homogeneous two-phase flow models and accurate steam-water table look-up method for fast transient simulations. Int. J. Multiph. Flow **95**, 199–219 (2017)
4. Chiapolino, A., Boivin, P., Saurel, R.: A simple phase transition relaxation solver for liquid-vapor flows. Int. J. Numer. Meth. Fluids **83**(7), 583–605 (2017)
5. Böttcher, N., Taron, J., Kolditz, O., Liedl, R., Park, C.-H.: Comparison of equations of state for carbon dioxide for numerical simulations, p. 9 (2012)
6. Bell, I.H., Wronski, J., Quoilin, S., Lemort, V.: Pure and Pseudo-pure Fluid Thermophysical Property Evaluation and the Open-Source Thermophysical Property Library CoolProp. Ind. Eng. Chem. Res. **53**(6), 2498–2508 (2014)

7. Daude, F., Berry, R.A., Galon, P.: A Finite-Volume method for compressible non-equilibrium two-phase flows in networks of elastic pipelines using the Baer-Nunziato model. Comput. Methods Appl. Mech. Eng. **354**, 820–849 (2019)
8. Wood, A.B.: A Textbook of Sound. G. Bell and Sons LTD, London (1930)
9. Daude, F., Galon, P.: Simulations of single- and two-phase shock tubes across abrupt changes of area and branched junctions. Nucl. Eng. Des. **365**, 110734 (2020)
10. Terashima, H., Kawai, S., Yamanishi, N.: High-Resolution Numerical Method for Supercritical Flows with Large Density Variations, p. 15 (2011)
11. Sod, G.A.: A survey of several finite difference methods for systems of nonlinear hyperbolic conservation laws. J. Comput. Phys. **27**(1), 1–31 (1978)
12. Mathias, P.M., Naheiri, T., Oh, E.M.: A density correction for the Peng-Robinson equation of state. Fluid Phase Equilib. **47**(1), 77–87 (1989)
13. Tiselj, I., Horvat, A., Gale, J.: Numerical scheme of the waha code. Multiph. Sci. Technol. **20**(3–4), 323–354 (2008)
14. Nakagawa, M., Berana, M.S., Kishine, A.: Supersonic two-phase flow of CO2 through converging-diverging nozzles for the ejector refrigeration cycle. Int. J. Refrig. **32**(6), 1195–1202 (2009)

Reacting Flows

Numerical Simulations of Real-Fluid Reacting Sprays at Transcritical Pressures Using Multiphase Thermodynamics

Mohamad Fathi(✉), Stefan Hickel, and Dirk Roekaerts

Delft University of Technology, Delft, South Holland, Netherlands
{m.fathi,s.hickel,d.j.e.m.roekaerts}@tudelft.nl

Abstract. Transcritical fuel sprays form an indispensable part of high-pressure energy-conversion systems. Modeling the complex real-fluid effects in the high-pressure multiphase regime of such sprays accurately, especially the hybrid subcritical-to-supercritical mode of evaporation during mixing fuel and oxidizer, is essential and challenging. This paper represents a novel numerical framework for accurate and efficient simulations of transcritical sprays. The spray is modeled using a diffuse interface method with multiphase thermodynamics, which couples real-fluid state equations with vapor-liquid equilibrium (VLE) calculations to compute thermo-transport properties. A physically consistent turbulence model for large-eddy simulations (LES) is used, with combustion being modeled via real finite-rate chemistry based on the fugacity of the species. The current method is accurate and free from semi-empirical drop break-up/evaporation models. LES results for the Engine Combustion Network (ECN) Spray-A benchmark demonstrate the potential of the proposed method and its advantages over traditional approaches.

Keywords: transcritical injection · turbulent combustion · multiphase thermodynamics · vapor-liquid equilibrium

1 Introduction

We consider cases where combustion is initiated by auto-ignition during or after the injection of a cold liquid fuel in a pressurized chamber filled with a hot gas oxidizer. The chamber pressure is greater than the critical pressure of the fuel but less than the cricondenbar pressure of the fuel-oxidizer mixture. This operating pressure causes the injectant to cross the two-phase region due to mixing with the environment. Such turbulent reacting sprays at transcritical pressures have drawn significant attention due to the complexity of the physical and numerical modeling. Besides real fluid effects due to the high pressure, the possibility of coexisting saturated liquid and vapor phases further complicates the modeling of transcritical fuel sprays.

The most common approach for turbulent spray modeling is primarily based on droplet breakup/collision and evaporation in the context of the Lagrangian particle tracking (LPT) method [5]. However, experimental evidence indicates that transcritical fuel injection is a mixing-controlled process instead of being controlled by droplet

M. White et al. (Eds.): NICFD 2022, ERCOFTAC 29, pp. 169–177, 2023.
https://doi.org/10.1007/978-3-031-30936-6_17

breakup or finite-rate evaporation on droplet surfaces [12]. For this reason, transcritical injectors are also often modeled in the context of the Eulerian dense-gas (DG) method [6], which assumes a single-phase working fluid with properties modeled via a suitable (cubic, e.g.) equation of state (EOS). While the DG method can be justified for combustion systems working close to the supercritical conditions, its utilization as an accurate numerical framework for transcritical sprays is questionable, since excluding the effect of the subcritical phase separation may lead to nonphysical or ill-defined states when a part of the fluid passes the meta-stable boundaries at transcritical conditions [8].

An approach with higher fidelity, which we refer to as multiphase thermodynamics (MT) [4,9,11], is to include vapor-liquid equilibrium (VLE) calculations for the multi-component real-fluid mixture, consistent with the applied EOS. In the MT method, rapid flash calculations (phase-splitting calculations) provide information on the composition and amount of saturated vapor and liquid phases, based on which the thermo-transport properties are determined by suitable mixing rules for the multiphase co-existence regions. In the case of a single-phase mixture, the MT method provides results identical to the DG method.

This paper presents the fundamentals of MT-based large-eddy simulations (LES) of transcritical sprays. Particular attention is paid to evaluating the thermo-transport properties of the working fluid in the two-phase coexistence regions. In addition to quasi-one-dimensional examples, the performance of the LES-MT method compared to conventional methods is assessed for the ECN Spray-A benchmark experiment, where transcritical n-dodecane is injected into the pressurized combustion chamber with and without oxygen.

2 Main Research Issues

In the MT method, a multi-component mixture is considered as the working fluid for which the Navier-Stokes equations (NSE) are solved. Due to the transcritical conditions during the mixing, the working fluid might be thermodynamically unstable when it passes the two-phase boundaries. In that case, the required thermo-transport properties of the working fluid are estimated based on the output of the rapid flash calculations. While one can use the amounts of saturated vapor and liquid to determine the overall mixture's extensive thermodynamic properties, estimating the dynamic viscosity and thermal conductivity requires a structure model.

3 Methodology

We solve the reacting compressible NSE free from any ideal-gas assumption in their conservative form for a working fluid with real-fluid thermo-transport and chemical properties. The working fluid can generally be a single-phase or two-phase mixture according to its local composition and thermodynamic state. In this section, the MT-based governing equations for the modeling of the real-fluid effects of such working fluid are presented.

3.1 Multiphase Thermodynamics

The NSE are closed by suitable volumetric and caloric EOSs. Additional VLE calculations must be conducted to account for the coexistence of vapor and liquid phases. This phase split information is required for calculating pressure and temperature from the mixture's density, internal energy, and mass composition.

3.1.1 State Equations

The most popular EOS is the ideal gas (IG) law, but its applicability is limited to gaseous media at relatively high temperatures and low pressures. A cubic EOS is usually selected as a suitable compromise between accuracy, complexity, and computational costs at transcritical conditions. The volumetric EOS in the general form of cubic EOS reads

$$p = RT/(\overline{\vartheta} - b) - a/[(\overline{\vartheta} + \delta_1 b)\,(\overline{\vartheta} + \delta_2 b)] \quad \text{with} \quad \overline{\vartheta} = \rho/W, \tag{1}$$

where ρ and W are the density and molecular weight of the mixture, a and b account for the non-ideal behavior resulting from inter-molecule attractive and repulsive forces, and δ_1 and δ_2 are two extra calibration parameters. The molar specific internal energy of the mixture, $\overline{e}$, is computed via the corresponding caloric EOS as follows:

$$\overline{e} = \sum_{i=1}^{N} X_i \overline{h}_i^{\circ}(T) - RT + (a - T\partial a/\partial T)/[(\delta_2 - \delta_1)\,b] \ln[(\overline{\vartheta} + \delta_1 b)/(\overline{\vartheta} + \delta_2 b)], \tag{2}$$

where X_i is the mole fraction of species i. The first two terms account for the absolute internal energy of the mixture at the actual temperature but at the standard reference pressure, and the last term accounts for the internal energy change via an isothermal thermodynamic path from the standard reference pressure to the actual one. The molar specific enthalpy of species at standard pressure, $\overline{h}_i^{\circ}$, is approximated via standard NASA polynomials.

The most widely used cubic EOSs are formulated based on two constant parameters and therefore have the intrinsic limitation of a unique and universal critical compressibility factor. This results in a systematic error for the specific volume (or density) at conditions close to the critical point. To overcome this limitation, Cismondi and Mollerup [2] proposed a three-parameter cubic EOS. This so-called RKPR EOS includes the effect of the actual critical compressibility factor in the evaluation of the third parameter. The RKPR EOS is employed for our computations, a detailed rationale is given in Ref. [4].

3.1.2 Flash Calculations

Mixing at transcritical pressure might result in the coexistence of saturated vapor and liquid phases. We, therefore, solve additional VLE equations to determine the composition and amount of each phase. The VLE equations for a mixture with N components comprise N equations for species mass conservation plus $N + 2$ equations for equality of temperature, pressure, and partial molar Gibbs free energy, $\overline{g}_i$, in liquid (L) and vapor (V) phases:

$$m_i^L + m_i^V = m_i, \quad T^L = T^V, \quad p^L = p^V, \quad \overline{g}_i^L = \overline{g}_i^V, \tag{3}$$

where m_i denotes the mass of species i in the mixture. This set of $2N+2$ equations must be supplemented with two more constraints to uniquely determine the specific volume, temperature, and composition of the liquid and vapor phases. These two constraints define the type of flash problem. As we solve the conservative form of the compressible NSE, that is, transport equations for energy and density, isoenergetic-isochoric phase-splitting calculations, also known as UV-flash calculations, must be carried out. The two additional constraints for UV-flash calculations are

$$\overline{\vartheta} = (1-\alpha)\overline{\vartheta}^L + \alpha\overline{\vartheta}^V, \quad \overline{e} = (1-\alpha)\overline{e}^L + \alpha\overline{e}^V, \tag{4}$$

with α as the vapor mole fraction. To evaluate the required specific internal energies of the liquid and vapor phases, we use Eq. (2) for each phase separately, and to compute α, we use the total mass balance rewritten as

$$\alpha = (W - W^L)/(W^V - W^L). \tag{5}$$

To solve the VLE equations, we utilize the newly introduced method of Fathi and Hickel [3] that performs UV-flash calculations very fast and robust via Newton iterations with the exact Jacobian based on an effective reduction method. We note that tabulation methods are a suitable alternative for non-reacting flows. When the number of transported species is large, however, both memory requirements and table query time grow unfavorably, while the cost of the Fathi-Hickel reduction method remains low and essentially independent of the number of species. For a comprehensive review and practical implementation guidelines, readers are referred to Ref. [3].

3.2 Multiphase Transport

According to Chung et al. [1], thermal conductivity λ and dynamic viscosity μ of dense fluids are a complex function of their temperature, specific molar volume, molecular weight, acentric factor ω, and critical point information:

$$[\mu, \lambda] = f(T, \overline{\vartheta}, W, T_c, \overline{\vartheta}_c, \omega). \tag{6}$$

For a multi-component mixture, the molecular weight, critical temperature, critical molar specific volume, and acentric factor are obtained according to the composition. It is essential to note that the underlying assumption here is that the multi-component mixture is a single (stable) phase. Therefore, it would be questionable to compute the overall dynamic viscosity and thermal conductivity of a two-phase mixture via these correlations by passing the overall composition of the working fluid in the coexisting region. To correctly estimate the nominal transport properties of the working fluid, we first compute the transport properties of the saturated vapor and liquid phases via Chung correlations separately and then combine these values with a structural model to find the overall transport properties of the mixture.

For evaluating the transport properties of two-phase materials, we assume a random spatial distribution of the vapor and liquid phases within a finite-volume cell. Using the EMT model [13], the overall thermal conductivity of the mixture can be computed from its saturated liquid and vapor phase via

$$(1-\theta)(\lambda^L - \lambda)/(\lambda^L + 2\lambda) + \theta(\lambda^V - \lambda)/(\lambda^V + 2\lambda) = 0, \tag{7}$$

in which $\theta = \alpha\overline{\vartheta}^V / [\alpha\overline{\vartheta}^V + (1 - \alpha)\overline{\vartheta}^L]$ is the fraction of the volume occupied by the vapor phase. Similarly, one can compute the overall dynamic viscosity of the multi-component mixture in the two-phase region.

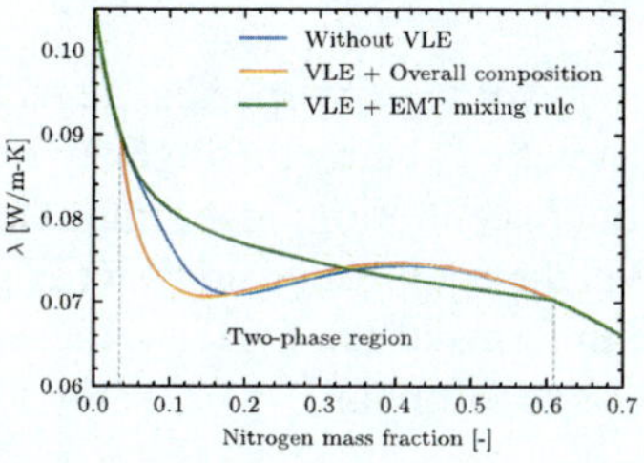

Fig. 1. Compositional functionality of the thermal conductivity (left) and dynamic viscosity (right) of n-dodecane and nitrogen mixture at 60 bar and 600 K using RKPR EOS with different inputs to Chung's method.

In Fig. 1, we compare different methods to compute the thermal conductivity and dynamic viscosity of binary mixtures of n-dodecane and nitrogen at a given temperature and pressure. All results are obtained with the RKPR EOS and Chung correlations, but with different inputs: form a dense-gas approach (no VLE calculations), from VLE calculations but using the overall (unstable in the two-phase region) composition, and from the proposed VLE-EMT model. The figure shows the proposed VLE-EMT model can reproduce transport properties free from any nonphysical oscillation of conductivity or undershooting of dynamic viscosity in the two-phase region.

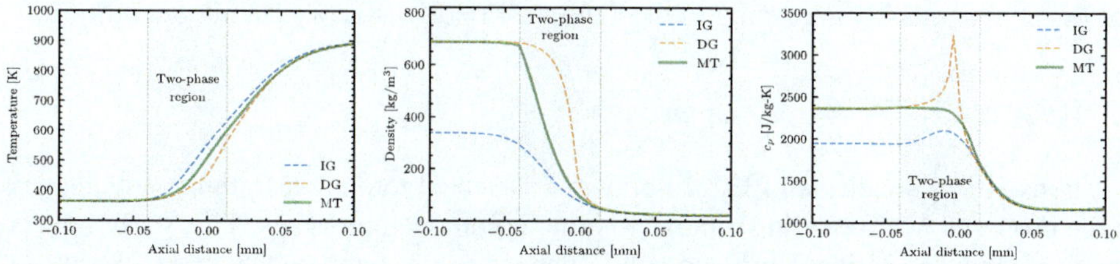

Fig. 2. The results of the MT method proposed in the current work are compared to conventional approaches of DG and GI approaches for a quasi-one-dimensional counter-flow example at ECN Spray A operating conditions.

3.3 Real-fluid Chemistry

Chemical reactions change the composition of a fluid mixture via the formation and destruction of species, which are expressed in the species mass balance equations by net mass production rates $\dot{\omega}_i$. At elevated pressures, real-fluid effects must be considered for the evaluation of this term. The species' production rates of N_r reactions can be expressed in a standard manner as

$$\dot{\omega}_i = W_i \sum_{r=1}^{N_r} (\nu_{ir}^P - \nu_{ir}^R)(k_{f,r} \prod_{j=1}^{N} C_j^{n'_{jr}} - k_{b,r} \prod_{j=1}^{N} C_j^{n''_{jr}}), \quad (8)$$

with C_j being for the concentration of species j. We use the species' fugacity for the evaluation of the concentrations $C_j \simeq f_j/(RT)$ with f_j being the fugacity of species j in the mixture. This fugacity-based evaluation of concentrations also simplifies the computation of the backward reaction coefficients via the equilibrium constant $K_c = k_f/k_b$. If the reaction rates are expressed as proposed in terms of the fugacity, the equilibrium constant can be computed similarly to the ideal-gas formula:

$$K_{c,r} = [p^\circ/(RT)]^{\nu_r} \exp[\Delta S_r^\circ/R - \Delta H_r^\circ/(RT)], \quad (9)$$

with $\nu_r = \sum_{i=1}^{N} (\nu_{ir}^P - \nu_{ir}^R)$ being the net change in the number of species in the reaction, and ΔH_r° and ΔS_r° being the reaction enthalpy and entropy net change at the standard atmospheric pressure.

3.4 Numerical Discretization

The NSEs are discretized in space by a conservative finite-volume scheme that uses a second-order central difference method for the diffusion terms and the adaptive local deconvolution method (ALDM) of Hickel et al. [7] for the inviscid fluxes. The van Albada limiter is utilized for the mass and energy flux reconstruction to avoid spurious under- or overshoots at sharp density gradients. A third-order SSP Runge-Kutta scheme is used for advection and diffusion, and a second-order Strang-splitting scheme is used to separate the stiff chemical reactions, see Fathi et al. [4] for details. The step size is set dynamically according to the Courant-Friedrichs-Lewy stability condition.

4 Results

The transcritical condition of ECN Spray-A is selected to demonstrate the advantages of the proposed MT modeling relative to other common approaches. This spray injects liquid n-dodecane (critical pressure of 18 bar) at a low temperature of 363 K into a preheated chamber with molar composition of 15 $\%O_2$, 6.23 $\%CO_2$, 3.62 $\%H_2O$, and 75.15 $\%N_2$, temperature 900 K and pressure 60bar. ECN Spray-A has a non-cavitating nozzle, which allows us to simplify the computational setup geometry. Note that the thermodynamic model can capture in-nozzle phase change and is applicable, e.g., to the C and D injectors.

We first demonstrate the crucial role of real-fluid and multiphase effects on the transcritical mixing of the fuel and oxidizer streams using a simple quasi-one-dimensional counter-flow example without chemical reactions. For a strain rate of 500 s$^-$1 close to the operating condition of Spray-A, Fig. 2 shows profiles of temperature, density, and heat capacity along the center line of the two opposite streams. The only difference between the DG and IG methods is the applied EOS; the RKPR EOS is used for the DG method. The MT method is based on the RKPR EOS with the aforementioned two-phase treatment. As our results indicate, the IG EOS yields an inaccurate density prediction and over-predicts the mixture's temperature. The deficiencies of the DG approach in the two-phase region manifest as a smoothed-out density solution and a non-smooth temperature profile due to the sharp nonphysical peak of isobaric heat capacity.

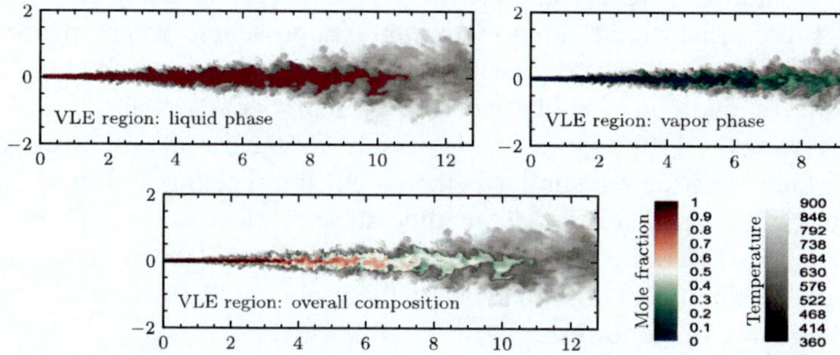

Fig. 3. LES-MT predictions of ECN Spray A. The background contour depicts the temperature field. In the two-phase region, the contours of the n-dodecane molar fraction of the saturated liquid, the saturated vapor, and the global mixture are shown.

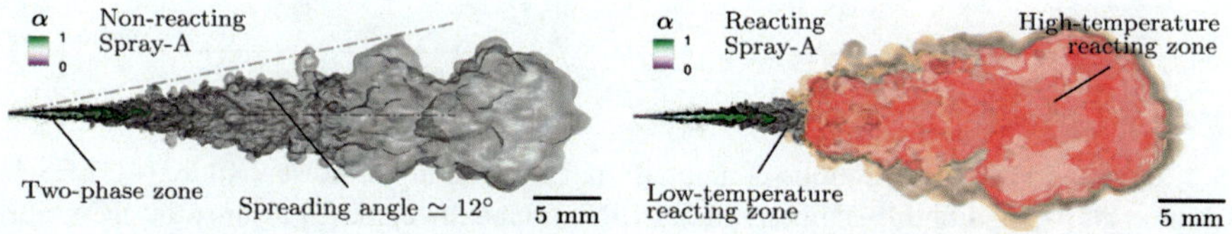

Fig. 4. LES-MT predictions of the inert (left) and reacting (right) ECN Spray A. Volume renderings at 800ţs after the start of injection.

For the LES-MT study of the ECN Spray-A, a multi-block structured grid consisting of 2864 blocks and 12.7 x 10^6 cells is generated using an adaptive mesh refinement technique. Because of the already high computational load of the multiphase thermodynamics method, a computationally efficient highly-calibrated 2-step reaction mechanism developed for the Spray-A is selected, following Hakim et al. [6].

The proposed MT method can evaluate the solubility of real-fluid mixtures at elevated pressure, as shown in Fig. 3. The figure highlights the two-phase region and shows the contours of the mole fraction of n-dodecane in the liquid, vapor, and overall mixture phases. The n-dodecane mole fraction is about 90 % in the liquid and 10 % in the vapor, which indicates the capability of the injectant to dissolve other species from the chamber gas.

Transcritical combustion typically initiates in low-temperature reacting zones, where the fluid state and transport properties deviate strongly from IG laws. Hence, making the assumption of the IG or even DG without accounting for local two-phase regions can lead to significant errors and uncertainties in transcritical combustion simulations. Volume renderings in Fig. 4 and quantitative data in Fig. 5 demonstrate that MT-LES yield accurate predictions for the spreading angle, liquid penetration lengths, vapor penetration lengths, ignition delay time, and flame lift-off length.

The LES-MT method is accurate but also has the advantage of being calibration-free. To judge its performance in relation to other approaches, we compare the vapor/liquid penetration trajectories via the LES-MT of the present work with LES-DG of Hakim et al. [6], and highly-calibrated LES-LPT of Gadalla et al. [5] with the experiments of Persson et al. [10]. Figure 5 shows results for the transient start-up interval of 0.2 ms in which we expect a similar behavior for the reacting and inert cases. The obtained results of the current LES-MT method are very close to the experimental measurements. The LES-LPT method was calibrated to match the experimental liquid length, which results in significant discrepancies in the vapor length prediction. Predictions of the DG method have the lowest accuracy in this case.

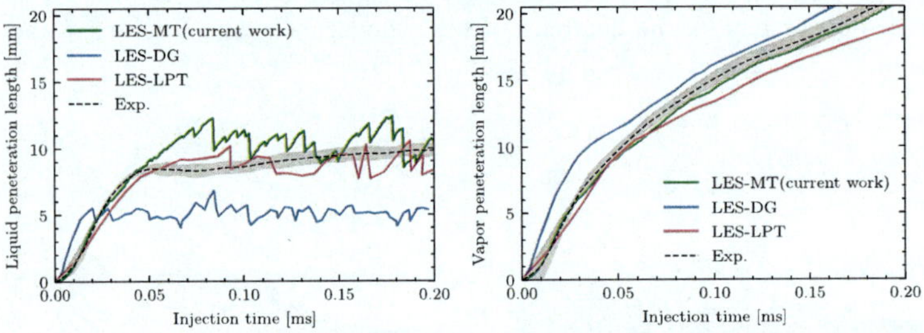

Fig. 5. Liquid and vapor penetrations computed via LES-MT in the present study [4], LES-DG of Hakim et al. [6], and highly-calibrated LES-LPT of Gadalla et al. [5] against the ECN Spray A experiment [10]

5 Conclusions and Future Work

A framework for large-eddy simulations with real-fluid multiphase thermodynamics (LES-MT) of transcritical sprays has been presented. This new methodology has been verified and validated for the transcritical ECN Spray-A at both inert and chemically reacting conditions. The predictions obtained with the proposed models are in better agreement with experimental data than results of traditional spray modeling methods from the literature.

References

1. Chung, T., Ajlan, M., Lee, L., Starling, K.: Generalized multiparameter correlation for non-polar and polar fluid transport properties. Ind. Eng. Chem. Res. **27**(4), 671–679 (1988)
2. Cismondi, M., Mollerup, J.: Development and application of a three-parameter RKPR equation of state. Fluid Phase Equilibria **232**(1–2), 74–89 (2005)
3. Fathi, M., Hickel, S.: Rapid multi-component phase-split calculations using volume functions and reduction methods. AIChE J. **67**(6), e17174 (2021)
4. Fathi, M., Hickel, S., Roekaerts, D.: Large eddy simulations of reacting and non-reacting transcritical fuel sprays using multiphase thermodynamics. Phys. Fluids **34**(8), 085131 (2022)
5. Gadalla, M., Kannan, J., Tekgül, B., Karimkashi, S., Kaario, O., Vuorinen, V.: Large-eddy simulation of ECN spray a: sensitivity study on modeling assumptions. Energies **13**(13), 3360 (2020)
6. Hakim, L., Lacaze, G., Oefelein, J.: Large eddy simulation of autoignition transients in a model diesel injector configuration. SAE Int. J. Fuels Lubricants **9**(1), 165–176 (2016)
7. Hickel, S., Egerer, C.P., Larsson, J.: Subgrid-scale modeling for implicit large eddy simulation of compressible flows and shock-turbulence interaction. Phys. Fluids **26**(10), 106101 (2014)
8. Ma, P.C., Wu, H., Banuti, D.T., Ihme, M.: On the numerical behavior of diffuse-interface methods for transcritical real-fluids simulations. Int. J. Multiphase Flow **113**, 231–249 (2019)
9. Matheis, J., Hickel, S.: Multi-component vapor-liquid equilibrium model for LES of high-pressure fuel injection and application to ECN Spray A. Int. J. Multiphase Flow **99**, 294–311 (2018)
10. Persson, H., Andersson, O., Kook, S., Lillo, P.M., Pickett, L.M.: Diesel spray ignition detection and spatial/temporal correction. SAE Int. J. Engines **5**(3), 1330–1346 (2012)
11. Qiu, L., Reitz, R.: An investigation of thermodynamic states during high-pressure fuel injection using equilibrium thermodynamics. Int. J. Multiphase Flow **72**, 24–38 (2015)
12. Siebers, D.L.: Scaling liquid-phase fuel penetration in diesel sprays based on mixing-limited vaporization. SAE Trans. **108**, 703–728 (1999)
13. Wang, J., Carson, J.K., North, M.F., Cleland, D.J.: A new approach to modelling the effective thermal conductivity of heterogeneous materials. Int. J. Heat Mass Transfer **49**(17–18), 3075–3083 (2006)

Experiments

Grid-Generated Decaying Turbulence in an Organic Vapour Flow

Leander Hake[1(✉)], Stefan aus der Wiesche[1], Stephan Sundermeier[1], Aurélien Bienner[2], Xavier Gloerfelt[2], and Paola Cinnella[3]

[1] Muenster University of Applied Sciences, 48565 Steinfurt, Germany
{leander.hake,wiesche,stephan.sundermeier}@fh-muenster.de
[2] Arts et Métiers ParisTech, 75013 Paris, France
{aurelien.bienner,xavier.gloerfelt}@ensam.eu
[3] Sorbonne Université, 75006 Paris, France
paola.cinnella@sorbonne-universite.fr

Abstract. An experimental and numerical investigation was carried out to study the quasi-homogeneous and isotropic decaying grid-generated turbulence in a subsonic organic vapour flow. The turbulent motion was measured by means of hot-wire probes which were located downstream of the grid and were operated in the constant-temperature mode. The working fluid was Novec 649, and the flow regime covered incompressible flow up to the inflow choking Mach number of the grid leading to downstream Mach numbers of order $M = 0.5$. High-fidelity compressible direct numerical simulations for Novec 649, modelled by real gas laws for typical CLOWT wind tunnel conditions (100 °C and 2 bar), were conducted to resolve details of the decay of homogeneous and isotropic turbulence (HIT) for Mach numbers 0.1 to 0.7. The numerical method using an in-house code were validated by the new experimental data. Exponents for the decay laws for turbulence intensity and length scales were determined.

Keywords: Isotropic Turbulence · Decay Laws · Compressible Flow · Organic Vapor · Choking Mach Number

1 Introduction

Following Davidson (2015), predicting the evolution of freely decaying, homogeneous turbulence is probably the most difficult problem in turbulence theory, because in that case only the "naked non-linearity" of the involved dynamics is present. Grid turbulence is the archetypal example of that theoretical ideal, and for essentially incompressible flows, a large body of literature has been established, see Kurian and Fransson (2009). However, there has been a longstanding debate as to whether the decay of the turbulent intensity *Tu* and the increase of the macro length Λ follow the Birkhoff-Saffman theory rather than the Loitsyansky-Batchelor theory. Whereas most of the literature works preferred Loitsyansky-Batchelor decay laws ($Tu \sim t^{-5/7}$ and $\Lambda \sim t^{2/7}$), there is theoretical and experimental evidence (Rotta (1972), Davidson (2015)) that grid turbulence might

M. White et al. (Eds.): NICFD 2022, ERCOFTAC 29, pp. 181–190, 2023.
https://doi.org/10.1007/978-3-031-30936-6_18

be also Birkhoff-Saffman turbulence ($Tu \sim t^{-3/5}$ and $\Lambda \sim t^{2/5}$). For engineering applications, the semi-empirical decay law $Tu \sim t^{-5/7}$ and $\Lambda \sim t^{1/2}$ was proposed by Roach (1986) on a large empirical database and is frequently considered for low-speed wind tunnel designing. In the scientific literature, a significant scattering of the experimentally obtained decay exponents is reported. That huge scattering can be traced back to the difficult problem of the correct determination of the virtual origin in grid experiments and/or to principal measurement uncertainties due to the need of hot-wire correction schemes, see Smolyakov and Tkachenko (1983).

Very little is known regarding compressibility effects and homogeneous isotropic turbulence (HIT) decay laws. The few available papers dealing with compressible flow of an ideal gas downstream of a turbulence grid, i.e., Zwart et al. (1997), Briassulis and Andreopoulos (1996), and Briassulis et al. (2001), reported contrary results and obtained, in parts, relatively small exponents for the decay laws. In the case of compressible flow, it is usually assumed that compressibility would substantially alter the turbulence dynamics only at very high turbulent Mach numbers, but "making measurements is made much more difficult by the presence of temperature or densities fluctuations (Bradshaw (1971))." For non-ideal gases like organic vapours consisting of complex molecules, only few direct numerical simulations have been carried out in the dense gas regime at very high turbulent Mach numbers (Sciacovelli et al. (2016), Giauque and Corre (2016)). Since our routinely employed turbulence models in computational fluid dynamics (CFD) analysis are often directly related to assumed HIT decay laws, their investigation is not only of academic but also of substantial practical engineering relevance. Since CFD methods has become a standard tool for the design and the optimization of organic Rankine cycle (ORC) power systems, the lack in our knowledge about HIT decay laws in compressible organic vapours is a severe one.

2 Experimental Setup

The present study employed a continuously running closed-loop organic vapor wind tunnel test facility which is shown in Fig. 1 and described in more detail by Reinker et al. (2019). After the settling chamber with flow straightener and four screens, the flow was accelerated in the first nozzle and entered then the test section tube where a second contraction zone and the turbulence grid were mounted. Roach (1986) provides a guideline for conducting grid turbulence studies which was considered here. This includes, among other items, the requirement of a low background turbulence intensity of the wind tunnel (which was here of order 0.1% without turbulence grid) and suitable large grid dimensions or a sufficiently small grid mesh size m in relation to the dimensions of the test section. The high-speed test section cross section was 50×100 mm, and a mesh size $m = 4$ mm with a grid rod diameter $d = 1$ mm were selected to achieve homogeneous flow fields downstream of the grid in the central part of the measurement zone.

A critical item in compressible flow experiments is the choking Mach number M_{ch} of the grid. The relevance of the choking effect on the results of compressible turbulence grid experiments was not considered by Zwart et al. (1997) or Briassulis et al. (2001). That issue opens partially their reported high-speed results to criticism. The choking

Mach number can be estimated by a relation (see for details aus der Wiesche et al. (2021))

$$\frac{A_{open}}{A_{total}} = \mathrm{M}_{ch}\left(\frac{2 + (\gamma_{pv} - 1)\mathrm{M}_{ch}^2}{\gamma_{pv} + 1}\right)^{-\frac{\gamma_{pv}+1}{2(\gamma_{pv}-1)}} \quad (1)$$

In the present case ($d = 1$ mm and $m = 4$ mm), the area ratio was $A_{open}/A_{total} = 9/16$ leading to a choking Mach number of about $M_{ch} = 0.25$ (for Novec 649 with a suitable isentropic exponent γ_{pv}). By applying higher pressure ratios with the compressor of the wind tunnel, it was possible to achieve higher downstream Mach numbers M of order $M = 0.5$, but when the choking limit was achieved, strong shock phenomena and additional peaks in the power density spectra were observed downstream of the grid. Due to the restrictions of the test section dimension, grids with higher porosity and higher choking Mach numbers could not be used without affecting the homogeneity of the downstream flow. Furthermore, the high density of the organic vapour led to high dynamics loads, and finer grids with smaller wire diameters and porosity were found to be structurally unstable or even seriously damaging. The working fluid was Novec 649 (at 100 °C and 2 bar), and some experiments were also conducted with atmospheric air for testing purposes. The compressibility factor Z during the Novec 649 tests was in the range of 0.90.

Due to the high dynamic loads, and the difficulty to conduct reliable organic vapour experiments at elevated temperature and pressure levels, fixed hot-wire probe positions were chosen in a staggered probe arrangement at three different distances $x/d = 80$, $x/d = 109$, and $x/d = 157$. That limited range of measurement positions was caused by the constraints of the basic test section and the use of a diffuser, see Fig. 1.

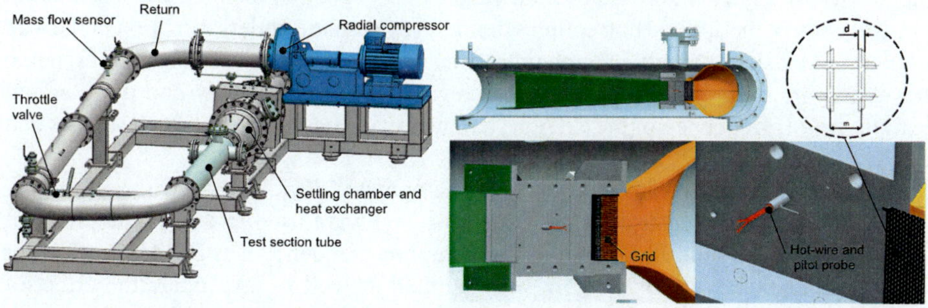

Fig. 1. Schematics of the wind tunnel test facility used for the present experiments and details regarding the test section and the turbulence grid including hot-wire probe instrumentation

The Mach number level was mainly controlled by the compressor running speed and by the filling level of the working fluid. During the individual in-situ calibration of the hot-wire probes, a miniaturized Pitot probe which was integrated in the probe stem was used. In the settling chamber, the static and the total pressure with a total uncertainty of $\Delta p/p = 0.7\%$) and the total temperature (with a total uncertainty of $\Delta T/T = 0.18\%$) were measured. The total temperature was varied by controlling the electrical heating power of the wind tunnel and by a variation of the coolant flow rate in the chiller. Due to the control system, it was possible to achieve a relatively steady state temperature level with a temperature drift of about 10^{-3}K/s. In the test section, both the static and total pressures were measured. The mass flow rate was measured independently through a flowmeter in the return of the wind tunnel (with maximum uncertainty level of 1.6–4.5% depending on the actual operation conditions). The thermodynamic properties were evaluated based on pressure and temperature measurements using REFPROP data base.

For the constant-temperature anemometer (CTA) measurements, two different hot-wire probes (single wire, gold-plated tungsten wires) were used for the investigations. Whereas the larger wire probes (wire diameter 10 μm, wire length 4 mm) were already successfully applied in recent studies (Reinker and aus der Wiesche (2021), Hake et al. (2022)), smaller probes with wire diameter 5 μm and wire length of 1 mm were employed in the present campaign, too. The use of two different wire sizes with the same electrical resistance enabled an independent test of the applied correction scheme for the finite wire effects. In high-Reynolds number flows, the dissipation length is typically much smaller than wire lengths, and hence a complex correction scheme must be used as discussed in detail by Smolyakov and Tkachenko (1983). Since the employed correction scheme based on certain assumptions (local isotropy and use of the Pao spectrum), the wire size study offered a chance to validate the correction scheme applied to the turbulent quantities. The large probes had a wire length of 4 mm which was of the same level as the macro length scale Λ or the mesh size m. Obtaining turbulent fluctuations with such large probes underestimate substantially the small-scale contributions making a large corrector factor necessary. The finer probes with 1 mm wire length represented an improvement regarding the resolution of small-scale effects. The ratio between the measured turbulence intensity Tu of the two probes was nearly constant, and its value (0.91 ± 0.02) was close to the theoretical prediction (0.88) based on the correction scheme (see Smolyakov and Tkachenko (1983)). This good agreement was considered as justification of the applied correction scheme. A systematic underestimation of the true turbulence intensity of about 7% (relative error) was predicted by the correction scheme for the finer hot-wire probes. A correction regarding high-frequency contribution and the finite cut-off frequency of the CTA system was made based on square-wave tests and investigations of the CTA transfer function.

For the electrical circuit, two different anemometer devices (one with an internal 5-kHz-low-pass filter and one without such a filter) were used to assess the effect of signal filtering due to the anemometer. The sampling frequency was 50 kHz, but higher sampling frequencies up to 200 kHz were applied as well to investigate the effect of aliasing for higher velocity levels. All samples were taken for 10 s; such a long measurement time in combination with a high sampling frequency ensured a negligible precision error. All data were digitalized and post-processed by means of MATLAB routines.

In the present work, a detailed hot-wire calibration process involving the determination of all relevant sensitivity coefficients was required for grid turbulence experiments with compressible organic vapours flows. The probes were calibrated in the empty test section, without a turbulence grid and a low background turbulence level, and in-situ downstream of the grid. The calibration process is described in detail by Hake et al. (2022) and based on a King equation with coefficients which are in fact functions of the

Mach number and the total temperature. Due to the high Reynolds number level and the high overheat ratio, the sensitivity coefficients for velocity and density were found to be essentially identical. Hence, the probes behaved like simple mass-flux sensors as in the case of incompressible flow. Since the isentropic exponent γ_{pv} was close to unity during the present experiments, the level of turbulent temperature and density fluctuations was negligible.

3 Numerical Method

The experimental investigation of small-scales in grid-generated turbulence is notoriously challenging, and high-fidelity numerical simulations offer opportunities to resolve details up to the smallest scales. Some general remarks about the use of numerical simulations for studying HIT can be found in Davidson (2015). The connection between the simulated decay in time and the kinetic energy decrease of the grid-turbulence with distance was established by the Taylor hypothesis. In the present study, the compressible homogeneous isotropic turbulence decay simulation was performed on a cubic computational domain of dimensions $[0; 2\pi L_{ref}]^3$ where L_{ref} was taken as 3 Λ with Λ being the integral length scale (see below). An in-house finite-difference code was used to solve the compressible Navier-Stokes equations. The inviscid fluxes were discretized by means of 10th-order standard centred differences whereas a 4th-order scheme was used for the viscous fluxes. This scheme was supplemented with a 10th-order selective filtering to eliminate grid-to-grid unresolved oscillations. The sub-scale motions are modelled using the Smagorinsky model, with Smagorinsky constants of $C_s = 0.18$ and $C_i = 0.11$. A four-stage Runge-Kutta algorithm was used for integration. Periodic boundary conditions were imposed in the three Cartesian directions of the computational domain. Concerning the fluid, the organic vapour Novec 649 was modelled through the Peng-Robinson-Stryjek-Vera (PRSV) equation of state (Stryjek et al. (1986)) and the Chung-Lee-Starling model (Chung et al. (1988)) for the transport properties. Indeed, the thermodynamic conditions selected ($P = 2.0$ bar,

$T = 373.15$ K and $\rho = 23.55$ kg/m^3) divided by the critical values leads to $P/P_c = 0.11$, $T/T_c = 0.84$ and $\rho/\rho_c = 0.04$, which corresponds to relatively dilute conditions. Therefore, in these conditions, the fluid is perfectly well modelled by the cubic equation of state.

In order to initialize the turbulence, one must provide a spectrum of isotropic turbulence. Here, the von Kàrmàn-Saffman spectrum

$$E(k) = \alpha \frac{u^2}{k_e} \frac{(\frac{k}{k_4})^4}{\left(1 + (\frac{k}{k_4})^2\right)^{17/6}} \exp\left(-a(\frac{k}{k_\eta})^b\right) \tag{2}$$

was used. In Eq. (2), the model constants were set to $a = 3C_k/2$ and $b = 2$ with $C_k = 1.613$ as the universal constant of the Kolmogorov law. Furthermore, $\alpha = 1.453$ and $k_e = 0.747/\Lambda$ were taken. To initialize the simulation, a stochastic turbulent field was generated in the physical space and then transposed into the Fourier space. The velocity field generated in the Fourier space was modified so that the energy spectrum associated to it fitted the reference spectrum of energy given. Thanks to an inverse Fourier-transform, the velocity field in the physical space was obtained. In the Fourier space, the Poisson problem was solved to initialize the pressure field.

4 Results and Discussion

In a first set of experiments, the reliability of the actual turbulence grid set-up and its instrumentation and data reduction method was assessed, and exponents for the decay laws were experimentally obtained and compared with available literature data. In a second set, the numerical method was validated.

The decay of the turbulence intensity *Tu* is shown in Fig. 2.a, and Fig. 2.b compares values for the decay exponent regarding the variance of the turbulent velocity field. The decay of *Tu* for experiments with air at essentially incompressible flow conditions was in excellent agreement with the correlation $Tu = 0.8\ (x/d)^{-5/7}$ proposed by Roach (1986). Systematically higher values for *Tu* and a somewhat weaker decay were observed for the organic vapour Novec 649 at higher Mach numbers (M about 0.2). Interestingly, the observed decay trend agreed reasonably well with compressible flow measurements from Zwart et al. (1997), see Fig. 2 (left). For comparing grid turbulence decay exponents, it is useful to consider the choking effect. Hence, the exponents are plotted against the normalized Mach number in Fig. 2 (right). The large scattering of literature data can be seen in Fig. 2. The Novec 649 result ($2n = -1.42 \pm 0.05$) for the lower Mach number agreed very well with the theoretical Batchelor turbulence value ($-10/7 \approx -1.4286$), and the values (-1.06 ± 0.05 and -1.12 ± 0.05, respectively) found at higher Mach numbers were very close to the value (-1.13 ± 0.02) reported by Krogstad and Davidson (2010) indicating Saffman turbulence (the systematic deviation to the theoretical value $-6/5 = 1.20$ can be explained by a weak temporal decay of the dimensionless energy dissipation coefficient). However, the present high-speed results are also compatible with a decay exponent $2n = -1$, which was obtained by Kistler and Vrebalovich (1963) at very high Reynolds numbers. Due to the high density of Novec 649 and the high velocity, similar high Reynolds numbers were achieved in the present experiments. The compilation of literature data shown in Fig. 2 (right), indicates a rough trend to lower decay exponent values with increasing normalized Mach number (the super-critical flow data ($\mathrm{M}/\mathrm{M}_{ch} = 1.24$) reported by Zwart et al. (1997) must be discussed separately).

To validate the numerical method, a first decay simulation for M = 0.10 was performed using for initialisation the measured integral length $\Lambda = 2.8$ mm and a turbulent Mach number of 0.003 (corresponding to the measured turbulence intensity of order 3% at $x/m = 20$). For that relatively low Mach number level, reliable experimental data, indicating a high degree of isotropy (skewness = 0.0623, kurtosis = 3.068), were available. For higher Mach numbers, the isotropy was destroyed by shock wave phenomena since the choking limit of the grid was achieved. Figure 3 shows an instantaneous view of the simulated velocity field in the numerical simulation. The comparison between the measured and simulated decrease of the turbulent intensity is shown in Fig. 3, too. The value at a distance $x/m = 20$ was selected as reference for the turbulence intensity to compare results for air and Novec 649 in the same diagram.

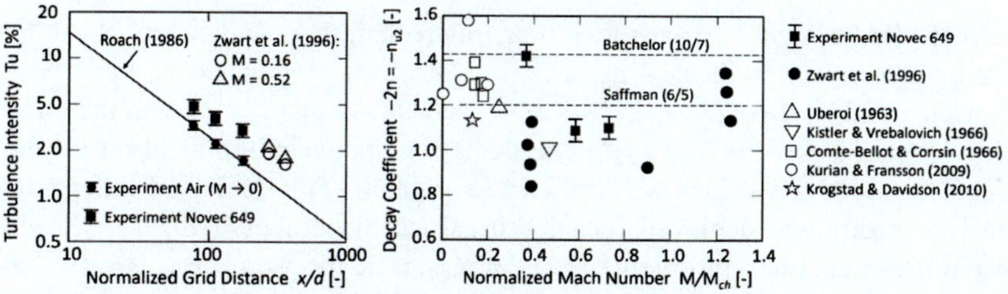

Fig. 2. Decay of the turbulent intensity Tu against grid distance (left) and decay exponents (right)

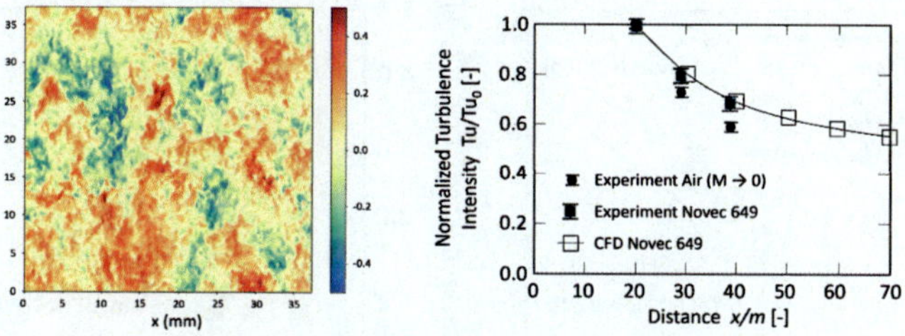

Fig. 3. Instantaneous view of the simulated velocity field (left) decay of the turbulence intensity (right)

The experimental data obtained for Novec 649 were in excellent agreement with the numerical prediction. For a sufficiently long time after the initialisation, the numerical method predicted a decay exponent of $n = -0.5$. This value was in a reasonable agreement with the data shown in Fig. 2 and correspond to the canonical exponent $2n = -1$ for the velocity variance. Interestingly, the decay for air (M → 0) was found to be slightly stronger, and the experimentally determined data points were systematically below the numerical curve in Fig. 3. This behaviour was already indicated by the decay exponent data shown in Fig. 2.

Assuming an invariant large-scale structure for the decaying turbulence, the exponent n for the turbulence intensity Tu is related to the exponent n_Λ for the macro length Λ by the conditions $n = -(\sigma + 1)/(\sigma + 3)$ and $n_\Lambda = 2/(\sigma + 3)$, Rotta (1972). Assuming $2n = 1$ would correspond to $\sigma = 1$ and this predicts $n_\Lambda = 1/2$ for the growth of the macro length. That exponent was already assumed by the correlation suggested by Roach (1986). In Fig. 4 (left), the experimentally observed increase of the macro length Λ is plotted and compared with the Roach correlation. The present high-speed results were systematically higher than the prediction of the Roach correlation, but the increase was compatible with the assumption of $\Lambda \sim (x/d)^{1/2}$. Interestingly, the same macro lengths were for moderate subsonic flow up to the choking Mach number. Using the value $\Lambda/d = 2.8$ for initialization of the numerical simulation, the relation between the macro and the Kolmogorov lengths were calculated and compared with experimentally obtained values in Fig. 4 (right). A good agreement was found, validating numerical method once again. The numerical simulations covered longer decay times than achievable by the present experimental test assuming the Taylor hypothesis, but the predicted trend, Fig. 4 (right), agrees well with the experimental data obtained for smaller lengths.

The three length scales (macro length Λ, micro length λ, and Kolmogorov length η) obeyed reasonably well the scaling laws for isotropic turbulence (i.e., $\lambda/\Lambda \sim 15^{1/2}$ $\mathrm{Re}^{-1/2}$, $\eta/\lambda \sim 15^{-1/2}\,\mathrm{Re}^{-1/4}$, $\eta/\Lambda \sim \mathrm{Re}^{-3/4}$, see Davidson (2015)) with the turbulent Reynolds number $\mathrm{Re} = Tu\ U\ \Lambda\rho/\mu$. The deviations between the ideal ratios and the experimentally obtained values was only a few percent. This indicated independently that nearly isotropy was achieved for subcritical Mach numbers ($M/M_{ch} \leq 0.98$) for the present grid-generated turbulence. For $M/M_{ch} > 1$, the isotropy was substantially disturbed downstream of the choked grid.

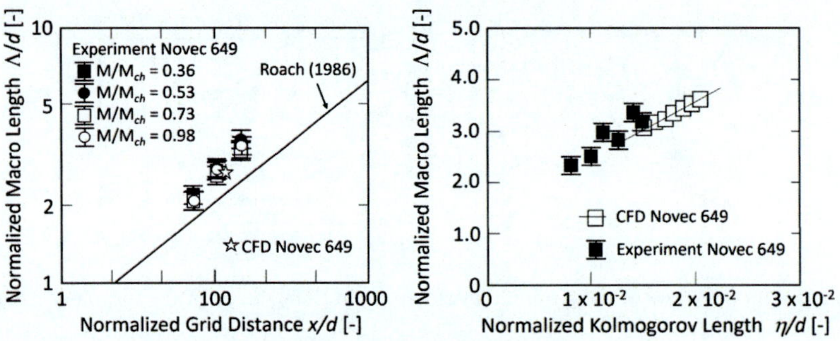

Fig. 4. Increase of the macro length Λ (left) and relation between macro length and Kolmogorov length (right)

5 Conclusions

Experimental and numerical investigations of the grid-generated turbulence decay indicated that by a decay exponent of $n = -1/2$ the results confirm each other. Furthermore, the turbulence decay in the Novec 649 atmosphere behaves differently than in air. Due to the smaller decay exponent, the turbulence intensity in Novec 649 decreases more slowly than in the air atmosphere. The measurements in the air atmosphere show a very good match with the literature results of Roach, who determined a decay exponent of $n = -5/7$. As already by the decaying turbulence findings, the comparison of the different length scales approaches also shows a good agreement with the results of the CFD simulation. Additionally, it is shown that the turbulence level Tu as well as the macro (integral) lengths Λ was higher in the Novec 649 atmosphere experiments.

For the higher flow speed ranges ($M/M_{ch} > 1$), it was observed that the isotropy was significantly disturbed downstream of the choked grid. The occurring disturbances due to reaching the grid choking Mach number showed up as high peaks in the power density spectra of the measurements. Finally, it can be said that the investigations regarding grid-generated isentropic turbulence decay in flow speed ranges above the choking Mach number of the used grid, is still a great challenge.

NOMENCLATURE

d	Rod diameter of the grid, m
E	Energy spectrum, -
k	Wave number, 1/m
m	Mesh size, m
t	Time, s
Tu	Streamwise turbulence intensity, -
U, u	Velocity, m/s
X	Distance, m
Z	Compressibility factor, -
η	Kolmogorov length, m
λ, Λ	Micro/macro (integral) lengths, m
σ	Decay law exponent parameter, -
γ	Isentropic exponent, -
M	Mach number, -
Re	Reynolds number (Re = $Tu\ U\ \Lambda\ \rho/\mu$), -

Subscript

ch	choking

Acknowledgement. This work was part of the Franco-German research project "REGAL-ORC" funded by ANR and DFG.

References

Aus der Wiesche, S., Hake, L., Cinnella, P.: Experimental and numerical assessment of an asymptotically correct choking Mach number expression for organic vapor flows through turbine cascades. Proceeding ORC 2021 Munich, Germany, paper-ID 83 (2021)

Bradshaw, P.: An Introduction to Turbulence and its Measurement. Pergamon, Oxford (1971)

Briassulis, G., Andreopoulos, J. (1996). Compressibility effects in grid generated turbulence. AIAA Meeting Papers, A9636509, NAGI-590, AIAA Paper 96–2055

Briassulis, G., Agui, J.H., Andreopoulos, J.: The structure of weakly compressible grid-generated turbulence. J. Fluid Mech. **432**, 219–283 (2001)

Chung, et al.: Origin of gases hydrocarbons in subsurface environments: theoretical considerations of carbon isotope distribution. Chem. Geol. **71**, 97–103 (1988)

Davidson, P.A.: Turbulence. 2nd edn. Oxford University Press, Oxford (2015)

Giauque, A., Corre, C.: Direct numerical simulations of homogeneous isotropic turbulence in a dense gas. In: Proceedings 1st Symposium NICFD, Varenna, Italy (2016)

Hake, L., et al.: Hot-wire anemometry in high subsonic organic vapor flows. In: Proceedings ASME TurboExpo, Rotterdam, The Netherlands, paper GT2022–81686 (2022)

Kurian, T., Fransson, J.H.M.: Grid-generated turbulence revisited. Fluid Dyn. Res. **41**, 021403 (32 pages) (2009)

Reinker, F., Kenig, E. Y., aus der Wiesche, S.: Closed loop organic vapor wind tunnel CLOWT: commissioning and operational experience. In: Proceeding ORC 2019 Athens, Greece, paper-ID 47 (2019)

Reinker, F., aus der Wiesche, S.: Application of hot-wire anemometry in the high subsonic organic vapor flow regime. In: Pini, M., De Servi, C., Spinelli, A., di Mare, F., Guardone, A. (eds.) NICFD 2020. ERCOFTAC Series, vol. 28, pp. 135–143. Springer, Cham (2021). https://doi.org/10.1007/978-3-030-69306-0_14

Roach, P.E.: The generation of nearly isotropic turbulence by means of grids. Heat Fluid Flow **8**(2), 82–92 (1986)

Rotta, J.C.: Turbulente Strömungen. Teubner, Stuttgart (1972)

Sciacovelli, et al.: DNS of turbulent flows of dense gases. In: Proceedings 1st Symposium NICFD, Varenna, Italy (2016)

Smolyakov, A.V., Tkachenko, V.M.: The Measurement of Turbulent Fluctuations. Springer, Berlin. (1983). https://doi.org/10.1007/978-3-642-81983-4

Stryjek, R., et al.: PRSV: an improved peng-Robinson equation of state for pure compounds and mixtures. Can. J. Chem. Eng. **64**(2), 323–333 (1986)

Zwart, P.J., Budwig, R., Tavoularis, S.: Grid turbulence in compressible flow. Exp. Fluids **23**, 520–522 (1997)

Study on the Operation of the LUTsCO2 Test Loop with Pure CO_2 and $CO_2 + SO_2$ Mixture Through Dynamic Modeling

Giuseppe Petruccelli(✉), Teemu Turunen-Saaresti, Aki Grönman, and Afonso Lugo

Lappeenranta-Lahti University of Technology LUT, Lappeenranta, Finland
{Giuseppe.Petruccelli,Teemu.Turunen-Saaresti,Aki.Gronman, Afonso.Lugo}@lut.fi

Abstract. Within the framework of the Horizon 2020 DESOLINATION (DEmonstration of concentrated SOLar power coupled wIth advaNced desAlinaTion system in the gulf regION) project, CO_2-based mixtures will be used as the working fluid of the power cycle coupled with the desalination plant. Desalination technologies require temperatures above 50 °C, therefore a fluid with a critical temperature above 70 °C is required to perform the compression step in the liquid phase, minimizing compression work and increasing cycle efficiency. Blending CO_2 with other fluids, such as SO_2, leads to an increase in the critical temperature of the mixture and makes it suitable for transcritical cycles in concentrated solar power applications, while preserving the advantages of pure CO_2 over steam cycles. Throughout the DESOLINATION project, CO_2 blends will be tested in the LUTsCO2 test loop to provide validation data for the design of heat exchangers. In this work, the adaptation of the test loop to the CO_2 blend is investigated. A dynamic model of the test loop is built in MATLAB-Simulink, which allows each component to be modeled independently and to realize a closed-loop cycle model coupled with a controller and real gas property tables. Steady-state simulations of the system are performed and the dynamic model of each component is verified with the design values. As a result, the study highlights the key performance and fluid dynamic differences which arise from the use of a CO_2 blend instead of pure CO_2.

Keywords: CO2 mixtures · Transcritical refrigeration cycle · Dynamic model · Simulink

1 Introduction

In recent years, supercritical carbon dioxide (sCO_2) power cycles have been considered one of the best alternatives to steam cycles, because of their higher efficiencies, compactness, and simpler layout of the system. Recent research has shown the advantages of s CO_2 cycles in terms of lower cost and better performance in waste heat recovery (Uusitalo et al. 2018), nuclear (Dostal et al. 2006), and concentrated power plants (CSP) (Crespi et al. 2020) applications. The development and demonstration of innovative

M. White et al. (Eds.): NICFD 2022, ERCOFTAC 29, pp. 191–200, 2023.
https://doi.org/10.1007/978-3-031-30936-6_19

CO_2 mixture power cycles for CSP plants, desalination and their coupling are some of the main objectives of the DESOLINATION project, an EU Horizon 2020 project. CO_2 blends are chosen as the working fluid for the innovative power cycle to improve the performance of the cycle (Manzolini et al., 2019; Crespi et al. 2022), turning the supercritical cycle into a transcritical one, and allowing an efficient coupling with desalination technologies that require a working fluid with a critical temperature above 70 °C. The key point of operating with a transcritical cycle is the compression of the fluid in liquid phase, where less compression work is required and the overall cycle efficiency increases. The high working temperatures of the power cycle (up to 700 °C) raise the issue of thermal stability of the mixture and material compatibility. Under these thermodynamic conditions, sulfur dioxide (SO_2) was selected as a promising working fluid to be adopted in power cycles (Morosini et al. 2022), and material and thermal stability tests are currently being conducted by the DESOLINATION project partners.

The LUTsCO2 test loop is a closed transcritical refrigeration cycle conceived and designed to study the behavior of pure CO_2 in the supercritical region, where fluid behavior is highly non-ideal, focusing on heat transfer across the Widom line (Petruccelli et al. 2021). The adaptation of the LUTsCO2 test loop to CO_2-SO_2 mixtures is investigated using dynamic modeling, which is a powerful tool for studying the steady-state and transient performance of sCO_2 cycles (Song et al. 2021), and MATLAB-Simulink was chosen to perform the dynamic simulations and assess the control of the system under different operating conditions and working fluids. In this paper, we first describe the experimental test loop and the mathematical models applied for each component, and then present the verification results based on the design conditions and the results obtained in off-design conditions for three SO_2 molar fractions: 0%, 25% and 40%.

2 LUTsCO2 Test Loop

The experimental test loop described in this paper consists of a closed-loop transcritical refrigeration cycle whose Simulink configuration is shown in Fig. 1, together with its controller, and the thermodynamic cycle in the T-s diagram in Fig. 2. The test rig includes two reciprocating compressors, that provide the required mass flow rate and compress the fluid up to 130 bar. The flow rate can be controlled by a frequency converter installed on one of the compressors. The hot gases leaving the compressors are cooled in an internal heat exchanger and then expanded by an expansion valve to an intermediate pressure close to the critical pressure, at which the fluid is first heated by an electric circulation heater and then cooled to the liquid-like state by a water-cooled gas cooler. The heat transfer occurring at the intermediate pressure is significantly affected by the fluid properties variations occurring in the critical point surroundings. Finally, the fluid is further expanded through a second expansion valve into the two-phase region at the desired pressure, corresponding to the compressor inlet pressure, before being completely vaporized and superheated in the evaporator. Cycle operation is controlled by PID controllers that keep the controlled variable close to the desired values.

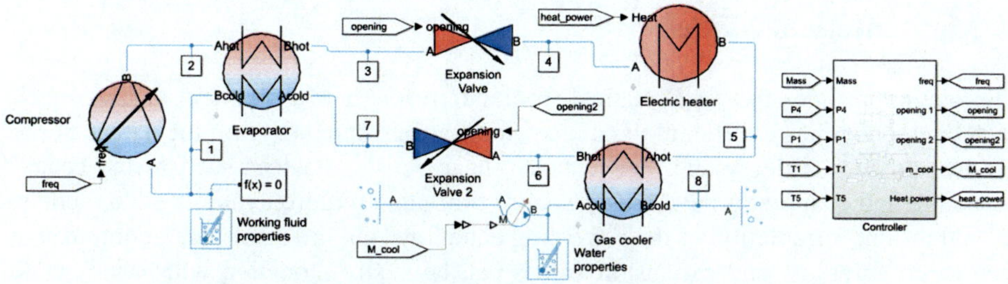

Fig. 1. LUTsCO2 test loop layout and Simulink model with controller

3 Thermodynamic Properties of CO_2-SO_2 Mixtures

The fluid is highly non-ideal near the critical point, and an equation of state for real gases is needed to properly define the fluid. Therefore, the thermodynamic and transport properties are calculated by the software using look-up tables extracted from the NIST REFPROP database (Lemmon et al., 2018). An in-house code was developed to generate MATLAB files suitable for implementation in Simulink. Look-up tables are generated for various molar fractions of SO_2, with a resolution of 0.26 bar in pressure and a variable resolution in temperature with an enhanced refinement near the critical point to properly capture the large gradients of properties in this region. Defining the properties of mixtures is one of the objectives of the DESOLINATION project and more accurate equations of state for mixtures, such as PC-SAFT, are still under development. The presence of SO_2 has a direct impact on the thermodynamic behavior of the mixture, especially on the location of the critical point. Figure 2 represents the critical point locus as a function of the SO_2 molar fraction in a pressure-temperature diagram. The values of interest for the DESOLINATION project are in the range of SO_2 molar fraction between 25% and 40%, which allows to reach a critical temperature above 70 °C, that is necessary to optimize the performance of the power cycle. In the following sections, only two SO_2 molar fractions are considered, namely 25% and 40%, which represent the limits of the range of interest of the DESOLINATION project.

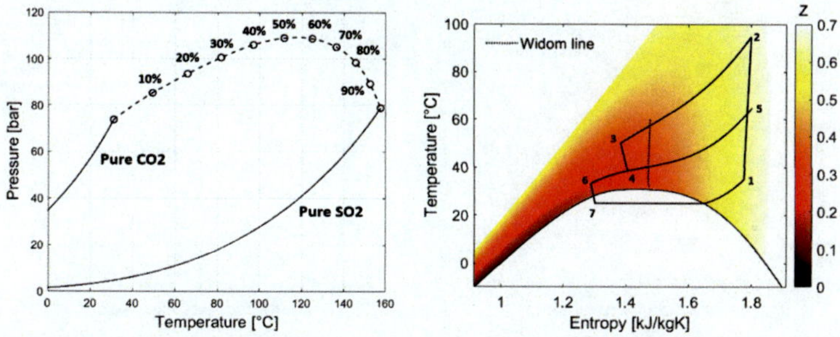

Fig. 2. Critical point locus for different SO_2 molar fractions (left) and temperature-specific entropy diagram of LUTsCO2 loop operating with pure CO_2 in design conditions superposed to the compressibility factor map (right)

4 Mathematical Models

This section outlines the mathematical formulation for each component of the model. Each component has a different dynamic response time depending on the nature of the physical process taking place. For instance, heat transfer devices such as the heater, gas cooler and evaporator have long response times and require dynamic-state models, i.e., a transient formulation of the governing equations. On the other hand, components such as compressors and expansion valves can be easily modelled with steady-state models due to their shorter response time. The overall model is simulated under the following assumptions: i) each component is insulated and heat transfer with the external environment is negligible; ii) pressure losses in the system are neglected; iii) flow in two-phase thermodynamic conditions is assumed to be in homogeneous equilibrium.

4.1 Compressor Model

Two semi-hermetic reciprocating compressors are used to compress the fluid and provide the required flow rate. For the sake of simplicity, the two compressors are modelled as a single component performing the same compression process, but with an overall mass flow rate that is the sum of the mass flow rates of both compressors. The compressor block is defined by the mass and energy conservation equations:

$$\dot{m}_{in} = \dot{m}_{out} = \dot{m}_{compressor\,map} \tag{1}$$

$$\Phi_{out} - \Phi_{in} = \dot{m}_{compressor\,map}\left(h_{T,out} - h_{T,in}\right) = \dot{m}_{compressor\,map}\Delta h_{compressor\,map} \tag{2}$$

where $\dot{m}$ is the mass flow rate, Φ is the energy flow rate and h_T is the total specific enthalpy. The values of mass flow rate and enthalpy rise are obtained from the compressor maps as a function of pressure ratio and corrected rotational frequency, as shown in Fig. 3. The compressor maps were generated using the online tool provided by the manufacturer (Bock GmbH 2022).

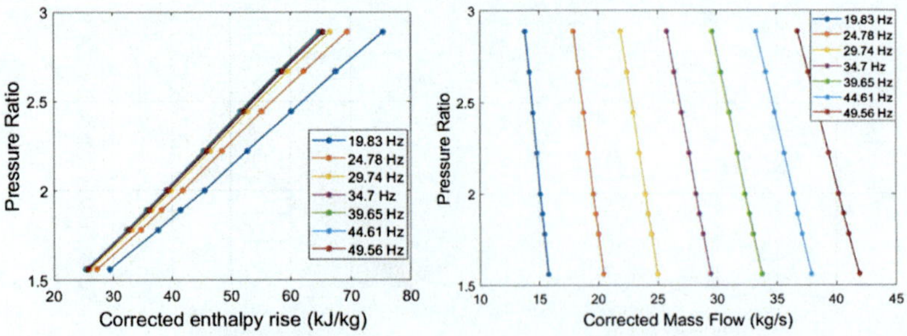

Fig. 3: Compressor maps

4.2 Heat Exchangers Model

The heat exchangers used in the test loop are the evaporator, the gas cooler and the electric heater. The evaporator and the cooler are printed circuit heat exchangers (PCHE) and have a great advantage in terms of dynamic modeling, since they have a simple layout for the fluid flows configuration. Instead, the electric circulation heater consists of a vessel in which the fluid flows and is heated by internal heating elements. These components are modelled in Simulink as circular tubes arranged in parallel and divided into five zones, assuming the same thermodynamic conditions of the flow in each zone. The PCHEs have a countercurrent flow configuration and heat is transferred from the hot side to the cold side through a thermal mass, instead the fluid in the electric heater is heated by a controlled uniform heat flux applied to the tube surface. The heat transfer between the fluids and the tube wall is calculated for each zone using the Gnielinski correlation (Gnielinski 1976):

$$Nu = \frac{\frac{f}{8}(Re - 1000)Pr}{1 + 12.7\sqrt{\frac{f}{8}}\left(Pr^{2/3} - 1\right)},\ h_{coeff} = \frac{kNu}{D_H},\ \dot{Q} = h_{coeff}S\left(T_w - T_f\right) \tag{3}$$

where Nu is the Nusselt number, f is the friction factor, Pr is the Prandtl number, h_{coeff} is the heat transfer coefficient, k is the thermal conductivity, D_H is the hydraulic diameter, $\dot{Q}$ is the thermal power exchanged between the fluid and the wall, S is the heat transfer surface area and T is the temperature. Equations (4)-(5) are the transient formulation of mass and energy conservation equations, respectively, which are solved for the fluids flowing through the heat exchangers:

$$\left(\frac{dp}{dt}\sum_{zones}\frac{\partial \rho}{\partial p} + \sum_{zones}\left(\frac{du}{dt}\frac{\partial \rho}{\partial u}\right)\right)\frac{V}{N_{zones}} = \dot{m}_{in} - \dot{m}_{out} \tag{4}$$

$$\frac{du}{dt}\frac{M}{N_{zones}} + u(\dot{m}_{in} - \dot{m}_{out}) = \dot{m}_{in}h_{T,in} - \dot{m}_{out}h_{T,out} + \dot{Q} \tag{5}$$

where p is the pressure, ρ is the density, u is the specific internal energy, V is the fluid volume and M is the fluid mass.

4.3 Expansion Valves Model

The expansion valves in the system perform adiabatic expansions to reduce the static pressure to the desired values. The pressure drop is achieved by introducing a variable local restriction into the system consisting of a contraction followed by a sudden expansion of the flow area. The total enthalpy remains constant along this component so that Eq. (6) holds:

$$u_{in} + \frac{p_{in}}{\rho_{in}} + \frac{U_{in}^2}{2} = u_R + \frac{p_R}{\rho_R} + \frac{U_R^2}{2},\ U_i = \frac{\dot{m}_{ideal}}{\rho_i A_i},\ \dot{m}_{ideal} = \frac{\dot{m}}{C_D} \tag{6}$$

where U is the ideal flow velocity, A is the cross-section area, C_D is the fluid discharge coefficient and subscript R indicates the restriction aperture. The pressure drop across the valve depends strictly on the mass flow rate through the restriction, which for turbulent flows can be calculated as follows:

$$\dot{m} = A_R(p_{in} - p_{out})\sqrt{\frac{2\rho_R}{|p_{in}-p_{out}|k_T}},\ k_T = \left(1 + \frac{A_R}{A}\right)\left(1 - \frac{\rho_{out}}{\rho_{in}}\frac{A_R}{A}\right) - 2\frac{A_R}{A}\left(1 - \frac{\rho_R}{\rho_{out}}\frac{A_R}{A}\right) \quad (7)$$

5 Verification of the Dynamic Model and Results for CO_2-SO_2 Mixtures Cycles

The overall dynamic model shown in Fig. 1 was simulated under steady-state conditions, and the results were compared with design values to verify the model. Subsequently, the same model was used to perform steady-state calculations in off-design conditions for pure CO_2 and CO_2-SO_2 mixtures with different SO_2 molar fractions. The simulated system is composed of the individual models described in Sect. 4, connected by the corresponding fluid networks according to the flow sequence. Based on the simulation results, the main differences in the operational capability of the test loop resulting from the use of CO_2 mixtures instead of pure CO_2 are highlighted and investigated.

5.1 Dynamic Model Verification in Design Conditions

The comparison between the results obtained with the steady-state Simulink model and the thermodynamic design points of the cycles is shown in Table 1. The model results are very accurate at each calculated point, with maximum relative errors of 0.19% for temperatures and 0.20% for pressures, showing that the dynamic model can be considered reliable. Table 2 compares the operating parameters of the cycle in design conditions with the results of the dynamic model. The largest relative errors occur in the calculation of compressor power, which is directly related to the pressure ratio across the compressor (cycle points 1 and 2), and in the cooling water outlet temperature. Nevertheless, the calculated errors are considered very small, confirming the accuracy of the model.

5.2 Off-Design Cycle Operation with Pure CO_2 and CO_2-SO_2 Mixtures

Three different cases of CO_2-SO_2 mixtures with different molar compositions were considered: (a) pure CO_2, (b) SO_2 molar fraction equal to 25%, and (c) SO_2 molar fraction equal to 40%. The cycle configuration for each of these cases was made setting the pressure levels at 120 bar for high pressure, 100 bar for intermediate pressure (110 bar for case (c) to avoid liquid phase in the heater and the gas cooler) and 55 bar for low pressure. In addition, the CO_2 mixture mass flow rates are kept the same in all three cases, with a value of 1.50 kg/s, and the superheat degree at the compressor inlet was set to 10 °C to prevent local gas condensation that would cause problems in the compressors operability. The water flowing in the open cooling circuit has an inlet temperature of 10 °C, and the mass flow rate is controlled for each configuration to achieve the desired superheat degree at the compressor inlet. The pressure-enthalpy diagrams obtained from

Table 1. Verification of the operating thermodynamic points of the cycle in design conditions

Cycle point	Temperature (°C)			Pressure (bar)		
	Design	Model	Error %	Design	Model	Error %
1	35.00	35.00	–	64.34	64.34	–
2	95.74	95.56	0.19	130	129.74	0.20
3	50.67	50.60	0.14	130	129.74	0.20
4	38.76	38.75	0.03	90	90	–
5	65.00	65.00	–	90	90	–
6	33.34	33.35	0.03	90	90	–
7	25.00	25.00	–	64.34	64.34	–

Table 2. Verification of the operating parameters of the cycle in design conditions

Parameter	Design	Model	Error %
CO_2 mass flow rate (kg/s)	1.60	1.60	–
Gas cooler thermal power (kW)	261.53	261.44	0.03
Evaporator thermal power (kW)	227.45	227.36	0.04
Electric heater thermal power (kW)	200.45	200.62	0.08
Compressor power (kW)	61.07	60.89	0.30
Cooling water mass flow rate (kg/s)	2.500	2.496	0.16
Cooling water outlet temperature (°C)	34.91	35.04	0.37

the simulation results of the three cases are shown in Fig. 4. The thermodynamic impact of the SO_2 molar fraction can be seen from the location of the two-phase region, which moves toward higher pressures and temperatures as the SO_2 molar fraction increases. The thermodynamic point at the inlet of the evaporator cold side (point 7) lies in the two-phase region, and this temperature corresponds to the saturation temperature at the current pressure for pure CO_2; on the other hand, if SO_2 is added to the mixture, the temperature increases as the fluid evaporates and is therefore not constant along the heat exchanger, as can be seen from the non-horizontal isotherm lines in the two-phase region in Fig. 4.

Table 3 shows the calculated temperatures and pressures at each cycle point for the three cases. Increasing the SO_2 molar fraction, while holding the same pressure levels, leads to higher temperatures in the system, which make the operation of the test setup with a SO_2 molar fraction of 40% unfeasible, since the maximum permissible temperature of the system is 150 °C. The closer the operating conditions are to the critical point, the greater the uncertainties in the fluid properties due to the large gradients in this region, leading to higher expected errors in the dynamic model and increased complexity in controlling the system. The main operating parameters are shown in Table 4. The three

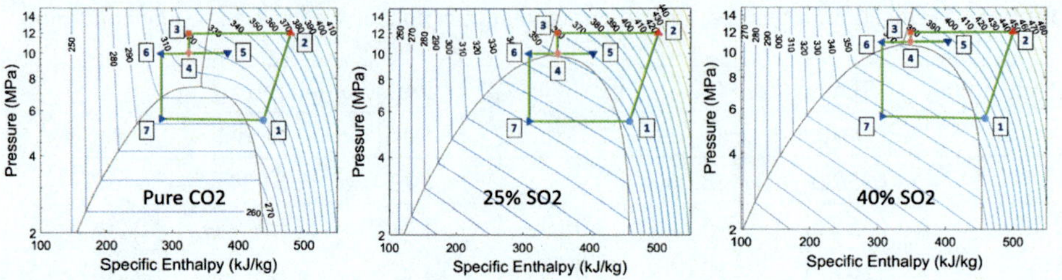

Fig. 4. p-h diagrams: pure CO_2 (left), 25% molar fraction SO_2 (center) and 40% molar fraction SO_2 (right)

Table 3. Operating thermodynamic points of the cycle in off-design conditions with pure CO_2 and CO_2-SO_2 mixtures

Cycle point	Pure CO_2		CO_2-SO_2 (25% molar SO_2)		CO_2-SO_2 (40% molar SO_2)	
	Temperature (°C)	Pressure (bar)	Temperature (°C)	Pressure (bar)	Temperature (°C)	Pressure (bar)
1	28.27	55	75.12	55	94.54	55
2	95.70	119.37	148.74	120.24	170.79	120.04
3	47.32	119.37	86.36	120.24	105.28	120.04
4	41.92	100	77.47	100	101.08	110
5	50.00	100	87.93	100	113.85	110
6	33.30	100	70.85	100	92.62	110
7	18.27	55	48.49	55	63.93	55

simulated thermodynamic cycle configurations do not present critical issues in terms of heat exchangers and compressors capacities. With respect to the design conditions at which the dynamic model was verified, the lower mass flow rates and, in particular, the higher fluid temperatures in the gas cooler (points 5–6), when SO_2 is added to the mixture, lead to a reduction of the cooling water flow rate and an increase in the water outlet temperature, which reaches 99.96 °C in the case with the SO_2 molar fraction of 40%, highlighting potential unwanted water evaporation in the heat exchanger during operation.

Table 4. Operating parameters of the cycle in off-design conditions with pure CO_2 and CO_2-SO_2 mixtures

Parameter	Pure CO_2	CO_2-SO_2 (25% molar SO_2)	CO_2-SO_2 (40% molar SO_2)
Working fluid mass flow rate (kg/s)	1.50	1.50	1.50
Gas cooler thermal power (kW)	151.55	140.97	145.76
Evaporator thermal power (kW)	233.12	222.72	227.58
Electric heater thermal power (kW)	88.72	77.69	82.71
Compressor power (kW)	62.90	63.30	63.07
Cooling water mass flow rate (kg/s)	1.348	0.507	0.354
Cooling water outlet temperature (°C)	36.87	76.43	99.96

6 Conclusions and Future Works

In this work, a dynamic model of the transcritical LUTsCO2 test loop was developed and simulated in MATLAB-Simulink. The model was verified by comparing the steady-state simulation results obtained for pure CO_2 with the thermodynamic design cycle and operating parameters. Using the verified dynamic model, the operability of the test loop with CO_2-SO_2 mixtures as working fluids was investigated. The main conclusions of the study are: (i) the dynamic model is able to accurately reproduce the design results with a maximum relative error of 0.30% at steady-state conditions; (ii) the heat exchangers and the compressors do not present any critical issues in their capacities in off-design conditions with CO_2-SO_2 mixtures; (iii) the high critical temperature of the CO_2 blends leads to higher temperatures in the system and lower cooling water mass flow rates in the gas cooler, resulting in higher water outlet temperatures that can reach values close to the saturation temperature during the operation of the test loop. The results obtained are very promising and future experimental data will allow to validate the dynamic model in steady-state and transient operations.

References

Bock GmbH. Bock VAP for stationary applications. Bock Website (2022). https://vap.bock.de/stationaryapplication Accessed 29 Jun 2022

Crespi, F., et al.: Thermal efficiency gains enabled by using CO2 mixtures in supercritical power cycles. Energy **238**(C), 121899 (2022). https://doi.org/10.1016/j.energy.2021.121899

Crespi, F., Sánchez, D., Rodríguez, J.M., Gavagnin, G.: A thermo-economic methodology to select sCO2 power cycles for CSP applications. Renew. Energy **147**(3), 2905–2912 (2020). https://doi.org/10.1016/j.renene.2018.08.023

Dostal, V., Driscoll, M.J., Hejzlar, P.: High-performance supercritical carbon dioxide cycle for next-generation nuclear reactors. Nuclear Technol. **154**(3), 265–282 (2006). https://doi.org/10.13182/NT154-265

Gnielinski, V.: New equations for heat and mass transfer in turbulent pipe and channel flow. Int. Chem. Eng. **16**(2), 359–368 (1976)

Manzolini, G., Binotti, M., Bonalumi, D., Invernizzi, C.M., Iora, P.: CO2 mixtures as innovative working fluid in power cycles applied to solar plants. Sol. Energy **181**, 530–544 (2019). https://doi.org/10.1016/j.solener.2019.01.015

Morosini, E., Ayub, A., Di Marcoberardino, G., Invernizzi, C.M., Iora, P., Manzolini, G.: Adoption of the $CO_2 + SO_2$ mixture as working fluid for transcritical cycles: a thermodynamic assessment with optimized equation of state. Energy Convers. Manage. **255**, 115263 (2022). https://doi.org/10.1016/j.enconman.2022.115263

Petruccelli, G., Uusitalo, A., Grönman, A., Turunen-Saaresti, T., Zocca, M.: Closed-Loop Supercritical Carbon Dioxide Wind Tunnel: Design and Components. In: 4th European sCO2 Conference for Energy Systems, Online Conference (2021). https://doi.org/10.17185/duepublico/73977

Song, P., et al.: Research on dynamic modeling of the supercritical carbon dioxide power cycle. Processes **9**(11), 1946 (2021). https://doi.org/10.3390/pr9111946

Uusitalo, A., Ameli, A., Turunen-Saaresti, T.: Thermodynamic and turbomachinery design analysis of supercritical brayton cycles for exhaust gas heat recovery. Energy **167**, 60–79 (2018). https://doi.org/10.1016/j.energy.2018.10.181

Preliminary Experiments in High Temperature Vapours of Organic Fluids in the Asymmetric Shock Tube for Experiments on Rarefaction Waves (ASTER)

Nitish Chandrasekaran[1(✉)], Theodoros Michelis[1], Bertrand Mercier[2], and Piero Colonna[1]

[1] Propulsion and Power, Delft University of Technology, Delft, The Netherlands
N.B.Chandrasekaran@tudelft.nl

[2] IMT Nord Europe, Institut Mines Télécom, University of Lille Center for Energy and Environment, Lille, France

Abstract. This paper describes the design, functioning, and preliminary experiments in the Asymmetric Shock Tube for Experiments on Rarefaction Waves (ASTER), a Ludwieg-type shock tube designed and realised to measure waves propagating in dense vapour flows of organic fluids. The setup is designed to operate at pressures and temperatures of up to 15 bar and 400 °C. The high and low-pressure sections of the tube are separated by a glass-disk barrier to ensure quasi-instantaneous opening. When the glass disk is broken, a rarefaction wave propagates into the tube. The wave speed is measured using a time-of-flight method with the help of four pressure transducers placed at known distances from each other. Leakage rates of 2.2×10^{-4} mbar·l·s^{-1} at vacuum and 5×10^{-4} mbar·l·s^{-1} at superatmospheric pressures were measured, which is considered sufficient for the conceived experiments. Preliminary rarefaction experiments in the dense vapours of dodecamethylcyclohexasiloxane, D_6, were successfully performed at various thermodynamic conditions. Also, a method for the estimation of sound speeds from the pressure sensor recordings is proposed. Results are found to be within 2.5% of the values predicted by the state-of-the-art thermodynamic model for D_6.

Keywords: ASTER · Ludwieg tube · Rarefaction wave · Time-of-flight · Speed of sound · Organic vapours

1 Introduction

Non-ideal Compressible Fluid Dynamic (NICFD) effects, and nonclassical gasdynamic effects within it, characterize the dense-vapour flows of high-molecular weight organic compounds, collectively known as Bethe-Zel'dovich-Thompson (BZT) fluids, in the critical point region and the vicinity of the vapour-liquid saturation curve [1]. The local fundamental derivative of gasdynamics Γ can exhibit values lower than 1 and can sometimes even become negative in this thermodynamic region giving rise to unconventional

M. White et al. (Eds.): NICFD 2022, ERCOFTAC 29, pp. 201–208, 2023.
https://doi.org/10.1007/978-3-031-30936-6_20

phenomena such as the admissibility of rarefaction shock waves (RSW), double sonic shocks, etc [2,3].

Though flows evolving in this region have been extensively studied theoretically, few experimental analyses have been carried out, and that too produced inconclusive results. Previous attempts to provide experimental evidence for the existence of RSWs in the vapour phase have been unsuccessful either due to the incorrect choice of working fluid or due to the presence of dissolved impurities such as air and moisture that are known catalysts for thermochemical decomposition [4,5]. The difficulty is further exacerbated by the fact that the $\Gamma < 0$ thermodynamic region spans a narrow range of pressures and temperatures. Though minor gradients in axial temperature have been shown to enhance nonlinear wave propagation in BZT fluids, large temperature fluctuations can cause mixed classical/nonclassical flows that are detrimental to the formation of RSWs, especially in a setup of limited length [6].

A series of shock tube experiments focussed on investigating NICFD flows in siloxane D_6 were pursued at TU Delft in the flexible asymmetric shock tube (FAST) facility [7]. The researchers performed preliminary speed of sound and wavespeed measurements in the non-ideal thermodynamic region of D_6 at temperatures up to 300°C and reported values within 8% and 1.6% respectively of the predictions by the state-of-the-art thermodynamic model. The authors also performed several rarefaction experiments at different initial conditions close to and in the predicted $\Gamma < 0$ region. However, they found no proof of either a RSW or a nonclassical isentropic expansion though they claimed that the measured value of Γ was close to zero [8].

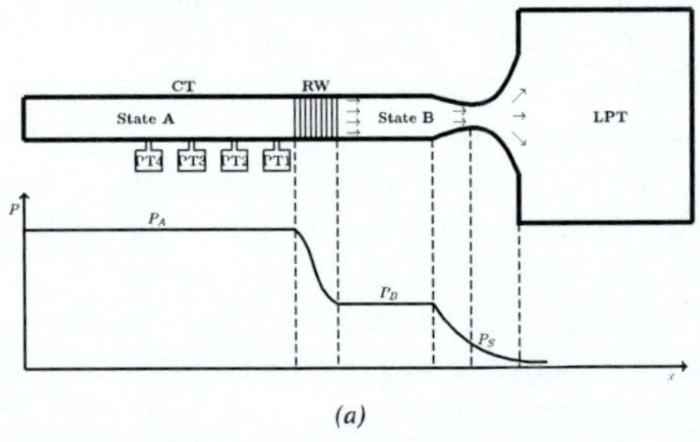

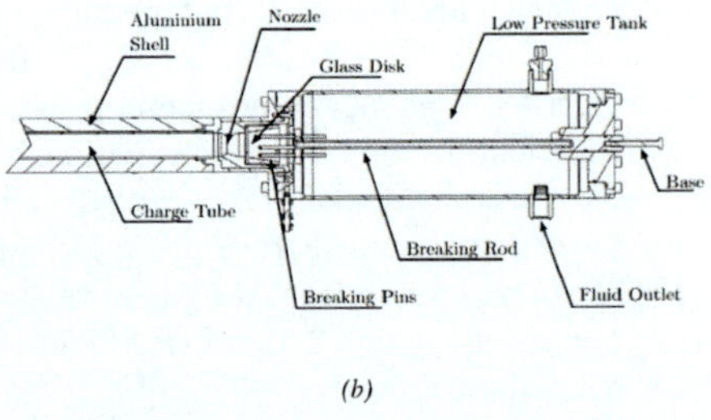

Fig. 1. (a) Schematic overview of a rarefaction wave experiment in the ASTER. The chart below shows the theoretical pressure profile in the charge tube after the breaking of the glass disk. The fluid is accelerated from rest at State A to State B and flows into the low-pressure tank through the nozzle, where sonic condition S is attained. The propagating rarefaction wave in the CT is captured using pressure transducers PT1 to PT4. (b) Cross-sectional drawing of the LPT showing the breaking pin mechanism and the glass disk

Recent measurements of the internal wall temperature of the FAST by the authors of this article showed large gradients in the axial direction. Since Γ is highly sensitive to the variation in the temperature, these measurements indicated that earlier experiments in the FAST might have suffered from mixed classical/nonclassical flow behaviour. Despite tight control and preventive measures that were already in place to control such gradients, the design of the FAST with multiple flanges and its large size made it unfeasible to maintain a constant temperature in the setup.

In this work, the design, development and commissioning of a new shock tube setup named the Asymmetric Shock Tube for Experiments on Rarefaction Waves (ASTER) are presented, along with preliminary measurements of rarefaction waves and local speeds of sound.

2 The ASTER Experimental Setup

The ASTER is designed based on the same principles of the FAST (see Fig. 1a) but is more compact and simple to operate. The working fluid is filled into the charge tube (CT), where it is heated to the desired temperature and pressure. A glass disk that separates the CT and the Low-Pressure Tank (LPT) (initially maintained at vacuum) is broken to initiate the experiment. Consequently, a rarefaction wave travels into the CT.

The CT is a 3 m long tube made of stainless steel (316Ti) with an outer radius of R = 17.5 mm and wall thickness of t = 2.5 mm. One end of the CT houses the inlet port through which the working fluid is injected with the help of a manual pump. The other end of the CT is connected to the LPT via a nozzle section. The nozzle creates a throat area to choke the flow, thereby preventing flow disturbances from travelling upstream into the CT. A 3 mm thick glass disk is placed at the exit of the nozzle. The nozzle opens into the LPT, a 1.1-litre cylindrical stainless steel tank of length 345 mm with an outer radius of 69.5 mm and a wall thickness of 4 mm.

Four dynamic high-temperature pressure transmitters PT1-PT4 (Kulite XTEH-10L-190SM-300PSI-A), each measuring pressure within 0.5% of the full range of 21 bar and with a sampling frequency of 250 kHz, are flush mounted along the CT at distances of 1.2, 1.4, 1.6 and 1.8 m from the nozzle throat respectively (at ambient temperature), for wavespeed measurements. These sensors are calibrated using a more accurate measurement from the Druck UNIK 5000 static pressure transmitter that is placed close to the inlet port of the CT.

The breaking of the glass disk is achieved using the mechanism shown in Fig. 1b. The breaking pins consist of six 8.5 mm long drill bits attached to a 320 mm long stainless steel rod that is mounted on a base of length 140 mm that protrudes outside of the LPT. Two Kalrez ® O-rings on the base ensure that the vacuum in the LPT is maintained and that the working fluid doesn't leak during experiments. To perform an experiment, the end of the breaking pin is hit with a hammer and the glass shatters in the least possible time. After every experiment, the LPT is opened to replace the glass disk and to evacuate the fluid from the setup.

The thermal energy required for heating the charge tube is supplied through two glass-silk heating jackets, each rated for 1250 W W, and having a 25 mm layer of insulation. The blankets are mounted on 17.5 mm thick aluminium shells to ensure sufficient temperature homogeneity along the CT. The temperature along the CT is measured and controlled using eight equidistantly placed K-type thermocouples that are bored into the aluminium shells to measure as close to the stainless steel wall as possible. In addition, a PT100 probe is placed into the CT via a thermowell to directly measure the fluid temperature. Both the thermocouples and the PT100 sensors are calibrated with the Fluke

9100 dry-well calibrator which has a rated accuracy of ±0.5 K. The LPT is heated using a 300 W heating jacket, the sole purpose of heating being to avoid temperature gradients at the nozzle section.

2.1 Choice of Working Fluid

Siloxane D_6 was chosen as the working fluid for experiments in the ASTER due to its high thermal stability, low levels of flammability and non-toxicity, and the large size of the non-classical thermodynamic region predicted by relatively accurate thermodynamic models [9]. The fluid was produced by Dow Corning and is certified to be 97.4% pure. The primary properties of the fluid are summarized in Table 1 [10]. Any moisture dissolved in the fluid must be removed before heating since its presence can adversely affect the thermal stability of the fluid. Dissolved water is removed by immersing 3Å molecular sieve desiccants in the fluid samples over several days.

Table 1. Relevant thermophysical properties of the working fluid D_6. MW is the molecular weight, P_c, T_c, and ν_c are the critical pressure, temperature and specific volume respectively, and ω is the accentric factor.

Fluid name	MW (kg/kmol)	P_c (bar)	T_c(oC)	ν_c (m^3/kg)	ω
D_6	444.92	9.61	372.6	0.0036	0.736

2.2 Facility Characterization

A series of leak tests were performed to assess the leak rate of the facility. The tightness of the setup is characterized using the average leak rate, defined as LR = $V\Delta P\ \Delta t^{-1}$, where V is the volume of the setup ($\sim 0.0032\,\mathrm{m}^3$), and Δp is the rise/drop in pressure measured after a time interval of Δt. At vacuum conditions, a LR of 2.2×10^{-4} mbar·l·s^{-1} was measured (air leaking into the system). Leakage tests at superatmospheric pressures conducted at different temperatures by filling the setup with air were consistently below 5×10^{-4} mbar·l·s^{-1} (air escaping the setup). These rates were deemed sufficient for the conceived tests.

The measuring principle of the ASTER is based on the time of flight method, which is a common technique employed to measure the speed of sound in gases. The propagating rarefaction wave in the CT is measured by the pressure sensors PT1-PT4 that are placed 20 cm apart from each other. The wave speed is then calculated as,

$$w = \frac{\Delta l}{\Delta t} \tag{1}$$

where Δl is the distance between a sensor pair, and Δt is the time difference of arrival of the wave at any pair of sensors. The distance Δl is calibrated with air, since it is a well-studied fluid for which accurate speed of sound data is available in literature [11]. The change in Δl due to temperature is accounted for using,

$$\Delta l_{p,T} = \Delta l_{p_0,T_0}(1 + \alpha(T - T_0)) \tag{2}$$

where $\Delta l_{p_0,T_0}$ is the length calibrated at p_0 = 8.74 bar and T_0 = 43 °C and α is the thermal expansion coefficient.

3 Results

3.1 Rarefaction Tests in D_6

A series of rarefaction tests were performed with D_6 at thermodynamic conditions in both the ideal-gas region and close to the saturation curve where NICFD effects can be expected. A summary of the measured values of the temperature and pressure and the corresponding thermodynamic properties are shown in Table 2.

Table 2. Thermodynamic conditions of the initial states of the rarefaction experiments. Γ and c for Tests 1 & 2 are estimated using the iPRSV EoS [12] while 3 & 4 are computed using the Helmholtz based EoS [13]

Test No	P_0 (bar)	T_0 (°C)	Γ_{model} (-)	Z (-)	c_{model}(m/s)	c_{exp}(m/s)	Δ c(%)
1	2.66	349.8	0.86	0.87	94.23	95.55	1.44
2	5.85	357.6	0.58	0.69	73.35	74.67	1.79
3	8.87	370.9	0.096	0.49	48.18	47.94	–0.48
4	8.96	369.6	–0.052	0.46	43.08	46.62	8.21

The speed of propagation, w, of the wave is estimated using a time-of-flight (ToF) method. Before its implementation, the signals from the pressure transducers are filtered using a zero-phase digital filter to limit the impact of noise. The ToF method is then implemented as follows: first, the head of the wave is identified as the point at which the pressure differs from the initial value P_0 by more than 10 mbar (see Fig. 2a). This threshold was found sufficient to eliminate the noise generated by the strike of the hammer. The tail of the wave is arbitrarily chosen such that only the smooth pressure drop across the wave is analysed and the wave reflections arising from the end-wall of the CT are neglected. This span is divided into several intervals and the ToF is applied to the corresponding subintervals of the different sensors to compute the local wave propagation speed.

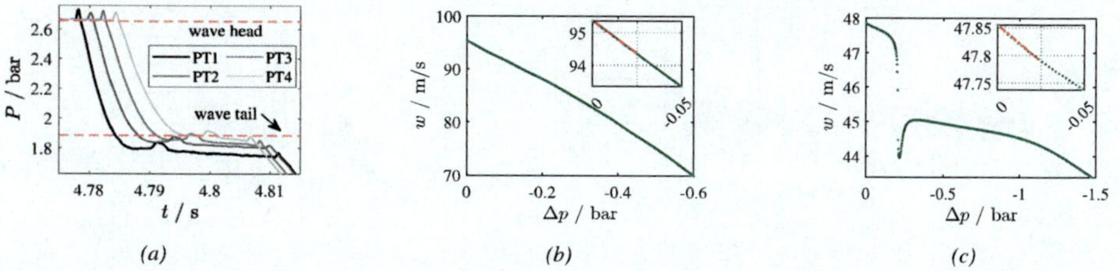

Fig. 2. (a) Recordings from sensors PT1 to PT4 for experiment 1. Red dotted lines show the limits of the relevant span of signals for ToF method; Wavespeed vs. Δp in D_6 for (b) Experiment 1 and (c) Experiment 3 (see Table 2). Figures in the inset show linear fit in dashed red line for soundspeed estimation. (Color figure online)

Figures 2b and 2c show the wavespeeds estimated using the ToF method for experiments 1 & 3 (see Table 2) respectively. For conditions in the ideal-gas region, a linear decrease in sound speed with pressure drop is expected. This is also observed in the wavespeed plot for experiment 1. The variation of the wavespeed with Δp for experiment 4, however, is not strictly linear as in the ideal case. This is an indication of the presence of NICFD effects since the wavespeed is no longer dependent only on the soundspeed and density, but also on Γ, whose value is lesser than 1 and variable in this thermodynamic region. In both cases, wavespeed monotonically decreases with Δp, indicating that Γ is always positive in these examples.

A linear fitting curve close to the head of the wave can give an estimate of the speed of sound in the undisturbed state since the wave propagation speed tends to the sound speed as $\Delta P \rightarrow 0$. The average error and the standard deviation with respect to the sound speed predicted by the thermodynamic model are $1.44\% \pm 1.63\%$ in the case of Test 1. This is very close to the model predictions, given that the model is valid only for pure D_6 while the one used in the experiments is only 97.4% pure. The predicted and the computed speeds of sound for the performed experiments are shown in Table 2. It is seen that as the experimental conditions get closer to and into the $\Gamma < 0$ region, the deviation between the model and the experimentally calculated sound speeds increases. This is expected since the thermodynamic model for D_6 is known to be inaccurate by as much as 30% in this region [7].

3.2 Alternate Speed of Sound Measurements

Though the speed of sound in D_6 can be extracted from rarefaction experiments, such runs don't allow for quick and repeated measurements at different conditions since the glass disk needs to be replaced between successive tests. This can be done only when the setup has cooled down, which increases the downtime between the experiments. To facilitate the rapid measurement of sound speeds at various thermodynamic states, an alternate approach to measuring the speed of sound is proposed here. Instead of using a glass disk, a 3 mm thick metallic disk capable of withstanding repeated strikes from the hammer is used as a barrier element. The metallic disk is actuated enough to

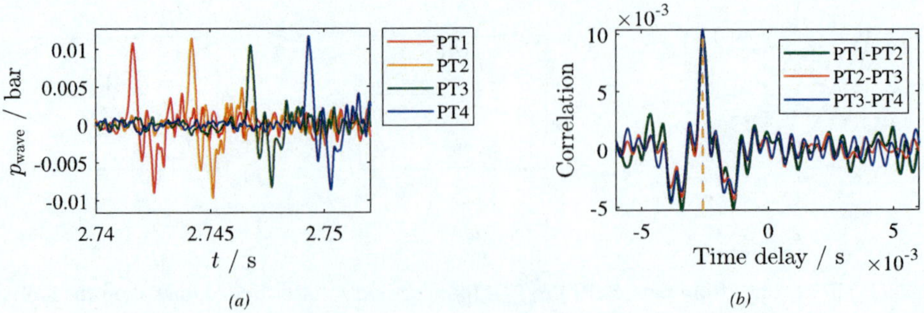

Fig. 3. (a) Over-pressure recording of a small-amplitude wave by sensors PT1 to PT4 (b) Output of cross-correlation technique showing a peak at time-delay between sensors

generate a small-amplitude wave in the CT. A cross-correlation (CC) technique for the ToF method is applied to the signals to estimate the sound speed. Cross-correlation is considered a more optimal ToF implementation since it is less sensitive to noise and uses all the information contained in the signals to produce a maximum value at the time delay between the two signals [14].

Figure 3a shows the wave amplitude recordings from the pressure sensors from a sound speed test in D_6 at 6.08 bar and 368.9 °C. The results from the CC of the different sensor pairs and the estimated time delay between the sensors are shown in Fig. 3b. Using Eq. 1, the sound speed in D_6 at these conditions is calculated to be 78.22 m/s, which is 2.8% higher than the model estimated value of 76.01 m/s. Figure 4 shows the measured sound speed values in D_6 at different temperatures and the corresponding values predicted by the thermodynamic model. The comparison shows a very good agreement in terms of the trend in c vs T. However, the model is seen to consistently underpredict the speed of sound by approximately 2.5%.

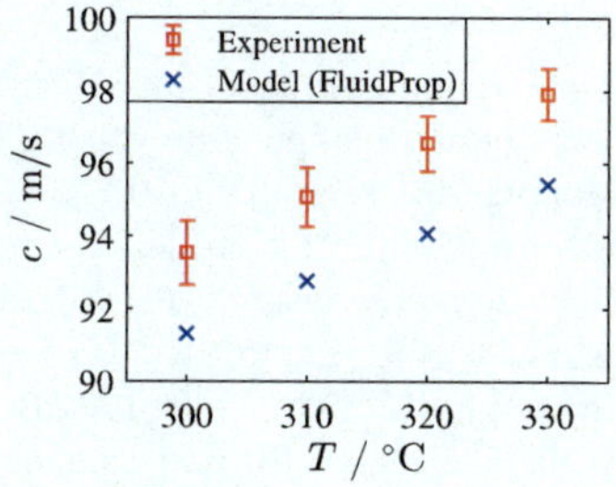

Fig. 4. Measurements of speeds of sound in D_6 at different temperatures and corresponding model predictions

4 Conclusion and Future Work

A new shock-tube facility for measurements on waves propagating in NICFD flows of organic vapours has been designed, developed and successfully tested in the Propulsion & Power Group at TU Delft. The facility is designed in particular to generate a nonclassical rarefaction shock wave in siloxane D_6 and is capable of withstanding up to 15 bar and 400 °C. Preliminary experiments on rarefaction waves in the non-ideal thermodynamic region of siloxane D_6 at temperatures and pressures up to 370 °C and 9 bar have been successfully performed. The speed of sound estimated from wavespeed measurements agrees within 2% of the state-of-the-are thermodynamic model predictions in the ideal-gas region but deviates at high P and T. A stand-alone approach to measure soundspeeds is also presented and preliminary measurements at temperatures up to 330 °C is reported, providing soundspeed values within 2.5% of the model predictions. These predicted values are however affected by high uncertainties of up to 30%. Further measurements are planned at thermodynamic conditions where nonclassical gasdynamic effects are expected. Sound speed measurements in this region can contribute towards improving the thermodynamic modelling of the fluids of interest.

Acknowledgements. This research has been supported by the Engineering Sciences Division (TTW) of the Dutch Organization for Scientific Research (NWO), Technology Program of the Ministry of Economic Affairs, grant number 15837.

References

1. Kluwick, A.: Non-ideal compressible fluid dynamics: a challenge for theory. J. Phys. Conf. Ser. **821**, 012001 (2017)
2. Nannan, N.R.: Advancements in nonclassical gas dynamics. Phd thesis, TU Delft, May 2009
3. Thompson, P.A.: A fundamental derivative in Gasdynamics. Phys. Fluids **14**(9), 1843 (1971)
4. Borisov, A.A., Borisov, A.A., Kutateladze, S.S., Nakoryakov, V.E.: Rarefaction shock wave near the critical liquid-vapour point. J. Fluid Mech. **126**, 59–73 (1983)
5. Fergason, S.H.: Dense gas shock tube: Design and analysis. Phd thesis, University of Colorado Boulder (2001)
6. Chandrasekaran, N.B., Mercier, B., Colonna, P.: Nonlinear wave propagation in dense vapor of Bethe-Zel'dovich-thompson fluids subjected to temperature gradients. Phys. Fluids **33**(10), 107109 (2021)
7. Mathijssen, T., et al.: The flexible asymmetric shock tube (FAST): a Ludwieg tube facility for wave propagation measurements in high-temperature vapours of organic fluids. Exp. Fluids **56**(10), 1–12 (2015). https://doi.org/10.1007/s00348-015-2060-1
8. Mathijssen, T.: Experimental observation of non-ideal compressible fluid dynamics with application in organic rankine cycle power systems. PHD thesis, TU Delft (2017)
9. Colonna, P., Guardone, A., Nannan, N.R., Zamfirescu, C.: Design of the dense gas flexible asymmetric shock tube. J. Fluids Eng. **130**(3), 034501 (2008)
10. Colonna, P., Guardone, A., Nannan, N.R.: Siloxanes: a new class of candidate Bethe-Zel'dovich-Thompson fluids. Phys. Fluids **19**(8), 086102 (2007)
11. Lemmon, E.W., Jacobsen, R.T., Penoncello, S.G., Friend, D.G.: Thermodynamic properties of air and mixtures of nitrogen, argon, and oxygen from 60 to 2000 k at pressures to 2000 MPa. J. Phys. Chem. Ref. Data **29**(3), 331–385 (2000)
12. van der Stelt, T.P., Nannan, N.R., Colonna, P.: The iPRSV equation of state. Fluid Phase Equilibria **330**, 24–35 (2012)
13. Lemmon, E.W., Bell, I.H., Huber, M.L., McLinden, M.O.: NIST standard reference database 23: reference fluid thermodynamic and transport properties-REFPROP, Version 10.0, National Institute of Standards and Technology (2018)
14. Khyam, M.O., Ge, S.S., Li, X., Pickering, M.R.: Highly accurate time-of-flight measurement technique based on phase-correlation for ultrasonic ranging. IEEE Sens. J. **17**(2), 434–443 (2017)

Experimental and Numerical Study of Transonic Flow of an Organic Vapor Past a Circular Cylinder

Stephan Sundermeier[1(✉)], Camille Matar[2], Paola Cinnella[2], Stefan aus der Wiesche[1], Leander Hake[1], and Xavier Gloerfelt[3]

[1] Muenster University of Applied Sciences, 48565 Steinfurt, Germany
{stephan.sundermeier,wiesche,leander.hake}@fh-muenster.de
[2] Sorbonne Université, 75006 Paris, France
camille.matar@etu.sorbonne-universite.fr,
paola.cinnella@sorbonne-universite.fr
[3] Arts et Métiers ParisTech, 75013 Paris, France
xavier.gloerfelt@ensam.eu

Abstract. The flow of an organic vapor at high subsonic Mach numbers past a circular cylinder was investigated using Background-oriented Schlieren (BOS) and computational fluid dynamics. The experiment was carried out in a closed-loop organic vapor wind tunnel (CLOWT) using Novec 649 as working-fluid which provides a compressibility factor of $Z = 0.91$ at the operating point investigated. For the first time, wake flow phenomena in organic vapor were optically resolved, including vortex streets and shock-wave interactions. In addition, a method for calculating the Strouhal number using Schlieren images is presented. Experiment and numerical investigation showed great agreement, demonstrating that BOS is ideally suited for visual analysis of flow phenomena in hot organic vapor.

Keywords: cylinder wake · Kármán vortex street · shock waves · organic vapor · Schlieren flow visualization · background-oriented Schlieren (BOS) · computatinal fluid dynamics (CFD) · real gas

1 Introduction

The experimental study was conducted in a closed-loop organic vapor wind tunnel (CLOWT) test section. For the organic vapor, Novec 649 by 3M, was used. Using Background-oriented Schlieren (BOS) optics occurring flow phenomena were visualized. The BOS approach offered advantages in comparison with conventional Schlieren optical methods regarding the demands on optical accessibility and mounting constraints of the cylinder in the test section. A maximum free-stream Mach number of 0.67 could be achieved at elevated pressure and temperature levels, leading to local supersonic flow regions and a cylinder based Reynolds number of $Re_{D,exp} = 5.12 \times 10^5$. Although the sensitivity of BOS is comparatively low, the significant density changes

M. White et al. (Eds.): NICFD 2022, ERCOFTAC 29, pp. 209–216, 2023.
https://doi.org/10.1007/978-3-031-30936-6_21

occurring in the organic vapor at high subsonic Mach numbers enabled detailed investigations of flow phenomena. A black anodized aluminum plate was used as the background for the BOS. A laser partially ablated the oxide layer of the plate to create a randomly arranged dot pattern. The combination of the illuminated background and the high-speed camera allowed an exposure time of up to 0.18 milliseconds corresponding to a recording frequency of $\alpha = 5586s^{-1}$. The selected background pattern was fine enough to capture flow phenomena downstream and upstream of the cylinder precisely. Furthermore, the high recording speed allowed the analysis of the emerging Karman vortex street. In addition to the vortex street, substantial moving shock waves were detected, similar to the phenomena close to the trailing edges of airfoils.
In addition to the experiments, an in-house finite difference flow solver is used to carry out Large Eddy Simulations (LES) at similar conditions as studied in the experiments. Due to the high Reynolds numbers investigated experimentally, the inlet Novec 649 conditions were fixed to a slightly lower but analogous cylinder based Reynolds number $Re_{D,num} = 2.0 \times 10^5$, and the Mach number is $M = 0.7$. These conditions are located just prior the cylinder drag crisis range. The presence of shock waves is investigated numerically and maps of density gradients are compared to the BOS images.

2 Experimental Setup

The experiment was carried out in a Closed Loop Organic Wind Tunnel (CLOWT). A test section with the circular cylinder within was installed inside the wind tunnel. For the experiment, the wind tunnel was first vacuumed and then heated to a temperature of 103°C. Subsequently the CLOWT was filled with a mass of 34 kg of Novec 649 from 3M, causing it to evaporate completely and thus serve as the organic vapour. Information about the CLOWT such as Design, Components and Control System can be found in [7]. For the Schlieren images a BOS system was used. In Fig. 1 the layout of the BOS is shown.

The system uses only a small amount of components and has the great advantage of optical accessibility, as only one side of the test section needs to be transparent. The BOS basically consist of camera, a light source and a patterned background. The latter was realised with a black anodised aluminium plate. A laser engraved a fine, randomly arranged pattern of dots with 0.2 mm in diameter on this plate. The resulting background with mounted circular cylinder is shown in Fig. 2(a). Areas of different density ρ, for example formed by different velocity ranges, causes certain areas of the pattern to shift. Software, such as Matlab in combination with PIVlab [10], can be used to evaluate these minimal displacements by means of image correlation. By using only horizontal displacements for image correlation, it is possible to produce synthetic Schlieren images that are comparable to Schlieren pictures of a z-type Schlieren optic. However, a significant disadvantage is the lower sensitivity compared to conventional systems, which can be attributed to the finiteness of the background pattern and the associated camera resolution required. The lower sensitivity ensures that flow phenomena only occur at higher density changes and thus only at higher Mach numbers. Therefore, lower flow velocities were not investigated in this work. More on Schlieren and shadowgraph techniques can be found in [9].

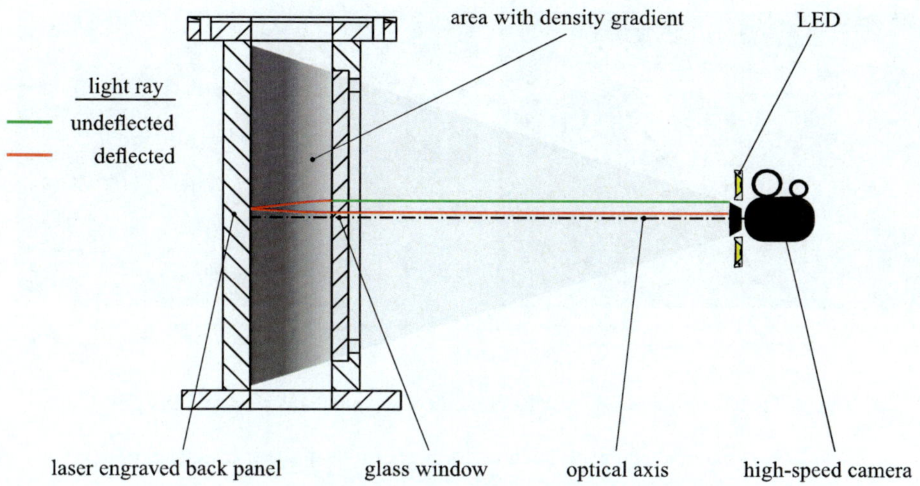

Fig. 1. Layout of the background-oriented Schlieren system.

3 Numerical Method

The thermodynamic behaviour of the gas is modelled by the Peng-Robinson-Stryjek-Vera (PRSV) equation of state. The dynamic viscosity is obtained with the Chung-Lee-Starling model. The compressible LES equations are solved with the in-house MUSI-CAA finite-difference solver. The inviscid fluxes are discretized by means of 10th-order standard centred differences whereas 4th-order is used for viscous fluxes. The scheme is supplemented with a 10th-order selective filtering to eliminate grid-to-grid unresolved oscillations. A four-stage Runge-Kutta algorithm is used for time integration and the IRS method [3] is implemented to enlarge the stability and allow the use of larger timesteps. The filter also acts as a regularization term draining energy at subgrid scales, so that no explicit subgrid-scale model is used (implicit LES). Due to the presence of shock waves a shock-capturing method is applied relying on the pressure gradient sensor introduced by Jameson [4]. The numerical domain consists in a structured curvilinear grid combining H- and O-type topologies, split in 1,600 sub-domains for parallel computation via the MPI library. It has a total of 200M points, with 250 2D–planes in the spanwise direction, extending over 1 diameter. Figure 2(b) shows some detail of the mesh. The grid resolution has been set based on previous experience and literature [1,2] and ensures a wall non-dimensional flow unit $y^+ \leq 1.5$ to fully resolve the boundary layer. Adiabatic no-slip conditions are applied at the wall, and non-reflecting Tam & Dong's conditions are imposed at the inlet, top, bottom and outflow boundaries. A sponge zone combining grid stretching and a Laplacian filter is added at the outlet. Periodicity is enforced in the spanwise direction. Because the IRS method allows for larger timestep due to increased numerical scheme stability, the CFL number reached based on the smallest cell is 3.5 which results in a computational timestep of $\Delta t_c = 2.66 \times 10^{-9}$s. The smallest length scales resolved are far larger than that of Kolmogorov η, as is the principle of LES. However, we provide an estimate of the latter's time scale indica-

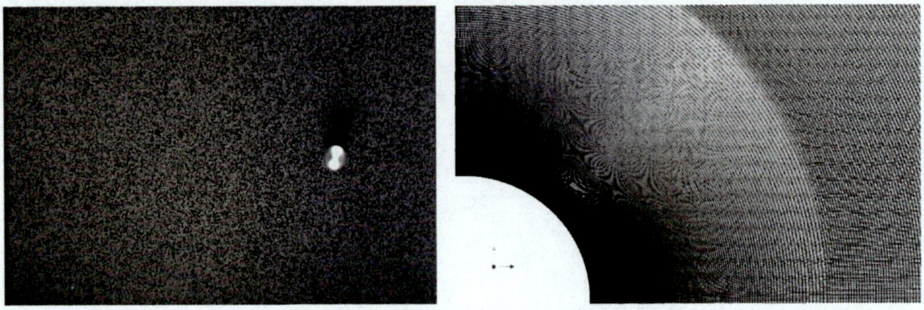

(a) Anodised and laser engraved aluminium background plate with randomly arranged dot pattern and circular cylinder attached.

(b) Computational domain details near the cylinder.

Fig. 2. The background used for BOS method and close-up view of CFD grid.

Table 1. Numerical and experimental flow conditions of the investigated case

	T_0	p_0	ρ_0	T_U	M_{in}	Γ_{in}	Z	Re_D	u
experimental	103 °C	260 kPa	23,08 kg/m^3	0.3%	0.67		0.91	5.12×10^5	62 ms^{-1}
numerical	97 °C	310 kPa	36,41 kg/m^3	0%	0.70	0.89	0.88	2×10^5	61.56 ms^{-1}

tive of our temporal resolution, especially at such CFL number. We can estimate the Kolmogorov time scale with a characteristic flow time scaled by the Reynolds number $\tau_\eta = \tau/Re^{1/2}$ where $\tau = D/U$, resulting in $\tau_\eta \approx 3.89 \times 10^{-8}$s which is one order of magnitude greater than what is imposed numerically.

4 Results

The flow conditions prevailed during the experiment as well as those for the numerical calculation can be found in Table 1.

To determine the theoretical frequency of shedding of the Kármán vortex street behind the circular cylinder, the Strouhal number shown in Eq. 1 is used.

$$S_t = \frac{f_s d}{u} \tag{1}$$

In this equation u describes the flow velocity, which equals 62m s^{-1} in the case investigated. A value of 0.2 is assumed for the Strouhal number S_t. f_s corresponds to the vortex shedding frequency, while D indicates the diameter of the cylinder, which is 5mm. After transforming the Eq. 1, the theoretical vortex shedding frequency f_s can be calculated as:

$$f_s = \frac{S_t u}{D} = \frac{0.2 \cdot 62ms^{-1}}{5 \times 10^{-3}} m = 2480s^{-1} \tag{2}$$

To estimate the actual Strouhal number a reference image with at least one significant vortex, shown in Fig. 5(a), is used. Starting from this image, the number n of images is determined, which elapse until another vortex is again present at the same position.

Multiplying $n-1$ by the recording frequency α of the high-speed camera yields a multiple of the actual vortex shedding frequency f_v. For the case shown in Fig. 5, this results in:

$$f_v = (n-1)\cdot\alpha = 4\cdot 5586 s^{-1} = 22344 s^{-1} \approx 9 f_s \tag{3}$$

This corresponds reasonably accurately to nine times f_s calculated in Eq. 2. Another way to obtain information about vortex shedding frequency is to adjust the recording speed of the camera to a multiple of the expected vortex shedding frequency. If the generated Schlieren images show vortices at the same location, the camera's recording frequency corresponds to a multiple of the real vortex shedding frequency. This method was not used in this work since the generation of the Schlieren images was not possible in real time.

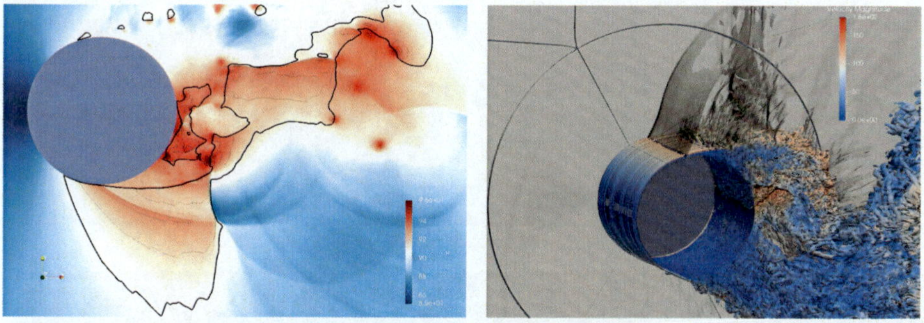

(a) Speed of sound c field with Mach number contours of a snapshot.

(b) Q criterion iso-surface colored by local velocity magnitude with the shock detection method from [6], and density gradient field in background.

Fig. 3. Snapshots from the two-dimensional LES.

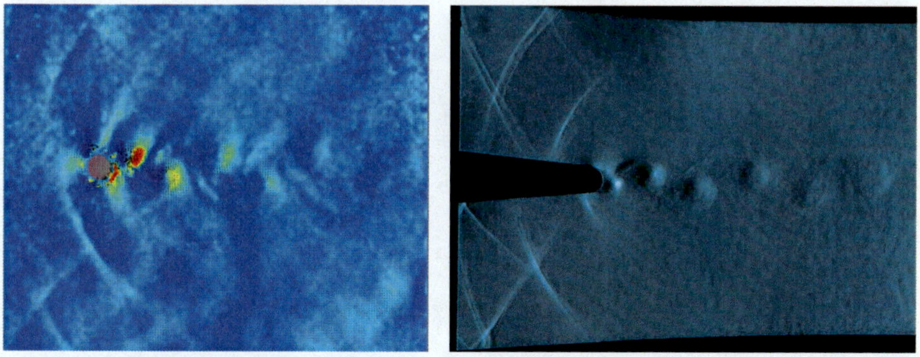

(a) Schlieren image of the flow around a circular cylinder in organic vapor at Mach 0.67.

(b) Schlieren image of vortex shedding behind a round trail- ing edge at Mach 0.65, adapted from [5].

Fig. 4. Comparison between a Schlieren image of the flow around a circular cylinder in the organic vapor Novec 649 captured by BOS and a Schlieren image of the flow past a round trailing edge.eps

Figure 4 shows the comparison of a Schlieren image of the flow around the cylinder at Mach 0.67 and a Schlieren image of vortex shedding behind a round trailing edge at Mach 0.65. Although the two images were taken under different flow conditions, except for the approximately equal inflow Mach number, as well as with different working fluids, there are strong similarities according to the vortex street. In addition both Schlieren images visualise clearly the substantial moving shock waves caused by the vortices.In Fig. 5, the Kármán vortex street occurring at a free-stream Mach number of 0.67 is shown with its dimensions. The distances between the vortices can be determined, for example, by the open source image analysis software ImageJ [8] and are listed in Table 2.

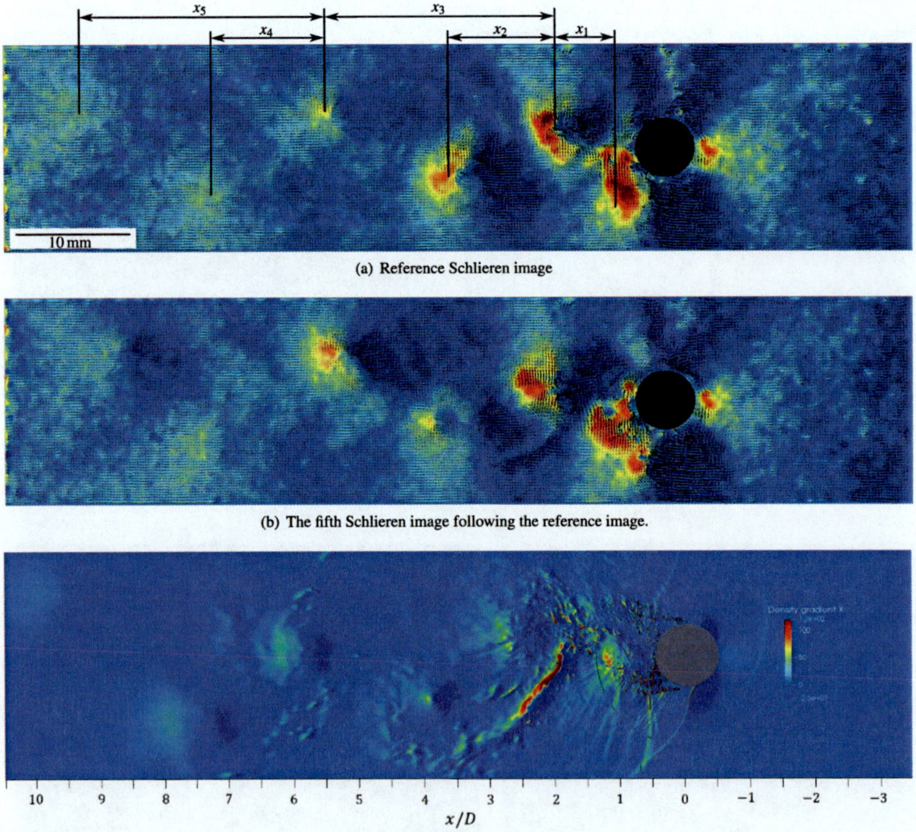

(a) Reference Schlieren image

(b) The fifth Schlieren image following the reference image.

(c) Numerical Schlieren image

Fig. 5. BOS images and a numerical Schlieren image with similar flow conditions.

Figure 5 shows the streamwise component of the density gradient field from the computation. The flow in the very near wake, less than one diameter behind the cylinder, is populated with small regions of strong density gradients. These correspond to

Table 2. Dimensions of the Kármán vortex street shown in Fig. 5

distance	x_1	x_2	x_3	x_4	x_5
x_i/D	1.0	1.9	3.9	2.0	4.9

the transitioning shear layers between the detached boundary layers and the free flow. Moreover, we clearly see a shock wave attached to the cylinder bottom wall where the boundary layer separates. For $x/D \in [1.5;3]$, a strong positive density gradient highlighting an elongated structure resembling a shock wave indicates that the flow can be locally supersonic in the near wake. Further downstream for $x/D \geq 3$ we clearly see how the shear layers have rolled-up to form the von Karman vortex street, which does not contain supersonic regions. Figure 3(a) shows real gas effects in the flow, as speed of sound increases during the initial expansion on the front section of the cylinder as it crosses the $M = 1$ contour line. The opposite is true for strong compressions, as can be seen in the near wake. This thermodynamic behaviour essentially reduces the maximum Mach number reached at the end of the expansion, weakening the shock waves which in turn modifies the near wake structures. In Fig. 3(b) a detailed three dimensional view of the transonic flow around the cylinder is shown. The breakdown of the shear layers is rapid, as shown by an iso-surface of the Q criterion, colored by the local velocity magnitude. The shock wave locations can be identified by applying the shock detection method proposed by Pagendarm [6], based on the density gradient projected on the streamlines everywhere in the flow. One weak shock exists at the boundary layer separation, followed by small compressibility regions due to the destabilised shear layer, ending with a strong normal shock. These intricate details of the instantaneous flow made available by the LES complete the experimental observations.

5 Conclusions

The BOS used was able to precisely capture flow phenomena that occur at higher Mach numbers. With the described method, the vortex shedding frequency could be determined, which is consistent with the expected shedding frequency. In addition, it could be shown that substantial moving shock waves emanate from the Kármán vortex street. This phenomenon is consistent with studies on airfoils with a rounded trailing edge [5]. This underlines the importance of basic flow experiments, such as the flow around a cylinder, in order to make application-specific predictions about complex flow phenomena. LES is used to help investigate intricate flow phenomena such as shock wave/wake interaction that are not accessible experimentally, and shows good promise with regards to the support it can bring to the Schlieren observations. The density gradient fields indicate the existence of locally supersonic regions present in the near wake, which vanish further downstream. These observations are in agreement with the BOS imaging. The role of the non-ideal thermodynamic behaviour of the gas is shown to be a crucial aspect for the formation of the near wakes structures, as this process is coupled with the presence of shock waves and supersonic expansions.

Acknowledgement. This work was part of the Franco-German research project "REGAL-ORC" funded by ANR and DFG.

References

1. Breuer, M.: A challenging test case for large eddy simulation: high Reynolds number circular cylinder flow. International Journal of Heat and Fluid Flow **21**(1), 648–654 (2000)
2. W. Cheng, D. I. Pullin, R. Samtaney, W. Zhang, and W. Gao. Large-eddy simulation of flow over a cylinder with Re_D from 3.9×10^3 to 8.5×10^5: a skin-friction perspective. *Journal of Fluid Mechanics*, 820(1):121–158, 2017
3. Cinnella, P., Content, C.: High-order implicit residual smoothing time scheme for direct and large eddy simulations of compressible flows. Journal of Computational Physics **326**, 1–29 (2016)
4. A. Jameson. Transonic flow calculations. *MAE report 1651, Princeton University*, 1983
5. A. P. Melzer and G. Pullan. The Role of Vortex Shedding in the Trailing Edge Loss of Transonic Turbine Blades. *Journal of Turbomachinery*, 141(4), 01 2019. 041001
6. H. G. Pagendarm and B. Seitz. An algorithm for detection and visual-ization of discontinuities in scientific data fields applied to flowdata with shock waves. *Scientific visualization: advanced software techniques.*, 1992
7. F. Reinker, E. Y. Kenig, M. Passmann, and S. aus der Wiesche.Closed loop organic wind tunnel (clowt): Design, components and control system.*Energy Procedia*, 129:200–207, 2017.4th International Seminar on ORC Power Systems, September 13-15th 2017, Politecnico di Milano, Italy.
8. Schneider, C.A., Rasband, W.S., Eliceiri, K.W.: NIH image to imageJ: 25 years of image analysis. Nature Methods **9**(7), 671–675 (2012)
9. G. S. Settles. *Schlieren and Shadowgraph Techniques: Visualizing Phenomena in Transparent Media*. Springer, Berlin, Heidelberg, 2001. OCLC: 851800772
10. W. Thielicke and E. J. Stamhuis. PIVlab - Towards User-friendly, Affordable and Accurate Digital Particle Image Velocimetry in MATLAB. *Journal of Open Research Software*, 2, 2014

Loss Measurement Strategy in ORC Supersonic Blade Cascades

Marco Manfredi, Giacomo Persico, Andrea Spinelli(✉), Paolo Gaetani, and Vincenzo Dossena

Department of Energy, Politecnico di Milano, Milano, Italy
{marco.manfredi,giacomo.persico,andrea.spinelli, paolo.gaetani,vincenzo.dossena}@polimi.it

Abstract. In the framework of Organic Rankine cycle (ORC) power systems, significant non-ideal gas effects typically take place in processes occurring within the cycle components. The flow non-ideality is of paramount importance for turbo-expander, whose design demands the application of advanced computational approaches, based on non-ideal compressible fluid dynamics (NICFD), to avoid misleading performance predictions. The validation of both low-fidelity and advanced aerodynamic methods based on state-of-the-art thermodynamic fluid models is therefore crucial but still limited, due to the lack of experimental benchmark data focused on ORC turbine losses. To fill this gap, an experimental campaign has been launched at Politecnico di Milano on a supersonic linear cascade representative of an ORC stator vane, with the aim of providing a reference for numerical simulations. This paper describes the devised measurement strategies conceived and implemented to investigate the flow-field established within the blade channels and downstream the planar row, particularly in the trailing edge region, and to retrieve the total pressure losses through the cascade with a relatively high spatial resolution, as they are the main outcomes of the experiments.

Keywords: Organic Rankine cycles · Experiments in non-ideal flow turbine blade cascades · Loss measurement

1 Introduction

Among non-ideal compressible flows in propulsion and power, organic Rankine cycle (ORC) turbine flows hold a relevant interest, since the expansion process occurs, at least initially, within a thermodynamic region close enough to the critical point for non-ideality to play a significant role. Also, small ORC units are often profitable if designed with a reduced number of stages [7] which typically results in supersonic stator design, at least for the first stage, which is also the one more severely affected by non-ideal effects. ORC turbine design/analysis tools typically rely on multi-fidelity models with increasing level of detail and complexity, including mean-line, through-flow, and computational fluid dynamic (CFD) based tools. CFD tools are not only at the basis of detailed blade design and analysis, but are also a reference for low fidelity tools validation, being experimental data, resolved at stage or interstage level, not available for ORC turboexpanders. It is worth recalling the relevance of reliable low fidelity

M. White et al. (Eds.): NICFD 2022, ERCOFTAC 29, pp. 217–228, 2023.
https://doi.org/10.1007/978-3-031-30936-6_22

tools in preliminary design and off-design analysis of turbines and whole ORC cycles [1,5,9,11]. The reliability of CFD simulations could be more critical when supersonic blade rows are considered, as complex shock structure originated at the blade trailing edge and having a relevant impact in loss production need to be properly caught.

With the aim of verifying the accuracy of CFD simulation, an experimental campaign on a supersonic linear cascade was designed, implemented and started on the Test Rig for Organic Vapors (TROVA) at Politecnico di Milano [8]. The cascade is designed to operate with siloxane MM at highly non-ideal conditions up to pressure and temperature of about 25 bar and 260°C respectively, with a corresponding compressibility factor $Z = P/\rho RT$ of 0.3, where P is the pressure, T the temperature, ρ the density and R the specific gas constant. This campaign is aimed at characterizing a converging-diverging planar cascade in terms of flow feature downstream the channel throat, including total pressure losses. Up to date, experimental data on typical organic fluid expansions were obtained for paradigmatic flow configurations [3,6,13,14,17] either isentropic or with simplified shock patterns, thus not representative of actual turbine blade rows, for instance missing any semi-bladed channel. Experiments on annular blade rows were also performed [2], but with a limited space resolution, a reduced level of non-ideality and by retrieving only numerically the total pressure distribution downstream the cascade. Beyond the challenges of measuring flow features at severe pressure and temperature conditions, in order to carry out the proposed experimental campaign, a strategy to measure cascade losses without resorting to fluid/thermodynamic conditions-specific calibration of employed instruments needs to be devised. The description and discussion concerning the implementation of this strategy is the main topic of this manuscript.

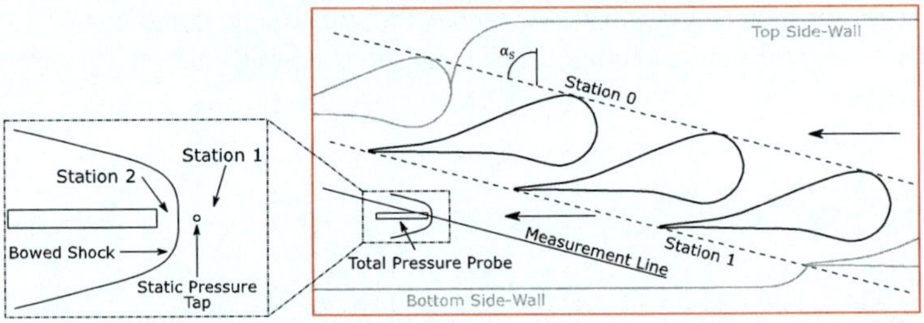

Fig. 1. Test section sketch.

The employed measuring techniques include double-passage schlieren to visualize density gradients and shock patterns in supersonic regions, endwall pressure taps at the blade channel diverging portion, and a miniaturized total pressure probe traversed downstream the blade trailing edge along two blade pitches. At each total pressure measuring point a wall static pressure tap is also located, which is needed to retrieve the flow total pressure upstream the probe-induced shock, namely the total pressure downstream the cascade, required to assess blade losses. The paper is organized as follows. Section 2 describes how the experimental campaign was motivated and conceived, while Sect. 3 reports the employed measurement techniques. Within Sect. 4 the method employed to assess cascade losses is presented and, finally, Sect. 5 draws the work conclusion and outlook.

2 Conception of the Experiment

Despite experimental data related to canonical compressible flows of organic fluids are available and mentioned in Sect. 1, experiments focused on the detailed investigation of non-ideal and supersonic flows through ORC turbine cascades, including total pressure loss estimation, are still missing in the open literature. The experimental campaign here described was conceived to fill this gap, aiming at matching multiple requirements. The linear cascade geometry was designed to reproduce the aerodynamics of supersonic stators implemented in high-temperature and high-expansion ratio ORC turbines, hence featuring a chocked converging-diverging bladed channel, a trailing-edge shock system, the shed and recovery of blade wakes, and the periodicity among different channels, so to reproduce a realistic loss generation through the blade row. Other crucial aspects were the selection of intake conditions relevant for state-of-the-art high-temperature ORC power plants, as well as the set-up of proper measuring techniques, capable of investigating the flow phenomena occurring in the blade trailing edge region and of retrieving the total pressure losses through the cascade with a relatively high spatial resolution. All these requirements needed to comply with the experimental facility constraints, such as the test section size and the transient operation typical of blow-down wind tunnels (a detailed description of the facility and of its operation can be found in [12,15,16]). The resulting experimental data may be exploited to develop and validate both low and high-fidelity CFD models implemented into ORC turbine design and analysis tools.

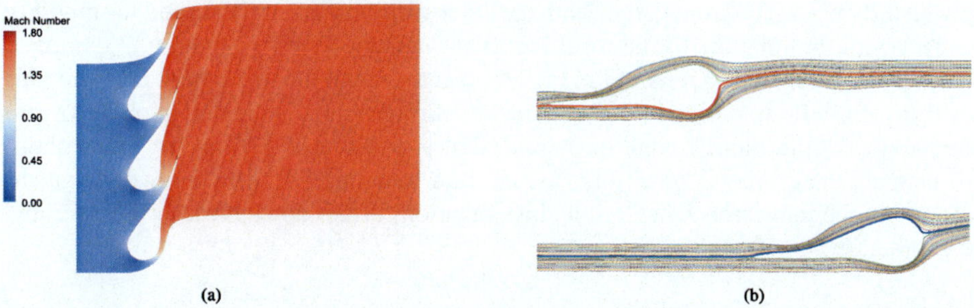

Fig. 2. Numerical results for the annular row configuration. (a) Mach number field (contour) with overlapped streamlines (grey solid lines). (b) Streamlines around a rotated blade profile: selected streamlines to design top (red) and bottom (blue) side-walls. (Color figure online)

The number of blades allocated into the cascade resulted from a trade-off between obtaining a good flow-field periodicity and complying with the reduced dimensions of the test section, a 93×180 mm rectangular plate. The test section, illustratively sketched in Fig. 1, features a linear cascaded with a geometric pitch of 45 mm, staggered at an angle $\alpha_s = 75°$, and composed by three complete blades, defining two central channels, and two side-walls integrating partial profiles to enclose the flow at the boundaries of the domain (reported for clarity in grey). As clarified in Fig. 1, the flow does not undergo any deflection expanding through the cascade, which is characterized by inlet and outlet parallel flows. This, on one hand, greatly simplifies the cascade integration within the TROVA test section; on the other hand, it makes the cascade representative of both radial-inflow and axial-flow supersonic nozzles. The former, indeed, typically receive swirled flow from the volute and make use of low-camber blades; the latter, instead, feature large flow turning but most of the turning occurs in the subsonic region upstream of the throat, while most of the entropy production occurs in the nearly-straight diverging section of the channel, where the flow is supersonic. This was also confirmed by the dedicated analysis concerning the inlet flow angle reported in [8], were results show that the flow rotation upstream the throat does not affect the downstream flow-field topology, as long as supersonic flows are established.

The detailed side-walls design was carried out exploiting dedicated CFD simulations, to attain an adequate downstream flow-field periodicity and reduced shock disturbances at the measuring station. For this analysis, a quasi-3D slice of an annular row representative of the cascade midspan section was considered, featuring the same blade profiles, pitch, and stagger angle. Simulations were performed using the ANSYS-CFX solver, integrating the RANS equations complemented by the $k - \omega$ SST turbulence model equations and by look-up tables to introduce the non-ideal thermodynamics of the fluid. Full details on (high-resolution) numerical schemes, mesh size, and boundary conditions can be found in [8]. The resulting flow-field is reported in Fig. 2a in terms of Mach number and velocity streamlines, while two focuses on the streamlines around a rotated blade profile are shown in the top and bottom part of Fig. 2b. The bold red and blue streamlines were extracted to define a tailored shape of the top and bottom cascade side-walls respectively, aiming at maximizing the cascade periodicity within the 2 central channels, avoiding upstream recirculating regions, and at minimizing downstream flow disturbances. The front part of both the side-walls was properly shaped to smoothly connect with the inlet duct. The resulting linear cascade layout, including 3 complete blades and 2 sidewalls, is reported in Fig. 3a, together with the in-channel pressure taps positions (in yellow) and the downstream measuring points (in red), which lie over a line parallel to the trailing edge locus placed at a distance of 17% of the blade chord downstream the trailing edges. Here cascade flow structure are still clearly distinguishable and probe interference negligible; investigation further downstream was prevented by geometrical constraints.

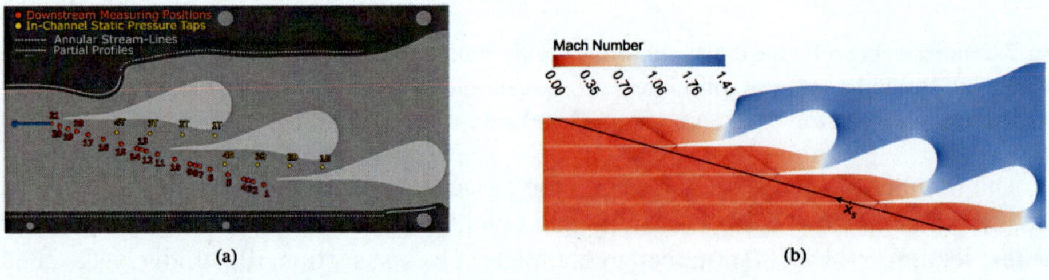

Fig. 3. Full linear cascade. (a) Final arrangement including side-walls design details, in-channel pressure taps position (in yellow), and downstream measurement points (in red). (b) 2D CFD simulation results: Mach Field. (Color figure online)

To assess the proposed side-wall design and define the spatial resolution and distribution of the measuring points, two-dimensional (2D) calculations of the flow at the midspan section of the actual cascade layout were considered. Structured multi-block meshes were created with ANSYS-ICEM CFD, which, differently from the physical domain, implement inlet and outlet mesh blocks extending upstream and downstream for about 0.75 and 5 blade chords respectively. The ANSYS-Fluent solver was applied since ANSYS-CFX does not accept two-dimensional geometries. This choice allows to avoid the implementation of fictitious layers in the span-wise direction, thus reducing the computational cost. Also in this case, the solver integrates the RANS equations, complemented by the $k-\omega$ SST turbulence model equations and by look-up tables to introduce the thermo-physical properties of the fluid. Since the component of flow Mach number perpendicular to the cross section is supersonic, calculations were performed by assigning at the inlet total pressure ($P_t = 25$ bar) and temperature ($T_t = 261.5°$C), flow direction, and turbulence. Further details concerning the numerical set-up can be found in [8]. To define the spatial resolution and position of the measuring points reported in Fig. 3a, the resulting Mach number field (reported in Fig. 3b) and the corresponding total pressure distribution were exploited. In agreement with the objectives of the experimental campaign, the downstream measuring grid results non-uniformly spaced, but conveniently refined in the vicinity of the wakes where higher gradients are expected, as shown in Fig. 3b.

The computed distributions over the black line in Fig. 3b (representing the measuring line) were processed to further assess the flow-field periodicity, also from a quantitative point of view. Figure 4a provides a qualitative proof of the attained periodicity, reporting as exemplifying flow property the Mach number distribution (orange line) versus the non-dimensional pitch-wise distance x_s from the bottom blade trailing edge, along with the position of the measuring points. To quantify the level of periodicity, the auto-correlation function was applied to the distributions over the measuring line of the most relevant flow properties, allowing to assess the degree of overlapping of the various signals with their delayed copies as function of the delay. The non-dimensional auto-correlation functions for entropy and Mach number are reported in Fig. 4a. As clearly shown, the peak distance, representing the fluid dynamic pitch, nearly coincides with the geometrical one (deviation lower than 0.6 mm, or $\approx 1.3\%$ of the pitch), and the two central auto-correlation peak values are very close to 1 (≈ 0.97). Both these outcomes certify the desired periodicity between the two central channels.

Further analyses were made by resorting to 3D numerical simulations of the full cascade layout, in order to keep into account three-dimensional effects, such as endwall boundary layers and their interaction with the shock pattern, and to assess they do not jeopardize the positive outcomes obtained through the 2D analyses. The meshing procedure for the 3D domain still relied on ANSYS-ICEM CFD but the number of elements greatly increased (9500000 cells) to guarantee the proper resolution in the endwall boundary layers. The numerical set up is identical to the one used for the 2D analyses except for inlet total pressure and temperature, which have been set equal to $P_t = 9$ bar and $T_t = 229°$C to numerically investigate another flow condition occurring during a typical test run. The resulting Mach number distributions at the cascade midpsan and over the measuring line for both 2D and 3D numerical domains are reported in

Fig. 4a. The results indicate, at first, a very a good flow periodicity also in off-design conditions, while P_t and T_t are decreasing as the high pressure vessel (HPV) is emptying. In addition, while strong 3D effects were not observed, very tiny differences between 2D and 3D Mach distributions can be seen along the measuring line in terms of wake position and depth, which are however sufficiently small not to challenge the validity of the experiments or to question the chosen potions of the measuring points.

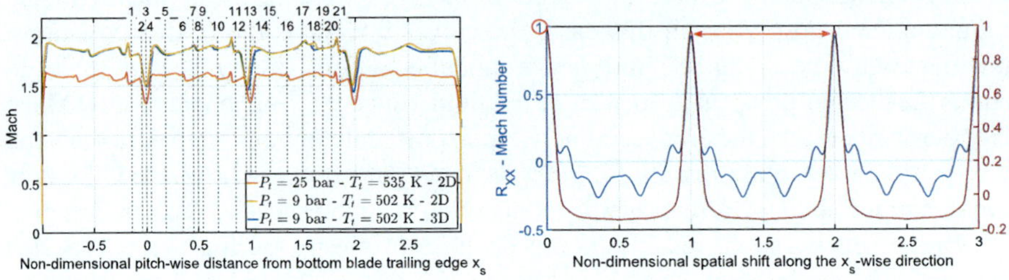

(a) (b)

Fig. 4. Periodicity analysis results - Distributions over the measuring line. (a) Mach number for different numerical domains and inlet conditions, and measuring points position (vertical dashed lines). (b) Auto-correlation function for entropy and Mach number: the unity value (marked with a circle) 1 corresponds to perfect periodicity.

3 Measurement Techniques

The detailed investigation of the flow field within the supersonic cascade, particularly in the blade trailing edge region, and the measurement of total pressure losses are the primary objectives of the experimental campaign. As described in Sect. 4, the direct measurement of total pressure downstream the cascade P_{t1} (refer to Fig. 1 for the subscript notation) is prevented by the formation of a bowed shock at the probe head. Four additional thermodynamic properties are then required to retrieve P_{t1} and, consequently, the total pressure loss. In particular, since the flow may be considered adiabatic due to the action of heating system and of insulation, the thermodynamic properties chosen in the present work are: total inlet conditions, namely P_{t0} and T_{t0}, total pressure downstream the probe-induced shock P_{t2} and static pressure upstream the shock P_1. Total conditions P_{t0} and T_{t0} are measured in the plenum, a settling chamber upstream the test section in which the flow is smoothly and moderately accelerated such that it can be considered uniform at the cascade inlet. The total temperature is provided through 2 thermocouples (of J and K type) with expanded uncertainty of 1°C. Due to the negligible kinetic energy in the settling chamber, P_{t0} is measured by a wall flush-mounted absolute piezo-resistive transducer for high-temperature applications with expanded uncertainty of approximately 0.1% of the full scale. The total pressure value is double-checked through another similar transducer placed at the end of a pneumatic line outgoing the test section and, therefore, not directly facing the flow. More generally, all the pressure measurements described in the following are performed using the same kind of piezoresistive transducer for high-temperature applications either flush-mounted or connected through a pneumatic-line, each one characterized by a different full scale value.

To characterize the pitch-wise distribution of loss coefficient a traversing system was devised, consisting of 21 measuring points not uniformly spaced along the two central blade pitches (see red dots in Fig. 3a). Along this line, P_{t2} is retrieved exploiting a pressure transducer connected by means of a pneumatic line to an ad-hoc designed total pressure probe (sketched in blue in Fig. 3a). Conversely, static pressure upstream the shock P_1 is provided by a locally-mounted sensor connected to the flow through a static pressure taps of 0.3 mm diameter (sketched in green in Fig. 3a). The probe head, pre-aligned with the flow, is 15 mm long, features a recessed stem, and is located at blade mid-span, while the taps are machined at the rear wall such that they are located exactly below the probe head tip. The motivations at the basis of this choice and the detailed description of the probe-tap configuration are reported in Sect. 4.

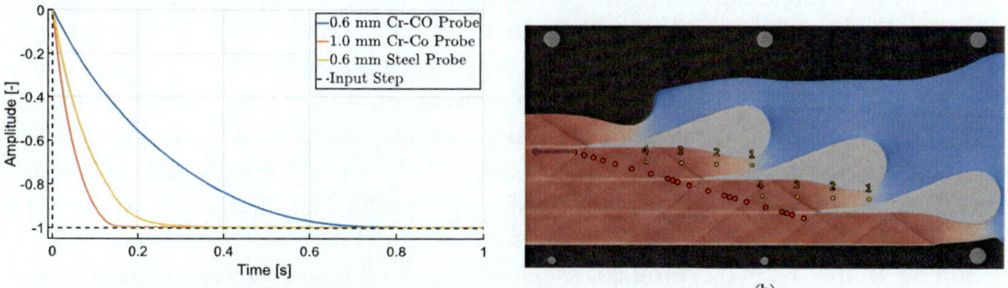

(a) (b)

Fig. 5. (a) Pneumatic lines dynamic responses to an input step for different probes. (b) Test section layout with overlapped Mach field and in-channel (yellow) and downstream (red) static pressure taps. (Color figure online)

The probe design presented several challenges related to four main requirements: necessity of local measurements, a dynamic response capable of following the (quasi-steady) HPV emptying, suitable structural stiffness to withstand high-pressure loads and avoid stem oscillations, a smart configuration to ease plug-in, orientation, and plug-out procedures. To guarantee resistance and rigidity, a cobalt-chrome alloy probe produced by means of stereolithography was considered. This material is characterised by a relatively high hardness and, in the first trials, the probe did not show any oscillations or deformations independently on the inlet conditions investigated (up to 25 bar). While probe external diameter was chosen equal to 1.6 mm, as a compromise between spatial resolution and instrument stiffness, the chosen value of 1 mm for the internal diameter resulted from a trade-off between dynamic response, integration area, and mechanical considerations. A dedicated dynamic analysis revealed that the 99.9% cut-off frequency of the pneumatic lines system implementing the 1 mm head probe resulted equal to about $f_{c,off} = 1.3$ Hz. These outcomes were considered acceptable, being $f_{c,off}$ larger than the typical frequency content of the HPV emptying dynamics, for which the 99.995% of the pressure signal energy content is always contained below 0.1 Hz. A further reduction of the probe head internal diameter would have produced an unacceptable drop in the cut-off frequency due an excessive shrinking at the 90° elbow region, caused by the inaccurate dimension tolerances of stereolithography. This is confirmed by the results shown in Fig. 5a, reporting the pneumatic line dynamic responses - experimentally characterized - to an input step for: two cobalt-chrome alloy probes with 1 mm and 0.6 mm internal diameter, and a steel probe with 0.6 mm internal diameter.

Since a limited number of access points are available, one single measuring position is investigated during a single test, making the probe traversing possible only by combining multiple test runs. This choice is justified by an already proven test repeatability [4] and, though extending the experimental campaign duration, it leads to two indubitable advantages. Firstly, the design and implementation of the sealing system was much easier, since no pitch-wise probe movement is required. The second one is the chance of exploring a wide range of operating conditions in a single test, from highly non-ideal to almost ideal, during the HPV emptying.

Beside the evaluation of loss coefficient, two main techniques were adopted to investigate the flow-field developing through the cascade, crucial for CFD development and validation. Four static pressure taps were manufactured on the end-wall plate, along the center-line of the diverging portion of the two central channels, complemented by an equal number of pressure transducer. The 8 taps are represented by the enumerated dots in Fig. 5b, in which the numerical (2D) Mach field is overlapped to the test section arrangement. The importance of these sensors is twofold: first, the processed pressure signals provide a quantitative estimation of the obtained periodicity between the flow developing within the two central channels; second, they contribute to characterize the flow expansion in the diverging portion of the channel and in the semi-bladed region, both upstream and downstream the fishtail shock stemming from the pressure side, as observable in Fig. 5b. Finally, a double-passage Schlieren equipment was also employed to visualize the flow-field density gradients. This allows to highlight with an almost continuous space resolution the pattern of fan/shock waves originated both within the diverging channels and at blade trailing edges, the morphology of the probe-induced shock structure, and the waves reflected at the bottom side-wall.

4 Loss Measurement Method

Since the cascade-exit flow is supersonic, a detached bowed shock sets up at the probe head, as shown in Fig. 1. This prevents from a direct measure of the total pressure downstream the cascade P_{t1} as the actual pressure retrieved through the probe is P_{t2}, the total pressure downstream the bowed probe-induced shock (obviously lower than P_{t1} due to shock losses). Under the assumptions of adiabatic conditions within the test section ($h_{t0} = h_{t1} = h_{t2}$) and of a normal shock in front of the probe head tap, 3 additional flow properties other than P_{t2} need to be concurrently measured to unambiguously retrieve P_{t1}. This is demonstrated by the three sets of equations reported in Equation (1), which specifies, from left to right, the thermodynamic model of the fluid (where the double subscript indicates that those equations are accounted for both in station 1 and 2), the definition of total enthalpy, and the integral balance equations through a normal shock.

$$\begin{cases} h_{1,2} = h(\rho_{1,2}, P_{1,2}) \\ s_{1,2} = s(\rho_{1,2}, P_{1,2}) \\ s_{1,2} = s(h_{t0}, P_{t1,t2}) \end{cases} , \quad \begin{cases} h_{t0} = h(T_{t0}, P_{t0}) \\ h_1 + \frac{V_1^2}{2} = h_{t0} \end{cases} , \quad \begin{cases} \rho_1 V_1 = \rho_2 V_2 \\ \rho_2 V_2^2 - \rho_1 V_1^2 = P_1 - P_2 \\ h_1 + \frac{V_1^2}{2} = h_2 + \frac{V_2^2}{2} \end{cases} . \tag{1}$$

In Equations (1) subscript t refers to total conditions, h and s stand for specific enthalpy and entropy per unit mass respectively, and V is the flow velocity, assumed fully normal to the central part of the bowed shock being the probe a priori flow-aligned. Equations 1 form a system of 11 mathematical relation in 15 unknowns, namely $h_{1,2}$, $P_{1,2}$, $\rho_{1,2}$, $s_{1,2}$, $P_{t1,t2}$ $V_{1,2}$, h_{t0}, P_{t0} and T_{t0}. Therefore, 4 independent flow properties need to be measured to attain a well-posed mathematical problem. As stated before, in the present work the inlet total conditions P_{t0} and T_{t0}, the static (P_1) and total pressure ($Pt2$) respectively upstream and downstream the probe-induced shock were chosen.

As introduced in Sect. 3, P_1 is measured through static pressure taps machined in the rear end-wall plate exactly below the total pressure probe head tip. This configuration is schematically represented in Fig. 6a, in which the abscissa x represents the axial distance from the probe head tip and the ordinate y is the span-wise distance from the rear end-wall, being the test section height equal to 18.7 mm. To retrieve the correct value of total pressure downstream the cascade (upstream the probe-induced shock) solving Eqs. (1), it is of primary importance to assess if the wall measured value is fully representative of the static pressure upstream the bowed shock. This choice is indeed theoretically supported by the boundary layer theory and by the fact that a planar cascade is involved, meaning that no span-wise pressure gradient are expected in the clean flow region. Moreover, the flow is supersonic, meaning that disturbances cannot be propagated upstream other then within boundary layers. However, the formation of a bowed shock at the probe head may affect the pressure measured at the wall if the shock-end wall boundary layer interaction happens sufficiently close to the wall tap, such that the distance travelled upstream by the disturbances is enough to modify the static pressure read by the static tap. The model developed by Moeckel [10] was firstly applied to estimate the bowed shock detachment distance from the probe head, its inclination in the oblique shock region, and, consequently, the intersection point between the shock and the rear end-wall. The results, reported in Fig. 6a for different values of the shock-upstream Mach number, seems to exclude the possibility of an interaction between the bowed shock and the retrieved wall static pressure. In fact, the Mach number at the cascade measuring station (i.e. upstream the bowed shock) is about 1.6 for the design conditions and increases while the inlet total pressure decreases as the test proceeds on (see also Fig. 4a).

To further asses the effectiveness of measured wall pressure in representing the one upstream the shock, a numerical investigation relying on CFD simulations were carried out. Some assumptions were needed to ease the meshing procedure and to investigate the physical problem without considering the full three-dimensional cascade layout with the total pressure probe as numerical domain. Therefore, a 3D prismatic domain was considered, characterized by a square cross section of 18.7 mm edge (equal to the cascade height) extruded in the axial, or stream-wise, direction for 120 mm. For sim-

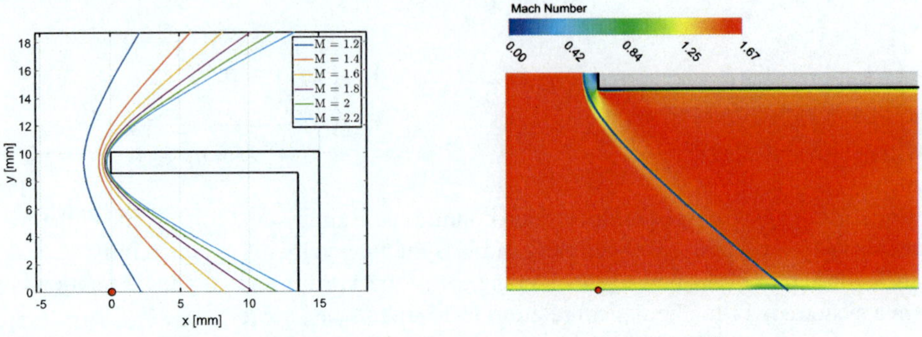

(a) (b)

Fig. 6. (a) Schematic representation of probe and tap (red dot) configuration. Bowed shock shape and location according to Moeckel's model [10] as a function of the upstream Mach. (b) Mach number distribution in the section defined by axial and span-wise direction and passing from the probe center. Comparison between the bowed shock shape resulting from CFD simulation and Moeckel's model. (Color figure online)

plicity the probe was schematized by a cylinder parallel to the prism, with an external diameter of 1.6 mm, and placed in middle of the domain. Exploiting the double symmetry of the problem and to lighten the computational effort, only a quarter of the numerical domain was actually meshed leading to a total number of cell elements of about 1 M. Calculations were performed using the ANSYS-Fluent solver and the same numerical set up described above. To set the boundary conditions (in terms of inlet total pressure and temperature, flow angle, and turbulence) the results of the planar cascade numerical simulations have been exploited, extracting the quantities of interest in one of the measuring points (green dots in Fig. 3a). The resulting Mach number field, in the section defined by axial and span-wise direction and passing from the probe center, is shown in Fig. 6b. As can be observed, the numerical results prove the lack of any interaction between the bowed shock and a pressure tap placed below the total pressure head tip (same axial position). In fact, the resulting pressure at the wall is equal to the one just upstream the probe-induced shock, confirming the possibility of using the measured pressure to retrieve P_{t1}. In Fig. 6b the bowed shock shape predicted by Moeckel theory for the exact value of the Mach number is also reported and compared with 3D high-fidelity results, highlighting a satisfactory agreement in terms of shock detachment distance from the probe head. On the contrary, some discrepancies concerning the bowed shock opening are observable moving far away from the normal shock region.

5 Conclusions and Future Works

This paper has presented a novel experiment on technically-relevant supersonic cascades operated with organic fluids. The experiment is aimed at investigating the shock systems originated at the blade trailing edge, the wake shed by the blades, and ultimately the total pressure loss caused by the aforementioned effects. Both flow visualizations and quantitative pressure measurements are considered, so to provide a comprehensive set of data for future validation of computational models. The key feature of the experiment is the linear arrangement of the blades, which are placed in the test section so to reproduce the actual periodicity of the annular cascade, but without imposing any flow deflection. This feature greatly simplifies the discharge of the flow from the test section and guarantees a high flow uniformity at the cascade inlet. Dedicated computational studies were carried out to define and assess the periodicity of the flow in the cascade, using both 2D and 3D flow models of the entire test section, in both design and off-design cascade conditions. Results show a noteworthy periodicity in the two complete channels reproduced in the cascade. In this paper, we have also proposed a methodology to retrieve the local total pressure value at the cascade exit using a total pressure probe operated in nulling mode (so that no aerodynamic calibration is required) combined with several pressure measurements on proper taps manufactured on the end-wall plate. In this way, the experimental pitch-wise profiles of cascade-exit total pressure and loss coefficient can be obtained. Preliminary trials recently performed in the facility have confirmed the flow configuration expected from CFD simualtions and have assessed the validity of the measurement concept. The experimental campaign is presently ongoing and the full set of experimental results will be presented in the near future.

References

1. Astolfi, M., Macchi, E.: Efficiency correlations for axial flow turbines working with non-conventional fluids. In: 3rd International Seminar on ORC Power Systems, pp. 12–14 (2015)
2. Baumgärtner, D., Otter, J.J., Wheeler, A.P.: The effect of isentropic exponent on transonic turbine performance. J. Turbomach. **142**(8), 081007 (2020). https://doi.org/10.1115/1.4046528
3. Beltrame, F., Head, A., De Servi, C., Pini, M., Schrijer, F., Colonna, P.: First experiments and commissioning of the ORCHID nozzle test section. ERCOFTAC Ser. **28**, 169–178 (2021)
4. Cammi, G., Conti, C.C., Spinelli, A., Guardone, A.: Experimental characterization of nozzle flow expansions of siloxane MM for ORC turbines applications. Energy **218**, 119249 (2021)
5. Da Lio, L., Manente, G., Lazzaretto, A.: A mean-line model to predict the design efficiency of radial inflow turbines in organic Rankine cycle (ORC) systems. Appl. Energy **205**, 187–209 (2017)
6. Gallarini, S., Cozzi, F., Spinelli, A., Guardone, A.: Direct velocity measurements in high-temperature non-ideal vapor flows. Exp. Fluids **62**(10), 1–18 (2021). https://doi.org/10.1007/s00348-021-03295-4
7. Macchi, E., Astolfi, M.: Organic Rankine Cycle (ORC) Power Systems. Technologies and applications. Woodhead Publishing Series in Energy: Number 107, Elsevier, New York (2017)
8. Manfredi, M., Persico, G., Spinelli, A., Gaetani, P., Dossena, V.: Design of experiments for supersonic orc nozzles in linear cascade configuration. In: 6th International Seminar on ORC Power Systems, pp. 1–12. Technical University of Munich (2021)
9. Manfredi, M., Spinelli, A., Astolfi, M.: Simultaneous optimization of cycle and expander design for truck on board waste heat recovery with orc technology. In: 6th International Seminar on ORC Power Systems, pp. 1–11. Technical University of Munich (2021)
10. Moeckel, W.E.: Approximate method for predicting form and location of detached shock waves ahead of plane or axially symmetric bodies. Technical report NACA-TN-1921 (1949)
11. Perdichizzi, A., Lozza, G.: Design criteria and efficiency prediction for radial inflow turbines. In: ASME 1987 International Gas Turbine Conference and Exhibition. American Society of Mechanical Engineers Digital Collection (1987)

12. Pini, M., Spinelli, A., Dossena, V., Gaetani, P., Casella, F.: Dynamic simulation of a test rig for organic vapours. In: ASME 2011 5th International Conference on Energy Sustainability, Parts A, B, and C. Energy Sustainability, pp. 1977–1988 (2011). https://doi.org/10.1115/ES2011-54212
13. Robertson, M.C., Newton, P.J., Chen, T., Martinez-Botas, R.F.: Development and commissioning of a blowdown facility for dense gas vapours. In: Turbo Expo: Power for Land, Sea, and Air, vol. 58608, p. V003T28A002. American Society of Mechanical Engineers (2019)
14. Spinelli, A., et al.: Experimental evidence of non-ideal compressible effects in expanding flow of a high molecular complexity vapor. Exp. Fluids **59**(8), 1–16 (2018). https://doi.org/10.1007/s00348-018-2578-0
15. Spinelli, A., Dossena, V., Gaetani, P., Osnaghi, C., Colombo, D.: Design of a test rig for organic vapours. In: Proceedings of ASME Turbo Expo, Glasgow, UK, June 2010
16. Spinelli, A., Pini, M., Dossena, V., Gaetani, P., Casella, F.: Design, simulation, and construction of a test rig for organic vapors. J. Eng. Gas Turbines Power **135**, 042304 (2013)
17. Zocca, M., Guardone, A., Cammi, G., Cozzi, F., Spinelli, A.: Experimental observation of oblique shock waves in steady non-ideal flows. Exp. Fluids **60**(6), 1–12 (2019). https://doi.org/10.1007/s00348-019-2746-x

Mach Number Estimation and Pressure Profile Measurements of Expanding Dense Organic Vapors

Adam J. Head[1(✉)], Theodoros Michelis[1], Fabio Beltrame[1], Blanca Fuentes-Monjas[1], Emiliano Casati[2], Carlo De Servi[1,3], and Piero Colonna[1]

[1] Propulsion and Power, Delft University of Technology, Delft, The Netherlands
{a.j.head,t.michelis,f.beltrame, c.m.deServi,p.colonna}@tudelft.nl

[2] Department of Mechanical and Process Engineering, ETHZ, Zürich, Switzerland
casatie@ethz.ch

[3] Energy Technology Unit, VITO, Mol, Belgium
carlo.deservi@vito.be

Abstract. This paper describes an experiment conducted within the nozzle test section of the Organic Rankine Cycle Hybrid Integrated Device (ORCHID) aimed at providing accurate data for the validation of NICFD flow solvers [5]. A supersonic flow of the dense vapor siloxane MM established in the nozzle of the setup was characterized by means of the schlieren technique and by pressure taps along the nozzle profile. The nozzle inlet conditions corresponded to a stagnation temperature and pressure of $T_0 = 253$ °C and $P_0 = 18.36$ bara. At these inlet conditions, the compressibility factor of the fluid is $Z_0 = 0.58$. The nozzle backpressure was equal to $P_b = 2.2$ bara. The experimental data-set includes: 1) the average mid-plane local Mach number, which was derived from the schlieren images by estimating the angle of the Mach waves originating from the roughness of the upper and lower nozzle surfaces, 2) the angle of a shock wave generated by a 5° wedge placed at the nozzle exit, also detectable in the schlieren images, and 3) the static pressure distribution along the flow expansion acquired with a *Scanivalve* DSA3218 pressure scanner device. The Mach number at the nozzle exit estimated based on the schlieren images is $M = 1.95 \pm 0.05$, very close to the expected value of $M = 2$ according to the design conditions of the experiment. The static pressure measurements have a maximum absolute uncertainty amounting to ± 1.80 kPa in the initial stages of the expansion. This information was used to assess the capability of the open-source SU2 flow solver in evaluating the NICFD effects in a supersonic flow of MM when the fluid thermodynamic properties are modeled with a cubic equation of state. For this purpose, two-dimensional Euler simulations were carried out with SU2 for the operating conditions achieved in the experiment. The numerical results are in good agreement with the experimental data. The largest deviation between the simulation and experiment is observed in the nozzle uniform region, where two dips in the Mach number occur due to a slight local decrease in flow velocity owing to two weak shock waves. The shock wave generated by the wedge located at the nozzle

M. White et al. (Eds.): NICFD 2022, ERCOFTAC 29, pp. 229–238, 2023.
https://doi.org/10.1007/978-3-031-30936-6_23

outlet propagates with two different angles, namely, $\beta_{above} = 37.6^\circ \pm 0.86$, and $\beta_{below} = 31.6^\circ \pm 0.64$, due to the axial misalignment of the wedge with respect to the flow.

Keywords: Schlieren measurements · data processing · error identification and uncertainty estimation

1 Introduction

The computer simulation of compressible flows is critical to the development of advanced technologies, like airplanes and aero-engines, rockets, and gas turbines, to name a few. Over the past few decades, CFD codes have reached a remarkable level of accuracy because of the extensive availability of accurate experimental data for their verification and validation [11]. However, these experimental data sets are limited to aerodynamic flows or classical and ideal gas dynamics. There is limited data available for flows typically encountered in the research subject of Non-Ideal Compressible Fluid Dynamics (NICFD), such as transcritical expansions close to the critical point and compressible two-phase flows. Although first experimental results on non-ideal compressible effects in supersonic flows of dense organic vapours have been recently published, see Ref. [2, 12, 14], the data in the literature remain insufficient in terms of quantity and quality for the validation of NICFD-capable flow solvers.

The Organic Rankine Cycle Hybrid Integrated Device (ORCHID) of the Propulsion and Power laboratory of the Delft University of Technology was built for this reason [5]. Notably, the setup is currently equipped with a nozzle test section to generate paradigmatic supersonic flow test cases similarly to what is done for classical and ideal gas dynamics, but for flows in the NICFD regime. The objective is to accurately measure quantities in the flow established in a convergent divergent nozzle, either empty or with an obstacle inserted to generate shock waves, for the validation of CFD codes capable of simulating NICFD flows.

This paper reports the results of an experimental campaign executed inside the ORCHID nozzle test section. In particular, datasets have been generated for: i) the total conditions at the inlet and outlet of a planar nozzle; ii) thanks to schlieren imaging, Mach lines over the entire length of the diverging part of the nozzle, and a shock-wave generated by a 5° wedge positioned at the exit of the nozzle in the testing channel; and, iii) static pressure measurements along the upper and lower surfaces of the planar nozzle. To ensure the reproducibility of the experimental test cases with a CFD model, a method for tracking the nozzle geometry, which changes position due to thermal expansion of the nozzle housing, was developed. Furthermore, uncertainties associated to the measured quantities have been calculated and reported.

The paper is structured as follows. Section 2 presents the layout of the nozzle test section together with the pressure measurement instrumentation. Furthermore, the schlieren layout and measurement procedure together with the specifications of the experiment are defined. In Sect. 3 the schlieren data post processing method and the

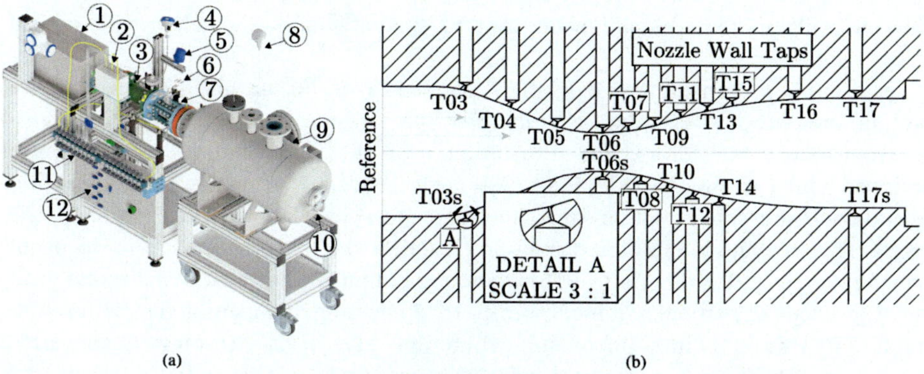

Fig. 1. The primary components and instrumentation of the nozzle TS. **(a).** ①: Flow Meter *KROHNE OPTIMASS 6400 F F* – Tag FT0004; ②: *Scanivalve DSA3218*; ③: Settling Chamber; ④: *WIKA UPT 20* – Tag PT011; ⑤: *WIKA TR10-C PT100* – Tag TT015; ⑥: *Taps*; ⑦: Nozzle Housing; ⑧: *WIKA UPT 20* – Tag PT014; ⑨: Receiver; ⑩: *WIKA TR10-C PT100* – Tag TT014; ⑪: Liquid Trap; ⑫: Scanivalve Control Panel. **(b).** A cross-section of the planar nozzle showing the positioning of the eighteen pressure taps. Three taps are symmetrically aligned and are indicated by the letter s. Detail A shows that the opening of each tap is perpendicular to the nozzle profile wall. Distances to taps from Reference in [mm]: T03(s) = 15.96; T04 = 24.72; T05 = 33.55; T06(s) = 42.4; T07 = 47.44; T08 = 49.9; T09 = 52.57; T10 = 55.01; T11 = 57.5; T12 = 59.91; T13 = 62.53; T14 = 64.89; T15 = 67.47; T16 = 79.35; T17(s) = 91.3. Two taps, T01 & T02, are located in the settling chamber at the first and second flange. Nozzle profile designed for the MM working fluid at the design conditions: $T_0 = 252$ °C, $P_0 = 18.4$ bar and a back pressure of 2.1 bar with a throat height $H_{th} = 7.5$ mm. See Nozzle B in [5, App. B] for the geometric points defining the profile. Fluid properties estimated with the multi-parameter equation of state available in the thermophysical property database of NIST [8].

uncertainty estimation of the experiment error sources is presented. Section 4 reports the average midline Mach number distribution in the diverging part of the planar nozzle together with the associated uncertainties. Furthermore, the pressure field obtained with a pressure scanner is also shown. The experimental results are also compared with the predictions of a CFD model. The conclusions and an outlook for future work are presented in Sect. 5.

2 The Nozzle Test Section: Layout and Measurement Procedure

Fig. 1a shows an isometric view of the ORCHID nozzle test section (TS) including the system instrumentation. A complete detailed overview of the main TS components, namely, the settling chamber ③, nozzle housing ⑦ and the receiver ⑨ can be found in [5, Ch. 6]. Figure 1b shows the converging diverging nozzle profile which has been designed with the method of characteristics [4]. The nozzle is made from two equal and removable profiles, which are inserted at the top and bottom of the nozzle housing.

The side walls of the nozzle are made from transparent windows which provide optical access.

Two pressure transmitters *Wika UPT 20*, see ④ & ⑧, are used to characterise the total pressure in the settling chamber and the pressure in the receiver. Both pressure transmitters have an expanded uncertainty of ± 0.1 bar. The test section is also equipped with a *Scanivalve* (SV) DSA3218 pressure scanner ② which is used, in combination with pressure taps, to measure the static pressures along the nozzle profile. The DSA3218 is a stand-alone digital pressure scanner which can accept up to 16 pneumatic inputs (ports $P_{SV}001$ to $P_{SV}016$) where each input is connected to a discrete pressure transducer. In particular, it incorporates 16 temperature compensated piezoresistive pressure transducers with a pneumatic calibration valve, RAM, 16 bit A/D converter, and a microprocessor in a compact self contained module. The pressure sensors are grouped according to their full scale specification, namely 34.5 barg ($P_{SV}001$–003), 17.2 barg ($P_{SV}004$–09), and 6.8 barg ($P_{SV}010$–016). Given an accuracy of $\pm 0.05\%$ of the full scale of the sensors, the expanded uncertainty for each sensor group is $\pm 0.0.0172$ bar, ± 0.0086 bar, and ± 0.0035 bar respectively. Figure 1b shows eighteen taps situated along the upper and lower nozzle profiles. Two extra taps are located on the flanges of the settling chamber. On the nozzle profile the taps were obtained by boring 0.2 mm diameter holes perpendicular to the profile surface, and afterwards, by drilling a 2.5 mm hole from the opposite side, see detail A. Eighteen 1/16 in. Swagelok tubes are braze-welded to the externally mounted surface of the nozzle profile and separately routed, together with the two tubes connected to the taps of the settling chamber, underneath the nozzle test section to small chambers equipped with sight glasses, i.e., the liquid traps (see ⑪ for the assembly and ⑫ for the valve control panel). This allows to visualise vapor condensation during pressure measurements. To avoid the accumulation of condensate in the lines connecting the pressure taps to the Scanivalve, the pressure scanner is equipped with a nitrogen purging system. The total temperature is measured at the nozzle inlet, see ⑤, with a *Wika TR10-C* PT100 transmitter. The expanded uncertainty is of Class A, i.e., $\pm 0.15 + 0.002\,|T|$ °C as per the EN 60751 standard. The same type of transmitter is also used to measure the temperature, see ⑩, in the receiver.

The adopted schlieren layout is a two-lens flat mirror z-type configuration. The exact same equipment, i.e. light source, pinhole, lenses and camera, employed in the experiment reported in [1] is adopted to visualize gradients of density. In the experiment described in this paper, an additional camera (*BOBCAT IGV-B1610* 16bit CCD) is used to image the nozzle profile geometry, allowing to track changes owing to thermal expansion and, hence, determine the nozzle throat size at various operating conditions.

The operating conditions targeted in the experiment, see Table 1, correspond to the design point of the ORCHID Balance of Plant (BoP) and nozzle test section. Figure 2 displays the thermodynamic states identifying the expansion process in the nozzle under the hypothesis of an isentropic adapted flow. The nozzle TS must be sufficiently prewarmed (at 180 °C for experiments with MM) in order to prevent excessive condensation during the initial stages of the TS operation. Furthermore, due to the inability of gate-valves at the inlet and outlet of the TS to perfectly seal against the flow direction, liquid tends to reside in the receiver during the ORCHID start-up sequence. Thus, when the TS is opened, it is necessary to wait for at least 20 min to allow all the liquid to

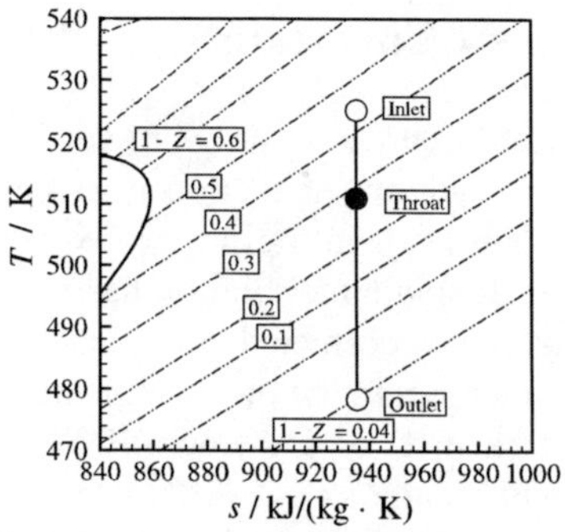

Fig. 2. $T - s$ diagram showing the isentropic expansion for the design conditions of the nozzle (also the target conditions for PR.028-NT.001) with contours of 1-Z (−··).

Table 1. The target boundary conditions of the experiment PR.028-NT.001.

Fluid	Tot. temp. TT015 / °C	Tot. press. PT011 / bara	Comp. F. Z_{Throat}
MM	252	18.4	0.56
Back p. PT004/bara	Outlet Mach. M_2	Mass flow FT001 /kg/s	
2.1	2	1.145	

evaporate and the operating conditions to stabilize around the values of total inlet pressure and temperature chosen for the experiment. During this settling period, the back pressure PT004 of the nozzle is set to the desired value. Once steady state is achieved, 1,000 schlieren image realisations are acquired at a frequency of 25 Hz, for a total time of 40 s. This ensures the statistical convergence of the measured quantities derived from the schlieren images. Spatial calibration of both cameras is performed by means of a calibration target located at the midspan of the test section, which bears circular markers of 1 mm spacing in both X and Y directions. A pinhole calibration model is employed which additionally accounts for lens distortion. A preliminary observation of the flow patterns in the schlieren images allow to verify if supersonic conditions have been reached in the nozzle and if the fluid is in a dry vapor state, before pressure measurements are carried out. A general outline of the pressure measurement procedure is as follows. The *DSA3218* pressure scanner is pre-warmed one hour before measuring in order to ensure the module becomes thermally stabilized. The pressure lines are purged with nitrogen in order to remove any condensate which may have accumulated during the nozzle starting phase, and which is therefore visible through the sight glass of the liquid trap, see Fig. 1a. Liquid purging ensures that the measuring ports are not blocked by, e.g., dust, metal filaments, tracer particles, and minimises the risk of damaging the transducers while measuring. Additionally, the presence of condensate in the lines is unwanted since it does not allow to correctly measure the flow pressure. The liquid is ejected back into the flow, and this can be checked because the liquid level in the trap must decrease. This procedure normally lasts approximately 10 s. A quick-zero adjustment is then performed and the recording of pressures can begin immediately. In this experiment the pressures are recorded at an acquisition frequency 1 Hz. Simultaneous flow visualization can be performed at this stage of the pressure measurements procedure. Periodic purging and zero-offset corrections must be made, depending on the duration of the pressure recording. For example, if the measurement lasts more than 3 min condensate starts to accumulate in the liquid traps.

3 Data Processing Method and Measurement Uncertainty

The acquired images have been pre-processed using LaVision Davis software version 8.3.0 [7] and subsequently post-processed by means of a Mach line extraction algorithm implemented in a general purpose programming software [13]. In the pre-processing phase, after correcting the dataset for shift and rotation, a spatial calibration has been performed. As a result, the raw distorted schlieren dataset is corrected such that any point in the image sequence corresponds to the correct location in the physical plane.

The Mach line extraction algorithm allows the evaluation of the slope of Mach waves with respect to the local flow velocity (Mach line angle μ) at different axial positions along the nozzle midplane ($Y = 0$) for each frame of the schlieren dataset. The uncertainties associated with both the Mach line angle value and the assigned position in the physical plane are also evaluated, see Ref. [5] for the details. From the Mach line angles, one can estimate the Mach number of the flow on the midplane provided that the flow in the nozzle is symmetric. The flow symmetry is verified by comparing Mach line angles above and below the midplane at different axial coordinates. It was observed that the average values of the angles of two concurring lines do not differ for more than their uncertainty band widths. In addition, the last curved Mach lines of the Kernel region originated both from the top and bottom profile coincide in a point which corresponds to the physical midplane of the nozzle, further proving flow symmetry. As a result, the extracted Mach line angles in the diverging part of the nozzle can be used to calculate, for each image, the corresponding Mach number distribution according to $M = \frac{1}{\sin \mu}$.

The algorithm consists of the following steps. First, the intensity of each image is normalised by dividing the intensity in each pixel by the standard deviation of this quantity calculated over a moving window of 5×5 pixels centered at the pixel being processed. This operation enhances the flow features captured by the schlieren technique. Second, the image is binarized and split in several overlapping interrogation windows, whose size is constant throughout the portion of the image corresponding to the uniform flow region, while it progressively decreases as the Mach lines of interest are closer to the throat. This gradual decrease of the interrogation window size is a function of the local height of the nozzle, and serves to ensure that the curvature of the detected lines does not exceed the local angular resolution. Furthermore, the step between an interrogation window and the next one is chosen to be one third of the width of the previous window. Given the imposed adaptation of the interrogation window size, it follows that the detection algorithm searches for more Mach lines in the kernel region. Third, the Hough transform is applied to each interrogation window [9] and the best line is selected. These operations are repeated for each frame. Finally, once the procedure has been applied to the whole image dataset, it is possible to compute the average and the standard deviation of the angle and length of all the detected Mach lines in each interrogation window. The Mach number distribution along the midplane and its total expanded uncertainty values can then be calculated. For more details on the line extraction algorithm and the uncertainty quantification procedure the reader is referred to [1,5]. Regarding the pressures acquired with the Scanivalve, the statistical averages and standard deviations are directly computed from the measurements obtained in steady conditions.

4 Mach Number Estimation and Static Pressure Measurements

The numerical solutions of the flow-field in the ORCHID nozzle are obtained using the open-source flow solver SU2 for a 2D half-nozzle with symmetry conditions along the midplane. The details about the numerical model, mesh convergence study and the specifications of the used workstation can be found in [3]. However, in this case, variations of specific heat capacity with temperature are considered and the $c_p(T)$ model reported in [10] was used. The critical point properties of MM were taken from [8]. The total inlet conditions assumed in the CFD model correspond to the average total pressure and temperature measured at the nozzle inlet during the experimental run once steady flow conditions established. Their values are $\bar{P}_0$ (PT011) = 18.36 ± 0.0229 bar, $\bar{T}_0$ (TT015) = 253.7 ± 0.66 °C. The pressure measured at the nozzle outlet is $\bar{P}_b$ (PT004) = 2.21 ± 0.0416 bar.[1] The nozzle profiles considered in the CFD model accounts for the thermal expansion of the nozzle housing. Notably, by analyzing with a dedicated camera the position of markers on the lateral surface of the nozzle profiles, the throat height was determined as $H_{th} = 7.73$ mm.

Figure 3a shows the first schlieren image from the recorded sequence, while Fig. 3b represents the same image after the moving filter has been applied to enhance the flow patterns. The average extracted Mach lines are superimposed on the midplane. A shock wave generated by a 5° wedge located at the nozzle outlet has an angle of $\beta_{above} = 37.6° \pm 0.86$, and $\beta_{below} = 31.6° \pm 0.64$ owing to the axial misalignment of the wedge which occurred during the nozzle TS start up. This will be prevented in the future by means of a more mechanically rigid support system. Figure 3c shows the comparison between the numerical prediction against the Mach number distribution along the nozzle mid-plane obtained experimentally. The experimental data is complemented with the estimated total expanded uncertainty. The axial coordinate x, which represents the distance along the nozzle mid-plane from the throat, is normalized with respect to the nozzle throat height H_{th}. The extracted Mach number ranges from $M = 1.4 \pm 0.04$ to 1.95 ± 0.05 at positions $x/H_{th} = 1.07$ and 7.0, respectively. The Euler simulation predicts higher values for the flow Mach number along the expansion process when compared to the measured Mach number. Nevertheless, the predicted Mach numbers match reasonably well with the experimental results, where the closest match is located inside the kernel region. The deviation tends to increase as the Mach number increases, i.e., towards the nozzle exit. In the uniform region, e.g., after the kernel-uniform transition point, there are two troughs located at $x/H_{th} = 4.5$ & $x/H_{th} = 6.4$. These troughs are the result of local decreases in the velocity caused by two weak oblique shock waves. This can be explained by the following. The nozzle glass panes have been mounted and sealed with a PTFE gasket giving a flow domain width of 20 mm. There was no loss of pretension on the bolts; the gasket retained its mechanical properties and the thermal expansion coefficient of glass is much lower than that of steel. This indicates

[1] The uncertainty sources associated with the measurements were classified and evaluated as Type A uncertainties (those dependent on measurement statistics) and Type B uncertainties (sensor characteristics) [6]. Type A and B uncertainty values were used to compute the total expanded uncertainty by means of a root sum square. The data acquisition uncertainty was quantified during the calibration of the NI9208 current modules and is deemed negligible.

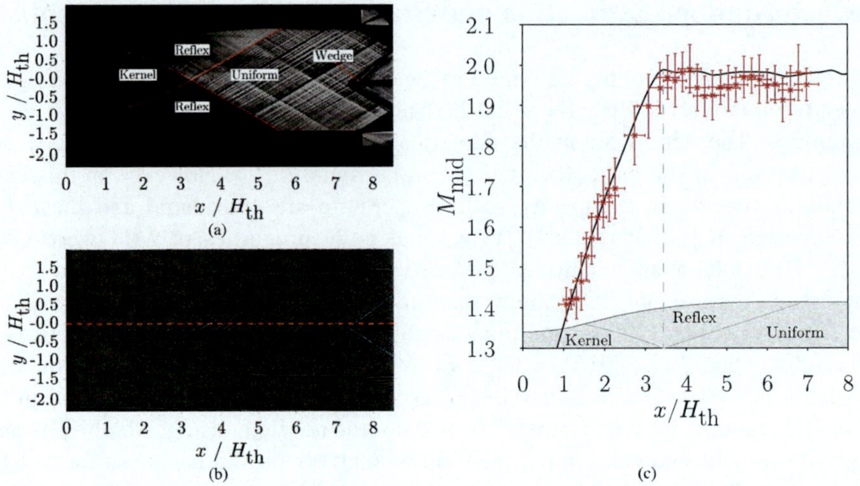

Fig. 3. A non-ideal expansion of MM at the conditions: $\overline{T}_0$ $(TT015) = 253.7$ °C, $\overline{P}_0$ $(PT011) =$ 18.36 bar and $\overline{P}_b$ $(PT004) = 2.21$ bar. Ref. = throat position. **(a).** A raw schlieren photograph showing Mach waves and an oblique shock wave at the nozzle exit. Knife-edge is horizontal. **(b).** Average extracted Mach lines superimposed on the first post-processed schlieren image. **(c).** Comparison between the measured flow Mach number with the corresponding total exp. uncertainty and the Euler solver numerical prediction.

that the width remained unchanged during the experiments but it will be checked with a confocal displacement sensor (CL-3000 WW) in future experiments. However, the measured nozzle height ratio $H_{\text{Exit,Measured}}/H_{\text{Throat,Measured}} = 2.68 \pm 0.059$ is lower with respect to the design height ratio of $H_{\text{Exit,Design}}/H_{\text{Throat,Design}} = 2.85$. Furthermore, the adapted pressure ratio $P_0/P_{\text{SV}}017 = 8.54$ is also lower than the design pressure ratio of 8.76. This results in a lower nozzle exit static pressure $P_{\text{SV}}017$ than the pressure in the receiver PT004. The mismatch leads to the generation of weak shock waves and the situation of overexpanded flow.

Figure 4 shows the measured static pressure distribution along the top and bottom nozzle profiles complemented with the expanded uncertainty listed in Table 2. Sixteen pressure values located at the indicated tap locations $T\#$ are compared against the results of the Euler simulation together with the computed compressibility factor plotted on the Y2 axis. The match is satisfactory with the greatest deviation inside the kernel and reflex regions; namely, at the Taps T06 = 1.2 %, T08 = 1.4 %, T09 = 2.7 %, T10 = 2.2 %, T13 = 1.7 % and T14 = 1.9%. A small discrepancy is observed at the tap location T17(s) and is most likely due to a small leakage of MM in the line connecting the Scanivalve to the liquid trap.

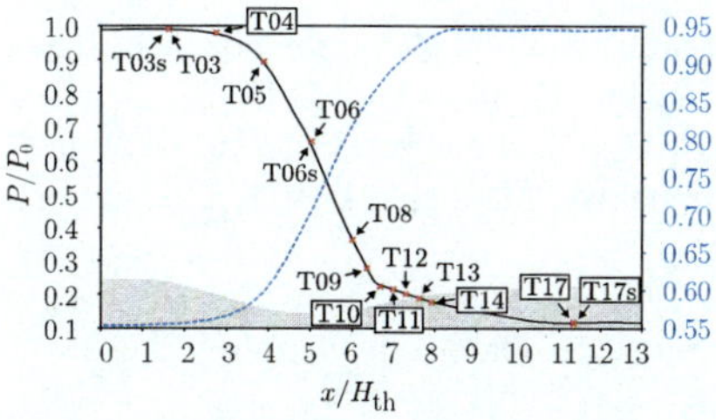

Fig. 4. Static pressures measured along the nozzle during the PR.028-NT.001 experimental run together with the related total expanded uncertainty bars, see Table 2. The pressure is non dimensionalized with the average value of the total pressure at the inlet $\overline{P}_0 = 18.36$ bara and the axial coordinate with the throat height $H_{\text{th}} = 7.73$ mm. Static pressures resulting from the Euler simulation (–) and exp. results (**x**). The centerline compressibility factor (- -) is calculated with the PR eqn. of state.

Table 2. Scanivalve experimental mean values and related uncertainties for the PR.028-NT.001 test. N.b. taps T01, T07, T15 and T16 not active during this experiment. The total expanded uncertainty includes Type A/B.

Tag	$\overline{P}$ / bara	U / bara *	Tag	$\overline{P}$	U / bara *
T02$^+$	18.33	0.0180	T09	5.11	0.0117
T03	18.15	0.0180	T10	4.12	0.0089
T03s	18.16	0.0180	T11	3.97	0.0037
T04	17.99	0.0179	T12	3.69	0.0037*
T05	16.43	0.0108	T13	3.47	0.0038
T06	12.07	0.0101	T14	3.26	0.0036
T06s	12.06	0.0110	T17	2.15	0.0037
T08	6.68	0.0111	T17s	2.08	0.0035

* Coverage factor $k = 2$. $^+$ T02 in same flange as PT011.

5 Conclusions and Future Work

The SU2 solver is able to accurately predict the flow Mach number and static pressure distribution in a supersonic vapor expansion of the fluid MM. The consistency of the experimental and numerical results indicate that the custom-designed Scanivalve purging and acquisition procedure is effective at measuring static pressures in hot organic vapor flows. Thermal expansion effects are significant and increase the throat height by approx. 3–4%. This decreases the nozzle height ratios ($H_{\text{exit}}/H_{\text{throat}}$) to such a degree that it is apparent in the Mach number results. Tracking the nozzle geometry is thus essential in these experiments. Initial shock wave experiments are promising but the mechanical rigidity of the model support system must be improved. Aligning the model along the axial plane is crucial for models of this size.

References

1. Beltrame, F., Head, A.J., De Servi, C., Pini, M., Schrijer, F., Colonna, P.: First experiments and commissioning of the ORCHID nozzle test section. In: Proceedings of the 3rd International Seminar on Non-Ideal Compressible Fluid Dynamics for Propulsion and Power, NICFD 2020. ERCOFTAC Series, vol. 28, pp. 169–178. Springer, Cham (2021). https://doi.org/10.1007/978-3-030-69306-0_18
2. Conti, C.C., Fusetti, A., Spinelli, A., Guardone, A.: Shock loss measurements in non-ideal supersonic flows of organic vapors. Experiments in Fluids **63**(117), 1114–1432 (2022)
3. Monjas, F., Blanca, H., Adam, J., De Servi, C., Pini, M.: Validation of the SU2 fluid dynamic solver for isentropic non-ideal compressible flows. In: Proceedings of the 4th International Seminar on Non-Ideal Compressible Fluid Dynamics for Propulsion and Power, NICFD 2022. ERCOFTAC Series, ERCO, vol. 28, p. IX, 180. Springer, Cham (2023)

4. Guardone, A., Spinelli, A., Dossena, V.: Influence of molecular complexity on nozzle design for an organic vapor wind tunnel. J. Eng. Gas Turb. Power **135**(4), 042307 (2013)
5. Head, A.J.: Novel experiments for the investigation of non-ideal compressible fluid dynamics: The ORCHID and first results of optical measurements. PhD thesis, Delft University of Technology (2021)
6. International Organization for Standardization (ISO) 1993 Guide to the expression of uncertainty in measurement. Guide 100:2008. Joint Committee for Guides in Metrology, JCGM. Corrected and reprinted (1995)
7. LaVision Inc. Davis, version 8.3.0 (2015)
8. Lemmon, E.W., Bell, I.H., Huber, M.L., McLinden, M.O.: 2018 NIST standard reference database 23: Reference fluid thermodynamic and transport properties-REFPROP, version 10.0, National Institute of Standards and Technology (2018)
9. Lo, R., Tsai, W.: Gray-scale hough transform for thick line detection in gray-scale images. Pattern Recognit. **28**(5), 647–661 (1995)
10. Nannan, N., Colonna, P., Guardone, A., Lemmon, E.: Multiparameter equations of state for selected siloxanes. Fluid Phase Equilibria **244**, 193–211 (2006)
11. Roache, P.J.: Verification and validation in computational science and engineering. Hermosa Publishers, Albuquerque (1998)
12. Robertson, M., Newton, P., Chen, T., Costall, A., Martinez-Botas, R.: 2020 Experimental and numerical study of supersonic non-ideal flows for organic Rankine cycle applications. J. Eng. Gas Turbines Power **142**(8), 081007 (2020)
13. The MathWorks, Inc. 2019 MATLAB version 9.7.0.1261785 (R2019b) Update 3
14. Zocca, M., Guardone, A., Cammi, G., Cozzi, F., Spinelli, A.: Experimental observation of oblique shock waves in steady non-ideal flows. Exp Fluids **60**(6), 101 (2019)

Author Index

M. White et al. (Eds.): NICFD 2022, ERCOFTAC 29, pp. 239–240, 2023.
https://doi.org/10.1007/978-3-031-30936-6